Green Energy and Technology

Climate change, environmental impact and the limited natural resources urge scientific research and novel technical solutions. The monograph series Green Energy and Technology serves as a publishing platform for scientific and technological approaches to “green”—i.e. environmentally friendly and sustainable—technologies. While a focus lies on energy and power supply, it also covers "green" solutions in industrial engineering and engineering design. Green Energy and Technology addresses researchers, advanced students, technical consultants as well as decision makers in industries and politics. Hence, the level of presentation spans from instructional to highly technical.

Indexed in Scopus.

Indexed in Ei Compendex.

Ataur Rahman · Sany Izan Ihsan ·
Fadly Jashi Darsivan Bin Ridhuan Siradj

Autonomous Driveline System for Electric Vehicle

Electric and Autonomous Electric Vehicle

Springer

Ataur Rahman
GameAbove College of Engineering
and Technology
Eastern Michigan University
Ypsilanti, MI, USA

Sany Izan Ihsan
Faculty of Engineering
International Islamic University Malaysia
Gombak, Kuala Lumpur, Malaysia

Fadly Jashi Darsivan Bin Ridhuan Siradj
Faculty of Engineering
International Islamic University Malaysia
Gombak, Kuala Lumpur, Malaysia

ISSN 1865-3529 ISSN 1865-3537 (electronic)
Green Energy and Technology
ISBN 978-981-96-8417-5 ISBN 978-981-96-8418-2 (eBook)
https://doi.org/10.1007/978-981-96-8418-2

This Springer imprint is published by the registered company Springer Nature Singapore Pte Ltd.
The registered company address is: 152 Beach Road, #21-01/04 Gateway East, Singapore 189721, Singapore

Preface

This is a comprehensive book on the development of Autonomous Driveline for Electric and Autonomous Vehicle focuses on technological details, mathematical relationships, and basic design principles in line with the development of green transportation systems. This book's good blend of theoretical review and technical design makes it an excellent platform for educating today's engineers, post graduate students, and industries personnel. It has a total of five parts with twenty one (21) chapters that focus on the development of electric and autonomous vehicle driveline.

Part I gives a brief review on adoption electric vehicle, inline with the development of sustainable environment. Part I also presents a case study on prototyping a retrofitted electric transport, which would enhance the core knowledge of the users to develop of alltrain electric off-road vehicle prototype.

Part II introduces an artificial intelligent control energy system which is built with organic solar supercapacitor (OSSC) as a structural frame of electric vehicles to meet the electrical load demand with solar renewable energy trapping. A solar supercapacitor prototype is developed by utilizing Carbon Fiber Reinforced Polymer, nano Zinc Oxide (ZnO), and Copper Oxide (CuO) fillers as the positive and negative electrodes and a dielectric layer sandwiched between the electrodes. The adaptive neuro-fuzzy intelligent system (ANFIS) controller has been used to control the power storage of the OSSC, power flow to meet the EV's auxiliary load demand, and battery charging. The test results of OSSC show potential of energy conversion efficiency (η_{ec}) 13.33%, open-circuit voltage (V_{oc}) 599 mV, current density (J_{sc}) 222.22 A/m^2, capacitance (C) 11.17 μF/cm^2, energy density (E_d) 120 Wh/kg, and power density (P_d) 29 kW/kg. Therefore, the OSSC can save about 3.4 kWh battery in 11.00 to 17.00 hours which would be able to make the EV about 15% energy efficient and contribute to emission control.

Part III presents an intelligent battery thermal management system for the EV. It presents a comprehensive review for the solutions of the limitation of the battery cooling thermal management system (BC-ThMS) for the electric or autonomous electric vehicle, intelligent cooling system design technique, and control. The performance study both in air cooling, liquid cooling, and phase change refrigerant based cooling system individually to study battery sustainability in any operational mode on maintaining the battery recommended temperature of 40ºC or below. An experimental study on an evaporative battery thermal management system has been conducted and found that it is potential to maintain the battery temperature 40 ºC or below. An intelligent system has been embedded with the EC-BThMS to make the system optimize in controlling the battery temperature in the range of 30–40 °C and warning send to the EV user if there is any chance to reach out the battery temperature beyond 40 °C. It may span the battery life, provide security and safety of the battery and saving the energy specially decreasing the frequency of battery charging.

Part IV presents an autonomous two-speed transmission. It presents the technique on the development of a fuzzy-logic-controlled servo-electromagnetic autonomous two-speed transmission (SEMA-2ST) model. The electromagnetic actuator is modeled in terms of electromagnetic capacity, number of coil turns, supply current, and electromagnetic force needed to shift gears. Experimental results of AEM-2SGB show on first-gear shift time 110 ms at 300 Nm motor torque and a second-gear shift time of 116 ms at 110 km/h vehicle speed, with a maximum current supply of 16 A using a 24 V lithium ion battery. The SEAM-2SGB reduces weight by 37–66%, transmission losses by 40%–90%, and battery life by 5%.

Part V presents an intelligent electromagnetic continuously variable transmission. It describes the development technique of Autonomous Electromagnetic Continuously Variable Transmission (EMACVT) for EV. It could be considered as the solution of EV future transmission. An analytical model of the EMACVT is developed by deep analysis on the kinematics of the CVT clamping forces and electromagnetic forces of the EMA. A fuzzy intelligent system has been introduced in the EMACVT to control the dynamic behavior of the CVT with controlling the current flow to the EMA actuation. There is a failsafe mode built into the CVT to prevent damage in case some of the sensors malfunction. During failsafe mode, the system shifted to manual mode for fixing the gear ratio equivalent to first gear and to save the CVT from damage by the central manual switching system. It is observed in experiment that time taken to accelerate car with 40–60 km/h with taking time in the range of 1.8–3.1 sec. The EMA develops the electromagnetic force in the range of 108–301 Nm for the supply current in the range of 6–9.33 amp.

Ypsilanti, Michigan, USA — Ataur Rahman
Gombak, Kuala Lumpur, Malaysia — Sany Izan Ihsan
Gombak, Kuala Lumpur, Malaysia — Fadly Jashi Darsivan Bin Ridhuan Siradj

About This Book

Artificial intelligence technologies are being applied around the world to solve transportation problems, mainly in the areas of traffic management, cyber security, road safety, public transport and urban mobility. A promising application of Autonomous system in EV/AEV aims to improve productivity by reducing accident rates of electric vehicles in close proximity and high-speed in-highway driving.

Autonomous drivetrain development focuses on the technological solution of an autonomous transportation system. It is an excellent platform to train today's engineers, graduate students and industry professionals to develop autonomous vehicle technologies. This book introduces an organic solar supercapacitor (OSSC) with artificial intelligence for storing solar energy in a car during EV operation, an autonomous battery temperature management system to control the battery temperature in the range of 30–35 °C, a two-speed servo-magnetic transmission to keep the car moving regardless of the road profile, and an IOT embedded emergency braking system for working in close proximity and at high-speed EV, and an autonomous electromagnetic CVT.

Neuro-Fuzzy Adaptive Intelligent System (ANFIS) controlled OSSC, the structural frame of autonomous vehicles, easily meets the auxiliary load requirements of the EV and reduces the battery charging frequency. Thus, OSSC can make an autonomous transport system approximately 5–15% energy efficient. The built-in intelligent battery temperature management system monitors the battery temperature between 30 and 40 °C and sends an alert to the EV user if therea is a possibility that the battery temperature will exceed 40 °C.

The Servo-Electric Magnetic Two-Speed Fuzzy Transmission (SEMA-2ST) provides electric drive regardless of the road profile. SEAM-2SGB reduces the weight of the EV by 25%, reduces losses by 90%, increases mileage by approximately 5%, and battery life span by 10–5%.

Autonomous electromagnetic continuously variable transmission (EMACVT) could be the powertrain solution for autonomous vehicles. Fuzzy intelligent control EMA controls the dynamic behavior of the CVT to ensure the vehicle propulsion in any road conditions. The CVT has a fail-safe mode built into it to prevent damage if certain sensors fail. In Safe Mode, the system went into manual mode to control the gear ratio equivalent to first gear and to protect the CVT from damage by the central manual shift system. EMACVT accelerates the electric vehicle from 0 to 60 km/h in 1.8 s.

Contents

About the Authors

Prof. Dr. Ataur Rahman Ph.D. is currently involved in the GameAbove College of Engineering and Technology, Eastern Michigan University, Michigan, USA. He was a senior Professor in the Department of Mechanical and Aerospace Engineering, International Islamic University Malaysia (IIUM). He was appointed as a Visiting Fellow for "Designing Automation System of Off-road Vehicle" in Mechanical Engineering, The University of Tokyo, Japan. He is the Chairman of IIUM Centre of Excellence of Electric Mobility and IIUM Flagship Project for IIUM Go-Green Campus. He has designed and prototyped IOT embedded High Tek Cooled IGBT INVERTER for EV/AEV, Advanced Electric Vehicles and Electric Coaster, Intelligent BTMS, Autonomous servo-electromagnetic (ASEM) 2-speed transmission, IOT ASEM braking system for high speed EV, Intelligent off-road vehicle, and Solar organic supercapacitor. He is the principal author of three books including two books in Nova Science Publication, Inc. New York, USA and IIUM Press, Gombak, Malaysia.

Assoc. Prof. Dr. Sany Izan Ihsan is an Associate Professor at the Mechanical Engineering Department, and currently the Dean of the Kulliyyah of Engineering, International Islamic University Malaysia (IIUM). He had his B.Sc. in Mechanical Engineering at the University of Wisconsin—Madison in 1995, M.Sc. in Aerospace Engineering at the University of Tennessee Space Institute (UTSI) in 1999 and Ph.D. in Engineering from IIUM in 2008. His research areas of interest include noise and vibration, ride quality, vehicle dynamic and control, semiactive control system, solar thermal, renewable energy, and engineering education.

Assoc. Prof. Dr. Fadly Jashi Darsivan Bin Ridhuan Siradj is an Associate Professor in the Department of Mechanical and Aerospace Engineering, Faculty of Engineering of the International Islamic University Malaysia. He is currently the Deputy Dean of Academic and Internationalisation. He previously served as the Head of Department of Mechanical and Aerospace Engineering. He completed his Ph.D. studies in the field of active vibration control in automotive applications. He received his Bachelor of Engineering and Master's degrees, both from the Faculty of Engineering, International Islamic University Malaysia in 1998 and 2002, respectively. At present, he is actively doing research in area of vehicle dynamics and in the area of sustainable energy and had previously received research grants from the Ministry of Education, Malaysia.

Part I
Adoption of Electric Vehicle

Chapter 1
Electric Vehicle and Environment

Abstract Around 14 million electric vehicles were sold globally, according to the IEA, a 35% increase from 2022. In 2023, there were 40 million electric vehicle, a 110% rise from 2022, and it looks like this trend will continue in 2024. The goal of the CEM's Electric Vehicles Initiative (EVI) was to achieve a 30% sales share for EVs by 2030 in order to minimize carbon dioxide emissions from the environment. The EV30@30 campaign reframed this target. According to estimates, a gasoline-powered automobile emits about 218 g of CO_2 per kilometer driven during its lifetime, compared to about 161 g for hybrid and plug-in hybrid cars and about 126 g for purely battery-electric cars. Power generation and the development of lithium batteries are necessary for this.

Keywords Electric vehicle · Environment · CO_2 emission · Market share

1.1 Introduction

In recent years, battery electric vehicles (BEVs) have drawn more interest due to their quickly developing technical capabilities. Although a lot of research has tried to investigate the social effects of BEV adoption, little is known about how much broad BEV adoption can influence economic and environmental factors at the same time. By performing a thorough evaluation of the effects of BEV deployment, this chapter seeks to close this research gap. A computable general equilibrium model is used to predict the impact of various scenarios based on the demand estimates from the discrete choice experiment. Three elements are considered when calculating the overall cost of ownership of BEVs: fuel prices, manufacturer financial incentives, and subsidy levels.

Additionally, increases in the productivity of BEV battery manufacturing are taken into consideration in light of current trends. This study demonstrates that while manufacturer incentives and fuel price changes have comparatively little effect on GDP growth, subsidies have a significant impact on GDP and BEV adoption because they stimulate household spending as well as industries related to vehicle manufacturing.

A. Rahman et al., *Autonomous Driveline System for Electric Vehicle*, Green Energy and Technology, https://doi.org/10.1007/978-981-96-8418-2_1

Productivity shocks only have an impact on BEV adoption in markets with competition, but they have a moderate effect on GDP. On the other hand, the effects on the ecosystem are less obvious. Even while BEVs reduce tailpipe emissions, part or all of the emissions savings may be offset by higher non-exhaust emissions brought on by increased manufacturing activity brought on by productivity increases or subsidies.

This implies that strategies to introduce subsidized BEVs must be combined with cleaner industrial and power generating measures in order to achieve the targeted emission reductions.

1.2 Future Mobility: Autonomous Electric Vehicle

The possible effects of autonomous vehicles on the road may require state and local governments to address the technology as it advances. The number of states contemplating legislation pertaining to autonomous vehicles has steadily grown every year.

Twenty-nine states in the United States: Alabama, Arkansas, California, Colorado, Connecticut, Florida, Georgia, Illinois, Indiana, Kentucky, Louisiana, Maine, Michigan, Mississippi, Nebraska, New York, Nevada, North Carolina, North Dakota, Oregon, Pennsylvania, South Carolina, Tennessee, Texas, Utah, Virginia, Vermont, Washington, and Wisconsin—as well as Washington, D.C., have passed laws pertaining to autonomous vehicles. Executive orders pertaining to autonomous vehicles have been issued by the governors of Arizona, Delaware, Hawaii, Idaho, Illinois, Maine, Massachusetts, Minnesota, Ohio, Washington, and Wisconsin.

High-performance computers, a growing number of advanced driver assistance system (ADAS) sensors, including high-resolution stereo and/or mono cameras, RADAR, and LIDAR, as well as future human–machine interfaces (HMIs), like head-up displays (HUDs) or large 4 K/8 K screens, will all be part of the extremely complex systems found in driverless cars.

Large volumes of data will be generated by these systems, necessitating advanced electrical assistance in the form of cables, assemblies, linkages, and high-speed data nodes. The interior of the autonomous vehicle will function as a true information highway, with data streams flowing in various directions or in parallel.

Continuous collection and interpretation of comprehensive environmental data is required for V2X communications. All of this data is necessary for fully autonomous cars to perform tasks like braking and accelerating accurately without the need for a driver.

Sensor technology and radio-based communication are both used in V2X communications. Vehicle systems can communicate with their local environment with the use of car sensors. Vehicles can communicate with traffic infrastructure including lights, signs, and tolls as well as with other vehicles thanks to radio systems. These environments actively inform the car about their condition and the shifting circumstances around them. For example, a sign tells you that the next turn-off is five miles away, a light signals that it is about to turn green, and a car a lane over signals that it is about to turn right. These signals all contribute to safer driving. Additionally, vehicles can read data and make autonomous judgments with the use of information about driving conditions, such as tolls, snowy roads ahead, or traffic slowdowns.

Data sensing and transmission must be 100% dependable and real-time. This implies that poor road conditions, such as fog, snow, or ice, or prolonged use, like thousands of hours of operation, cannot cause connectivity to fail. Using extremely reliable automotive-grade communication systems from skilled and knowledgeable vendors with engineering and materials science backgrounds, automakers may accomplish this goal. Preferred suppliers for OEMs must be knowledgeable about contact physics, have developed the precise production procedures and equipment necessary to create them, and know which materials are optimal for strong electromechanical connections.

Autonomous driving is transforming our perception of mobility and automobiles. New automobiles will improve passenger safety, comfort, and entertainment in addition to making transportation possible. Imagine sending emails, doing video conferences, resting in your car while you drive, and completing a project before you even get to work.

These days, getting about is a necessary but frequently unpleasant task. Fully autonomous driving will allow passengers to focus on their own objectives, transforming idle travel time into productive times for engagement, learning, working, and unwinding.

1.3 Global Electric Vehicle

Future automotive innovation will be shaped by four major technologies: automation, connectivity, electric power, and the shared economy. Perhaps the most obvious and significant change affecting the automobile sector is the rise of autonomous vehicles. However, electric vehicles (EVs), which are most suited to embrace automation and connected vehicles, are currently the focus of the automobile industry's future. EVs are already the primary mode of transportation in many nations worldwide. In recent years, a number of high-tech electric automobiles, such as Tesla and Porsche, have been produced with amazing characteristics and sold in large quantities.

Around the world, the number has risen to hundreds of thousands. This suggests that the automobile industry's future lies in electric vehicles. Furthermore, a large number of major automakers have participated in the electric vehicle market. Toyota, Honda, Hyundai, Mercedes-Benz, Mitsubishi, Tesla, Renault, BMW, Chevrolet, Ford, and Citroen are a few examples. They all have at least one highway-capable model. They have a strong commitment to the development of electric vehicles.

In 2018, there were more than 5.1 million electric cars worldwide, an increase of 2 million since 2017. This nearly doubled the record number of new registrations in 2017. With about 1.1 million electric cars sold in 2018 and 2.3 million units sold, the People's Republic of China (henceforth referred to as "China") continues to hold the top spot in the global electric vehicle market. By the end of 2018, there were 1.2

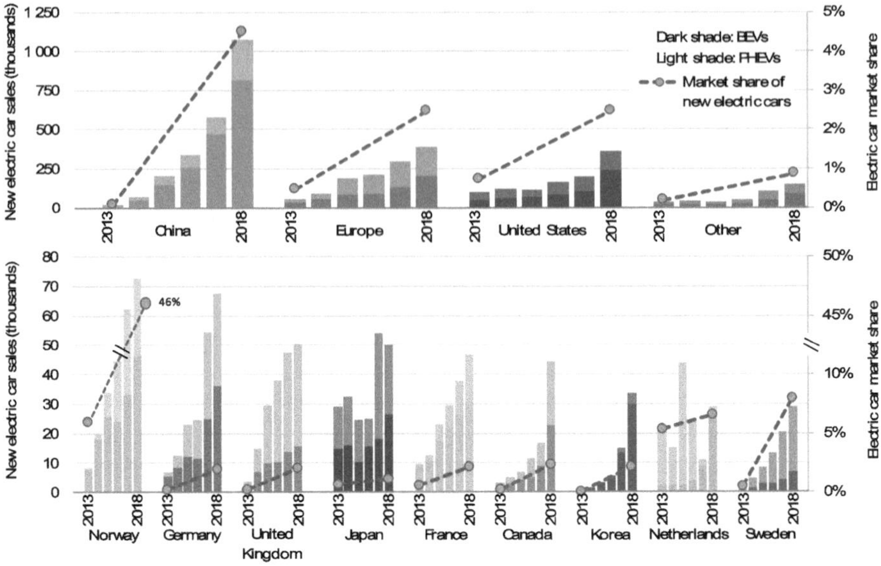

Fig. 1.1 Electric vehicle scenario (Global EV Outlook, 2019)

million electric cars in Europe and 1.1 million in the US, with a 385,000 and 361,000 increase in market share, respectively, from the year before (Fig. 1.1).

In terms of market share for electric cars, Norway continued to lead the world in 2018 with 46% of new sales, more than doubling Iceland's second-largest share of 17% and six times larger than Sweden's third-largest share of 8%. Along with traditional passenger cars, an estimated 5 million low-speed electric vehicles (LSEVs) were produced in 2018, an increase of about 700,000 from the previous year. China was the location of every LSEV. In 2018 and the first few months of 2019, shared "free floating" electric foot scooters grew significantly in major cities worldwide. Approximately 129 cities in the US, 30 in Europe, 7 in Asia, and 6 in Australia and New Zealand currently have foot scooter programs in place.

Electric cars (EVs) were mostly used in freight transportation as light-commercial vehicles (LCVs), with 250,000 units produced in 2018—an increase of 80,000 from 2017. In 2018, sales of medium trucks were between 1,000 and 2,000, with China accounting for the majority of these purchases. 5.2 million light-duty vehicle (LDV) chargers, 540,000 of which are open to the public, and 157,000 bus rapid chargers served the world's EV stock in 2018.

In 2018, there were an estimated 5.2 million EV chargers for light-duty vehicles (LDVs) globally, continuing the upward trend. Nearly 540,000 publicly available chargers, including 150,000 fast chargers, 78% of which are in China, supplement the majority, which are slow chargers (levels 1 and 2) at homes and offices. By the end of 2018, there were roughly 300,000 fast chargers deployed worldwide, including 156,000 bus fast chargers.

According to estimates, the world's EV fleet used 58 terawatt-hours (TWh) of electricity in 2018, which is comparable to Switzerland's total electricity use in 2017. The highest percentage of EV energy demand (55%) was still accounted for by two-wheelers, although LDVs grew at the fastest rate of any mode of transportation in 2017–18. In 2018, 80% of the global demand for EV electricity came from China. In 2018, the global EV stock released approximately 38 million tons of carbon dioxide.

1.4 EVs Energy Saving

The anticipated rise in EV usage across all modes will have an effect on the rise in oil consumption for road transportation. According to the New Policies Scenario, the anticipated demand for oil products in 2030 would be avoided by 127 million tons of oil equivalent (Mtoe), or around 2.5 million barrels per day [mb/d]. According to the EV30@30 Scenario (the EV30@30 Campaign aims to achieve a 30% sales share for EVs by 2030), the EV stock will lower the demand for oil products by 215 Mtoe (4.3 mb/d) in that year.

However, there will likely be a considerable increase in the demand for electricity to power EVs. The global EV fleet is expected to generate over 640 TWh of electricity in 2030, according to the New Policies Scenario of EV30@30 (Fig. 1.2). When compared to 2018 levels (58 TWh), this represents a more than ten-fold increase and is equal to the total final power consumption of France and Spain in 2016. According to the EV30@30 Scenario, the worldwide EV fleet would grow to a size of 1110 TWh in 2030, which is almost twice as much as the New Policies Scenario. In 2020, light-duty vehicles overtake two- and three-wheelers as the road modes with the highest electricity consumption under the New Policies Scenario.

About 60% of all vehicles in 2030 will be LDVs, with buses coming in second (26%), two- and three-wheelers in third (12%), and trucks in fourth (2%). According to the EV30@30 Scenario, buses will account for 20% of the electricity consumption from EVs in 2030, while LDVs will account for the majority (65%). According to projections of electricity consumption in both scenarios, EVs will likely play a far larger role in power systems than they have in the past. Uncontrolled charging of EVs may result in increased demands for transmission capacity and peak power generation. The total yearly electricity demand of EVs, the effect of daily charging patterns on load profiles, and the location power levels used for charging all affect how much power systems can be affected.

1.5 EVs GHG Emissions Reduction

The combined evolution of EV energy use and electricity generation's carbon intensity determines the well-to-wheel (WTW) GHG emissions from the EV fleet. WTW emissions from an EV are currently lower than those from an ICE vehicle fueled by

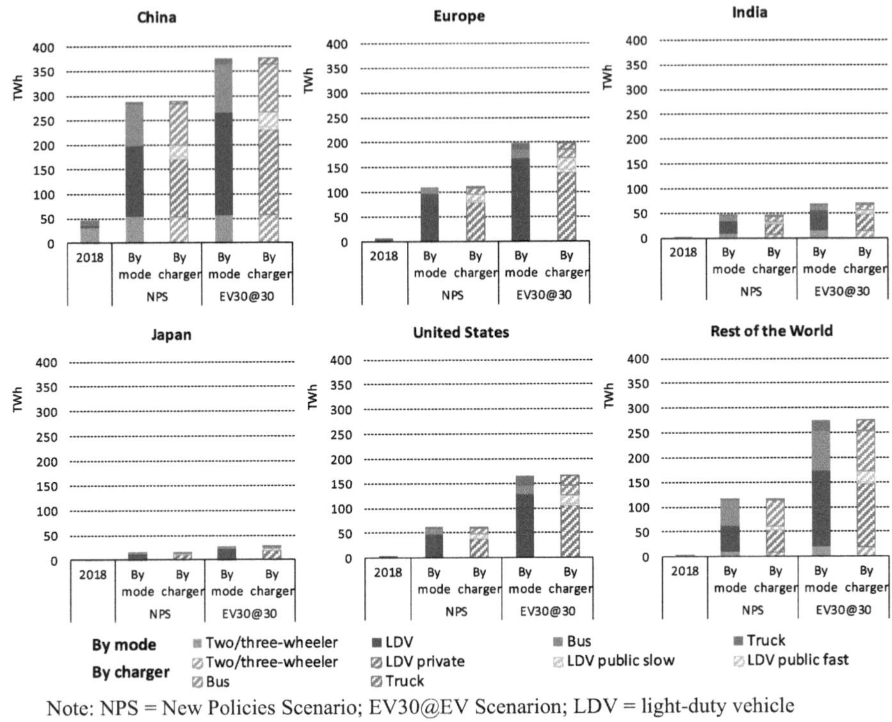

Fig. 1.2 Energy saving (Global EV outlook, 2019). *Note* NPS = New policies scenario; EV30@EV scenarion; LDV = light-duty vehicle

a blend of liquid and gaseous fuels, according to the global average carbon intensity of power generation. According to the New Policies Scenario (NPS), greenhouse gas emissions from the EV fleet are expected to be approximately 230 Mt CO_2-eq in 2030; however, if the same fleet of vehicles were powered by internal combustion engines (ICEs), these emissions would nearly quadruple to 450 Mt CO_2.

An equivalent fleet of ICE vehicles would emit roughly 770 Mt CO_2 in 2030, compared to the projected 230 Mt CO_2-eq emissions of the EV30@30 Scenario, which combines the accelerated deployment of EVs with a trajectory for power generation de-carbonization consistent with the International Energy Agency's (IEA) Sustainable Development Scenario (SDS). For the EV30@30 Scenario to limit the rise in GHG emissions due to the rapid expansion of the EV stock, it is crucial that electricity generation rapidly decarbonize. In the absence of these steps, the EV fleet's WTW GHG emissions in the EV30@30 Scenario would be approximately 340.00 Mt CO_2 by 2030. The overall value chain, or the life cycle of EVs in comparison to alternative options, determines whether or not EVs result in net advantages in terms of GHG emissions savings.

Overall, a mid-sized global average BEV and a plug-in hybrid electric car emit less than an average global ICE vehicle using gasoline over their lifecycles, when

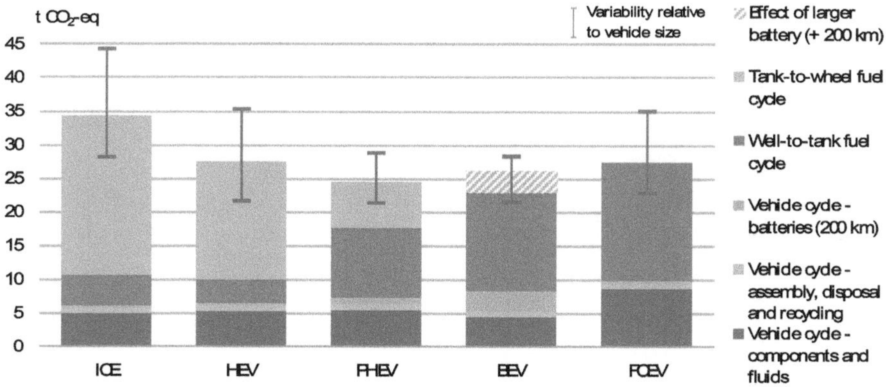

Fig. 1.3 Emission of vehicle (IEA, 2019)

taking into account the scale-up of battery manufacturing facilities (compatible with state-of-the-art plants, and in line with those that are assumed to be scaled up in the New Policies and EV30@30 scenarios), as well as the current global average carbon intensity of power generation (518 g CO_2/kWh, including losses). The impact's magnitude varies according to the ICE vehicle's size. For BEV vehicles with smaller batteries and a shorter operating range, net savings are higher.

The greenhouse gas emissions of BEVs powered by electricity, as measured by the current global average carbon intensity, are comparable to those of FCEVs powered by hydrogen.

The decarbonization of power generation systems, as illustrated in Fig. 1.3, offers the greatest potential for reducing emissions over the course of an EV's vehicle life cycle. In nations with low carbon intensity of the generation mix, net savings are higher today. If electricity generation rapidly decarbonizes, this could be a major advantage for BEVs and PHEVs over alternative powertrain technologies in the future. The ability of EVs to provide appreciable net GHG savings over competing technologies varies globally, however, due to regional and power system variations in carbon intensity. To achieve GHG savings from the electrification of road transportation, areas that primarily rely on coal for electricity production must shift to a lower carbon generation mix.

1.6 Sales and Market Share of EV

The number of electric cars sold worldwide topped one million in 2017 and around two million in 2018 (Fig. 1.4). After two years of slower growth, the electric car market grew by 68% year over year between 2017 and 2018, a robust rate similar to that of 2015.

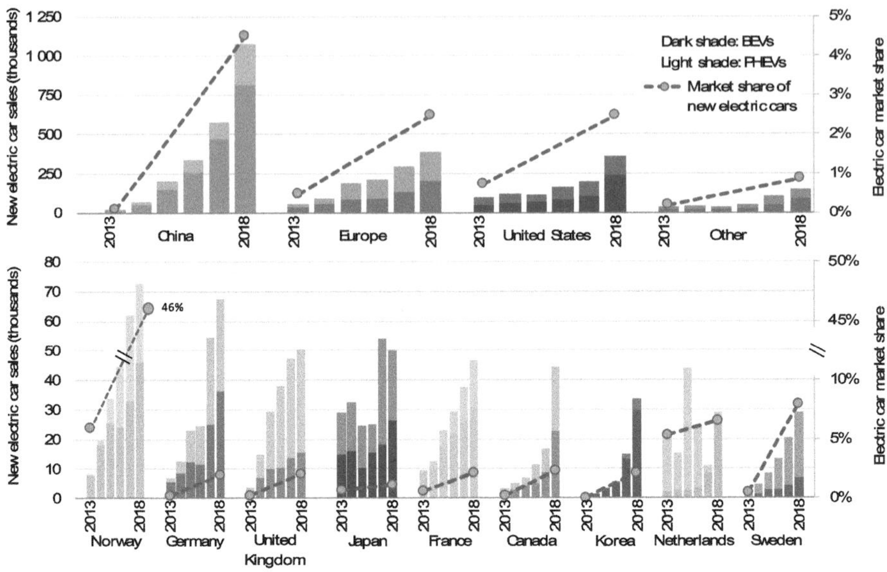

Fig. 1.4 Sales of EV worldwide (EV outlooks, 2019)

1.7 EVs Sales in China

With nearly 1.1 million sold in 2018, up from nearly 600,000 in 2017, and holding a 55% share of the global electric car industry, China continues to be the largest market for electric vehicles worldwide. The dynamism of China's electric car market is further shown by the fact that this gain contrasts with the general reduction in all passenger car sales that occurred in China relative to 2017. Europe sold 385,000 electric cars in 2018, making it the second-largest market for these vehicles. With 361,000 sales, the US is the third-largest market for electric vehicles.

1.8 EVs Sales in Europe

The growth rate of electric car sales in Europe in 2018 was 31% higher than in 2017, which is less than the global average and lower than the growth rate of 41% in 2016. Europe is home to the nations where sales of electric cars are most prevalent. Norway's 2018 percentage was close to 50%, more than 2.5 times higher than Iceland's (17.2%), the next-highest country, and six times higher than Sweden's (7.9%). Norway's sales volumes are followed by those of Germany, the United Kingdom, and France. 2018 saw a robust recovery in sales in Denmark and the Netherlands, after they had fallen in 2017.

1.9 EVs Sales in USA

In contrast to the previous year, when sales grew by only 24%, U.S. sales increased by 82% in 2018, outpacing the global market's growth rate. One factor contributing to this trend is the introduction of the Tesla Model 3, whose sales in 2018 completely offset the 134,000 more BEVs sold in 2018 (compared to 2017).

1.10 Conclusion

Sales of EVs grew by about 80–82% in 2018, which is a significant rise above the 20–24% growth in 2017. In nations where the generation mix's carbon intensity is low, net savings are greater. If power generation decarbonizes quickly, this might provide BEVs and PHEVs a major edge over alternative powertrain technologies.

Acknowledgements The authors would like to express their gratitude to the Kulliyyah (Faculty) of Engineering at International Islamic University for providing financial support in order to recruit a research assistant for this project.

References

1. https://www.google.com/search?q=ev30%4030&rlz=1C5CHFA_enMY841MY841&oq=ev30%4030&aqs=chrome..69i57.10518j1j7&sourceid=chrome&ie=UTF-8
2. 1https://www.google.com/search?q=Global+EV+Outlook+2019&rlz=1C5CHFA_enMY841MY841&oq=Global+EV+Outlook+2019&aqs=chrome..69i57j0l4.3391j0j7&sourceid=chrome&ie=UTF-8.
3. https://www.google.com/search?q=IEA%2C+2019&rlz=1C5CHFA_enMY841MY841&oq=IEA%2C+2019&aqs=chrome..69i57j0l5.7383j0j9&sourceid=chrome&ie=UTF-8

Chapter 2
Challenges and Solutions for Electric Vehicle Deployment

Abstract The decision to buy an electric car has always been heavily influenced by capital expenditure; 63% of consumers think that buying an electric vehicle is out of their price range. However, the emphasis is now on the difficulty of expanding the required infrastructure and raw material supply to allow the widespread adoption of EVs as battery prices decline and cost parity between EVs and ICE vehicles is reached by 2026. The following issues need to be solved in the EV market: market penetration, customer awareness, charging infrastructure, safety and energy security, and battery recycling and monitoring. Demand for EVs. Smart charging stations can improve the sustainability, economy, and efficiency of charging. However, there are interoperability issues and the cost of installing public charging stations can be high. An essential part of an EV is the battery, which stores energy and supplies power for cruising and acceleration. The traction motor needs power supply, which raises the battery's temperature. To regulate the temperature and ensure safety, an effective battery thermal management system is necessary.

Keywords Capitalisation on EV purchasing · Customer perception · Energy security · Battery recycling · Policy on EV adoption

2.1 Introduction

Every market has factors that influence growth as well as obstacles. Germany has said that it would stop selling new cars with internal combustion engines (ICEs) in 2030. Despite having one of the biggest ICE car fleets in the world and being home to well-known brands like Mercedes-Benz, Audi, and Porsche, the industry was not surprised by the move. Germany has merely joined the global race to cut emissions and electrify transportation, as more than 40 nations have pledged to phase out internal combustion engines by 2050. With firms like Toyota and Volkswagen committing $170 billion in electrification initiatives, sales of electric vehicles (EVs) are up 80% globally in 2021.

A. Rahman et al., *Autonomous Driveline System for Electric Vehicle*, Green Energy and Technology, https://doi.org/10.1007/978-981-96-8418-2_2

Apart from removing exhaust emissions and mitigating a portion of the transportation sector's 23% worldwide CO2 emissions, electric vehicles also offer crucial grid flexibility as we transition to a higher proportion of renewable energy (RES). But even with this worldwide push, just 7.2% of all vehicles sold in 2021 were electric. One of the main drivers of the market's expansion over the projection period has been the rise in demand for electric vehicles, advancements in battery technology, supportive laws and regulations, and the introduction of new plug-in EV models. The market for electric vehicles has also been expanding at an exponential rate during the anticipated period.

This is because businesses have been developing business models that enable consumers to swap or replace EV batteries after they are drained, such as battery-as-a-service (BaaS) and battery swapping. By reducing the amount of time users must spend charging their batteries, this enhances customer happiness and tackles a major deterrent for EV adoption. Moreover, it is anticipated that the e-mobility trend will accelerate during the forecast period in tandem with the growth in EV sales. Over the projected period, the market has grown further thanks to the ongoing installation of integrated charging stations and the development of green power-generating capabilities.

2.2 Challenges of EVs Adoption

Considering that 63% of consumers believe that an electric car is beyond their means, capital expenditures have always been a significant consideration when making an electric vehicle purchase. The difficulty of expanding the required infrastructure and raw material supply to support the widespread adoption of EVs is becoming more prominent, nevertheless, as battery prices decline and cost parity between EVs and ICE vehicles is reached by 2026. The following four issues are identified:

Inadequate Charging Infrastructure

Charging stations are harder to locate than conventional filling stations, typically due to high infrastructure construction costs and investment requirements. Charging stations are an expensive investment due to the installation costs, which range from $2,500 for a slower charger to $35,800 for a fast charger, as well as additional fees like permits and regulations. Additionally, letting individuals charge where they typically park—at home or at work—presents its own set of difficulties, including managing network access, multi-family buildings, and charging slot availability. This discourages customers from converting to electric vehicles and results in a decrease in the network of operational charging stations. A typical AC slower electric car charging station is depicted in Fig. 2.1.

Risk of Grid Overload

Power systems are already under stress due to the increased share of renewable energy sources and the problem of more erratic energy supply. A greater demand

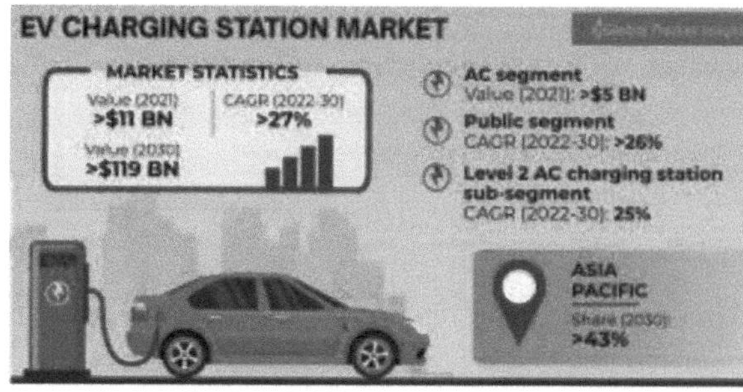

Fig. 2.1 Conventional low capacity AC charging system [1]

for electricity is created by the growing number of EVs, which may necessitate additional grid infrastructure investments. Utility companies and power generators are also having difficulty understanding the fast expanding EV market by predicting when and where this power will be needed. However, if EVs are charged in off-peak times, such as late at night or early in the morning, the chance of grid overload is reduced.

High Carbon Grid

Gray networks, which heavily rely on fossil fuels, are making electric vehicles less effective as a means for businesses and consumers to cut emissions. To persuade consumers that switching to electric vehicles is beneficial and lowers carbon emissions, it is imperative to decarbonize the grid as much as feasible.

Critical Minerals and Rare Earth Materials

Compared to ICE vehicles, electric vehicles consume around six times as many mineral resources. By 2040, the IEA predicts that there will be 70 million electric cars on the road, which would result in a 30-fold rise in the need for minerals. These subterranean resources are plentiful; the only question is whether they will be harvested responsibly, in line with social responsibility management principles, and in time to satisfy demand. It is anticipated that nickel shortages and issues with the expansion of lithium manufacturing would occur. Additionally, producers may employ inferior minerals as a result of this shortage, which will impair battery performance.

Crisis of Electrical Power

Compared to gasoline or diesel vehicles, electric vehicles have a far lower pollution impact. Compared to gasoline or diesel cars, which can only convert 17%–21% of the energy stored in the fuel to the wheels, electric vehicles can convert about 60% of the electrical energy from the grid to power the wheels. However, the adoption of electric vehicles is hindered by the electrical energy crisis in many nations. More than 1.5 billion of the 800 million people who lack access to energy for a sustainable way of life today reside in Asia. This deficiency is a personal tragedy as well as having broad ramifications for national security and energy security (ASIA NIKKEI, 2022). India and China are experiencing blackouts.

2.3 Public Perception and Attitude

To achieve substantial improvements in vehicle energy intensity and zero emissions, advanced power systems and lower vehicle road loads are required. Only fuel cell-powered automobiles, electric vehicles, and low-emission hybrid vehicles have so far been able to comply with CARB standards. Based on an analysis of EU Directives, optimal electrical and hybrid electric vehicle design has so gained popularity recently as a practical means of satisfying environmental protection and vehicle performance standards.

Developing a dynamic modeling and simulation on the discharging of the battery current for vehicle different load conditions, simulating heat generation into the battery module for different discharging currents, and developing a battery cooling thermal management system to maintain the battery temperature in the range of 25 to 35 °C are all part of this study's goal to develop a green transportation system for energy savings and an emission-free environment. In this chapter, public perception and attitude have been thoroughly discussed.

The authors Rahman et al. [2] and Rafia et al. [3] used a personal survey to gauge respondents' awareness, concern, and perception of green transportation. August 2013 is when the survey is carried out. Selayang, Cheras, Ampang, Taman Jaya, and other residential districts of Kuala Lumpur comprise the 200-person random sample. Investigating the rationale behind the growth of green transportation is the goal of this survey as, should the government wish to carry out any development initiative, it is critical to understand public knowledge, awareness, and perception. Participants in the study are asked if they would prefer to buy green, eco-friendly transportation. 54% of respondents said they would genuinely like to buy environmentally friendly green transportation, according to the survey's results. This is a positive approach to Malaysia's electric vehicle (EV) development.

A question concerning their concerns regarding green transportation is also posed to study participants. The findings of the survey are shown in Fig. 2.2. It shows that 36% of respondents think that battery range is the biggest issue, followed by high cost (23.6%). Reliability, safety, performance, and charging infrastructure are the top concerns for 12.2%, 10%, and 8.1% of respondents, respectively. The study asks participants if they are aware of green transportation. According to Figs. 2.3 and 2.4, the majority of respondents (42.1%) are familiar with plug-in hybrid electric vehicles (PHEVs), whereas 40.5 and 17% of respondents, respectively, are familiar with hybrid electric vehicles (HEVs) and electric cars (EVs).

Questions concerning the respondents' opinions of EV sustainability are posed. 38.4% of respondents "somewhat agree," according to the study's findings, that EVs are more environmentally friendly than conventional gasoline-powered cars and other options. Additionally, it shows that, when compared to traditional modes

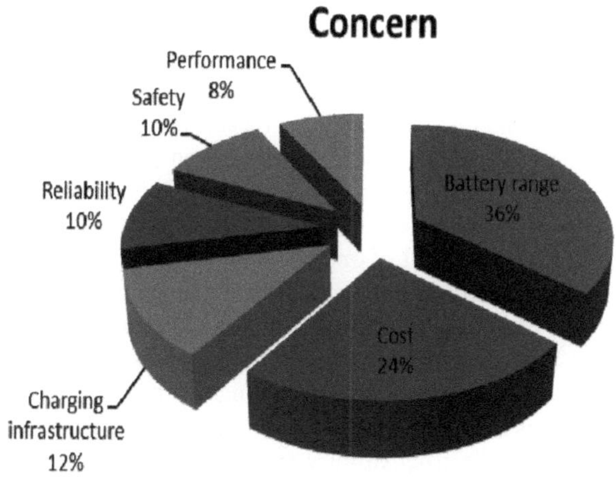

Fig. 2.2 Concern about green transport [2]

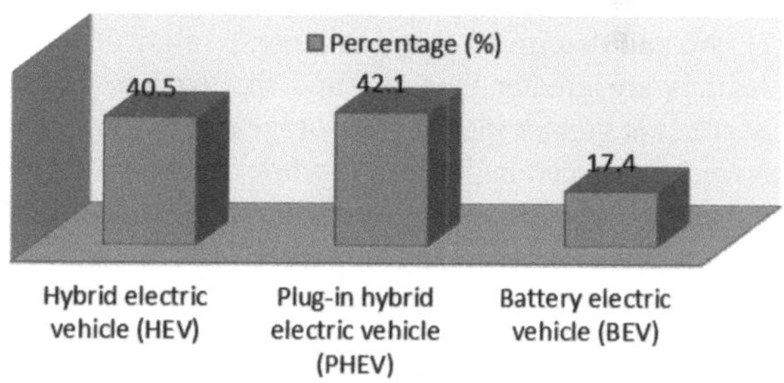

Fig. 2.3 Awareness about green transport [3]

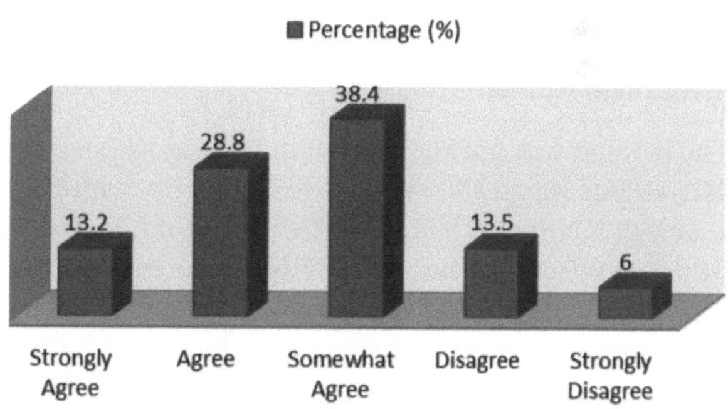

Fig. 2.4 Perception of the respondents toward sustainability of green transport [3]

of transportation, EVs are more environmentally benign, with 28.8% of respondents agreeing and 13.2% strongly agreeing.

Confirmatory Factor Analysis (CFA)

The goal of a measurement model, according to Kline [4], is to indicate if the observable signs that reflect a latent variable are suitable as a measuring tool. Estimating the correlation between the observed and underlying latent variables is the aim of a measurement model [5]. Four fit indices are used to assess a measurement model's adequacy: the comparative fit index (CFI), root mean square approximation (RMSEA), normed chi-square, and chi-square statistic. General recommendations recommend cut-off values for these indices in order to get a satisfactory model fit: While CFI values should be greater than 0.9, the normative chi-squared and RMSEA values should be fewer than 5 and 0.088, respectively [6].

Prior to testing the structural equation model, CFA was carried out on all of the measurement items at once. A standardized factor loading of 0.40 or above, ideally 0.70 or higher, offers compelling evidence of convergent validity, according to Hair et al. All of the items in this study had significant factor loadings, with the majority of them being higher than 0.60, indicating sufficient convergent validity. The CFA tests showed that all seven dimensions had a sufficient model-to-data fit: RMSEA value less than 0.080, CFI value above 0.95, and normed chi-square value below 2.41. The construct validity and reliability were also assessed by these tests.

The reliability coefficient, which shows the consistency of the entire scale or the overall dependability of the questionnaire, was determined by Cronbach's Alpha [7]. We can determine that convergent validity has been established if all indicator loadings are greater than 0.7 and the average variance extracted (AVE) for each construct is greater than 0.5, as suggested by Fornell and Larcker. All item loadings above 0.7, as seen in Table 2.1, while Table 2.2's AVE is greater than 0.5.

Additionally, it was suggested that when a construct's AVE is higher than the correlation it shares with other components in the model, sufficient discriminant validity should be proven [7]. Convergent discrimination is identified when the diagonal components (square root of the extracted variance) are larger than the off-diagonal elements (correlations among the constructs), according to a study by Chiu and Wang [8] (Table 2.2). We can infer from this study that the scales have a high degree of construct validity.

Structural Model

The p-value was not significant, indicating satisfactory model fit, and the CFA analysis's $\chi 2$/df was 1.380 ($\chi 2 = 4.139$, df $= 3$), with GFI $= 0.951$, AGFI $= 0.917$, CFI $= 0.980$, NNFI $= 0.971$, and RMSEA $= 0.048$. With p-values less than 0.1 and R2 values between 0.44 and 0.33, five of the nine routes were significant. This means that while 51% of the variance in intention to use was explained, the variance in environmental repercussions ranged from 44 to 33%. Multiple regression analysis outcomes using structural equation modeling methods. There is a significant and negative correlation ($p < 0.01$) between the environmental consequences (ECN) and the buying intention of EFCs (PIN). Having an R2 of 0.56 means that sentiment toward the environmental consequences of purchasing EFCs can explain about 56% of the variance in intention to purchase the same car. Therefore H1a is not supported, and the results contradict those of Follows and Jobber [9], who discovered a positive correlation between ecologically conscious purchase intention and environmental repercussions.

In summary, the purchase intention of EFCs (PIN) is strongly and adversely correlated with individual consequences (ICN) at $p < 0.01$. About 67% of the variation

Table 2.1 Construct validity of confirmatory factory analysis [6, 7]

Items	Stand. loadings	Reliability
***Self-transcendence value (SVN)**: (Normed* $x^2 = 1.818 = 1.818 = 1.818$, *, CFI = 0.996, RMSEA = 0.064)*		
• Helpfulness is important to me	0.77	0.87
• Forgiveness is important to me	0.81	
• Love and affection are important to me	0.75	
• Friendship and generosity are important to me	0.85	
***Conservation values (CVN)**: (Normed,* $x^2 = 1.536 = 1.536 = 1.536$, *, CFI = 0. 995, RMSEA = 0.052)*		
• Respectfulness and politeness are important to me	0.77	0.76
• Self-control such as restraint and self-discipline are important to me	0.84	
• A clean and tidy environment is important to me	0.82	
***Self-enhancement values (SVE)**: (Normed,* $x^2 = 1.46 = 1.46 = 1.46$, *, CFI = 0.984, RMSEA = 0.046)*		
• Success: For the achievement of my life, working hard and aspiring are important to me	0.56	0.62
• Self-satisfaction: An enjoyable and leisurely life is important to me	0.67	
***Environmental consequences (ECN)**: Normed,* $x^2 = 1.19\ \chi^2 = 1.19\chi^2 = 1.19$, *, CFI = 0.999, RMSEA = 0.030)*		
• Conventional cars can create air pollution	0.77	0.82
• Conventional car can create smog in large cities	0.84	
• Conventional car produce greenhouse gases such CO_2, N_2O that contribute to global warming and climate change	0.69	
• Conventional cars affect water quality because oil and particles get washed into lakes and rivers	0.59	
***Individual consequences (ICN)**: Normed* $x^2 = 2.246\ x^2 = 2.246\ x^2 = 2.246$, *CFI = 0.991, RMSEA = 0.079)*		
• Environmentally friendly car reduces CO_2 emission	0.76	0.81
• Environmentally friendly car can decrease the use of petroleum	0.76	
• Environmentally friendly car can reduce greenhouse gas emission	0.85	
• Environmentally friendly car is comfortable to use	0.52	
***Purchase intention (PIN)**: Normed* $x^2 = 1.335x^2 = 1.335$, *, CFI = 0.999, RMSEA = 0.041)*		
• I would buy an environmentally friendly car if it's quality is lower than a conventional car	0.77	0.88
• I would buy an environmentally friendly car even if it's performance is lower than a conventional car	.85	
• I would buy an environmentally friendly car even if it has a less appealing design	.89	

Table 2.2 Discriminate validity coefficients

	AVE	CVN	SEVN	SVN	ICN	ECN	PIN
CVN	. 0.56	**0.754**					
SEVN	0.54	0.287	**0.735**				
SVN	0.68	0.527	0.354	**0.827**			
ICN	0.61	0.232	0.305	0.311	**0.782**		
ECN	0.52	0.271	0.419	0.374	0.225	**0.722**	
PIN	0.57	0.135	0.071	0.185	0.533	0.056	**0.761**

Note: Diagonal elements in bold are the square root of the average variance extracted (AVE) Off-diagonal elements are the correlations among the constructs. For discriminate validity, diagonal elements should be larger than off-diagonal elements[24]

in the intention to buy the EFCs can be explained by individual effects, according to an R2 of 0.67. H1b is therefore supported. This result is in line with that of Follows and Jobber [9], who discovered a negative relationship between purchase intention and individual consequences. The environmental repercussions (ECN) of buying EFCs were found to be positively and significantly correlated with self-transcendence values (SVN).SVN and ECN's positive relationship validated H2a's positive relationship hypothesis. This aligns with a previous discovery [1] that was significant at $p < 0.01$. A significant and positive correlation was observed between self-transcendence values (SVN) and the individual consequences (ICN) of buying a "green" car ($p < 0.05$). The positive correlation between SVN and ICN refuted the negative association hypothesized by H2b. This contradicts a previous discovery [8] that was significant at $p < 0.05$. The association between conservation values (CVN) and ECN and ICN is positive but negligible. Since it was assumed that there would be a negative link, H3was not supported. Given the significant correlation between CVN and ICN, H3b was supported.

Self-enhancement values (SEVN) were found to be adversely related to PIN ($p < 0.01$) and positively and significantly related to ECN and ICN ($p < 0.01$); as a result, H4b and H4c were supported. Since it was assumed that there was a negative link, H4a was not encouraged.

2.4 Policy on the Adoption of EV

A vision statement and a set of goals are usually the first steps in policy efforts to encourage the deployment of EVs. Adoption of electric vehicles and charging standards is a first step. In addition to stimulating demand and encouraging automakers to expand the number of EVs available on the market, procurement programs also serve as a catalyst for the early rollout of publicly accessible charging infrastructure. Offering financial incentives is another beneficial policy tool, especially to close the price difference between EVs and less costly internal combustion engine (ICE) vehicles and to encourage the early installation of charging infrastructure.

Because EVs perform better in terms of local air pollution, economic incentives are frequently combined with other policy measures that improve the value proposition of EVs (such as reduced tolls or parking costs, or waivers of access restrictions). The implementation of fuel economy requirements, zero-emission vehicle mandates, and increased ambition in public procurement programs are measures that offer vital incentives to increase the availability of vehicles with low and zero tailpipe emissions. Minimum standards to guarantee "EV readiness" in newly constructed or renovated buildings and parking lots, the installation of publicly accessible chargers in urban areas and on highway networks, and requirements for interoperability and minimum availability levels for publicly accessible charging are all examples of regulatory actions pertaining to charging infrastructure.

As the markets and infrastructure develop, certain policy measures may need to be modified as the deployment of EVs and charging infrastructure changes, as has only been seen in Norway thus far. The adjustment of gasoline and car taxes and their impact on government revenue are two examples.

Key Policy Developments in 2018/19:

- European Union: A number of important policy tools were authorized in the European Union. The Clean Vehicles Directive, which allows for the public purchase of electric buses, and fuel economy requirements for cars and trucks are two examples. For new and refurbished buildings, the Energy Performance Buildings Directive establishes minimum standards for charging infrastructure. In several European countries, incentives are popular to encourage the deployment of EVs and chargers.
- China: China's policy changes include limiting investment in new ICE vehicle factories and proposing to increase the average fuel economy of the country's fleet of passenger light-duty vehicles (PLDVs) by 2025 (updating the 2015 limitations). Employing distinct incentives for automobiles according to their battery properties (such as zero-emission car credits and subsidies under the New Energy car mandate).
- Japan: By 2050, Japan's automotive strategy, which involves collaboration among industrial stakeholders, intends to use a mix of fuel cell electric vehicles (FCEVs), BEVs, PHEVs, and hybrid electric vehicles (HEVs) to reduce greenhouse gas (GHG) emissions from domestic automakers by 80% (90%) including exported vehicles. There was an announcement of an update to the fuel economy standards for vehicles and a revision to the rules for trucks.
- Canada: Canada presented a plan for future EV adoption that included quite aggressive measures in some provinces, such Quebec's obligation to use zero-emission cars (ZEVs), which is comparable to California's. The most strict ZEV mandate in the world, 30% ZEV sales by 2030 and 100% by 2040, was declared by British Columbia. This puts Canada in a comparable situation to the ten US states that have enacted ZEV laws.

2.5 Advanced Technology: Mitigate the Challenges

Typically, cars sit still 95% of the time. As illustrated in Fig. 2.5, smart and flexible charging technology takes advantage of unused car battery power to supply more electricity to the grid during times of high demand or, in certain situations, just automatically pauses or lowers charging power. On the other hand, it enables users to charge during off-peak hours for a price that is at least one-third that of charging during peak hours, which lowers consumer costs and eases grid congestion during peak hours. In order to better predict unexpected spikes in electricity demand, the charging system enables EV owners to schedule charging according to power limits, price, and priority. Additionally, unused power can be sold back to the grid.

Additionally, the technology allows the grid to expand its capacity, meet the growing demand from electric vehicles at a lower cost to consumers, lessen the strain on the grid system, and prevent energy prices.

Charging Infrastructure in California

The California Energy Commission is making investments in the technologies and charging infrastructure through the Clean Transportation Program, which is accelerating the state's shift to clean, zero-emission electric vehicles. These investments are essential to achieving the state's target of having 1.5 million zero-emission vehicles on California roads. Strategic regional planning to encourage the use of these vehicles is also being supported by the Energy Commission.

The Energy Commission is currently constructing a network of interconnected, easily accessible direct current fast chargers to enable electric car drivers to freely move about the state. This is the biggest network of charging stations in the US that are open to the public.

This network enables battery electric vehicles to be charged quickly on all of the Golden State's major highways and thoroughfares. In the shortest amount of time, electric vehicle drivers can travel throughout California and conveniently charge their vehicles by focusing on transit routes.

Fig. 2.5 A fast charging point [9]

2.6 Advanced Zero-Carbon Fast Charging System

As seen in Fig. 2.6, an ultrafast charging station is being developed for the zero-carbon IIUM electric vehicle as well as any other electric vehicle. A 96-kWh battery should be charged for 15 min to reach 60% capacity and 30 min at most to reach 100% capacity. Lead-gel electrochemistry batteries with a nominal voltage of 150 V and 300Ah will power the station. One unit of a 5 kW horizontal axis wind turbine with an inverter set on an 18 m tower and 25 kW solar photovoltaic panels with a 100 A inverter will provide the energy needed to charge the batteries.

Any electric vehicle, including e-motorcycles, needs energy for speedy battery charging, which the inverters will supply. There won't be any carbon emissions at all. Tenaga Nasional Sdn Bhd is to receive the excess electricity from the ZCUCS.

Energy Source: Renewable Energy

Future energy must be accessible, safe, and equitable, and the World Economic Forum is dedicated to helping keep global warming to less than 1.5 °C over pre-industrial levels. In Palm Spring, Indio, a large wind turbine is used to produce megawatts of electricity (Fig. 2.7).

Energy Efficiency Building

When comparing California's 2016 Building Energy Efficiency Standards to worldwide standards established in 2015, it was discovered that the state's requirements saved 29% more energy annually than the international standards. The annual electricity savings from California's rules would power 300,000 more households. An illustration of solar panels on a building is shown in Fig. 2.8.

Battery Thermal Management System

Figure 2.9 illustrates the development of an evaporative cooling battery thermal management system (EC-BThMS) for the Proton Saga EV's battery pack. Based

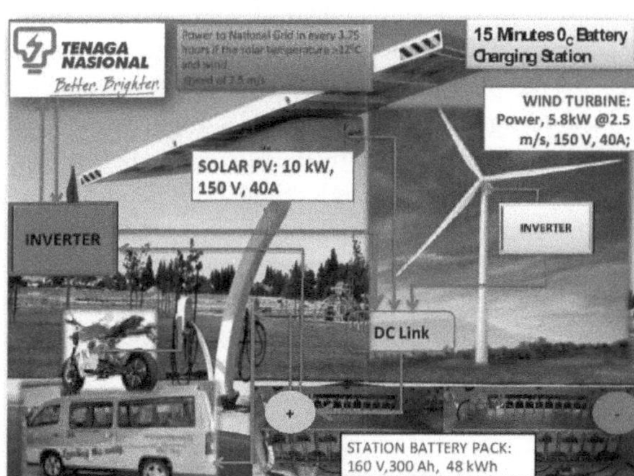

Fig. 2.6 Renewable energy powered ultrafast DC charging station (ZCUCS) for EV [10]

Fig. 2.7 Palm Spring, Indio, Dessrt, California, USA for Mwatt power [9]

Fig. 2.8 Energy from energy efficient building [11]

on the "Refrigerant Cycle" concept, this cooling system was created. An expansion valve, compressor, condenser, independent cooling ducts, and a set of thermocouples are all features of the EC-BThMS. For the EC-BThMS, the evaporator consists of 14 cooling ducts. It has been assumed that the cooling ducts will develop as follows: (i) the flow is homogeneous, two-phase (saturated liquid and vapor), one-dimensional steady state; (ii) the duct cross-sectional area is constant and comparable to all ducts; and (iii) the pressure at each inlet and outlet is the same.

According to the EC-BThMS's basic idea, heat produced within the battery modules will be absorbed by the refrigerant before being released into the surrounding air. Copper's high heat conductivity of 386 W/m0K led to its selection as the material. Each cooling duct was positioned between the battery modules, sandwiched side by side, and quite close to the modules (for thermal contact). Additionally, every evaporative duct has an intake port that permits low-pressure refrigerant (80% saturated liquid and 20% vaporized) to enter the ducts and exit as low-pressure refrigerant (roughly evaporated) via outlet pipes. The EC-BThMS that was created is displayed in Fig. 2.9.

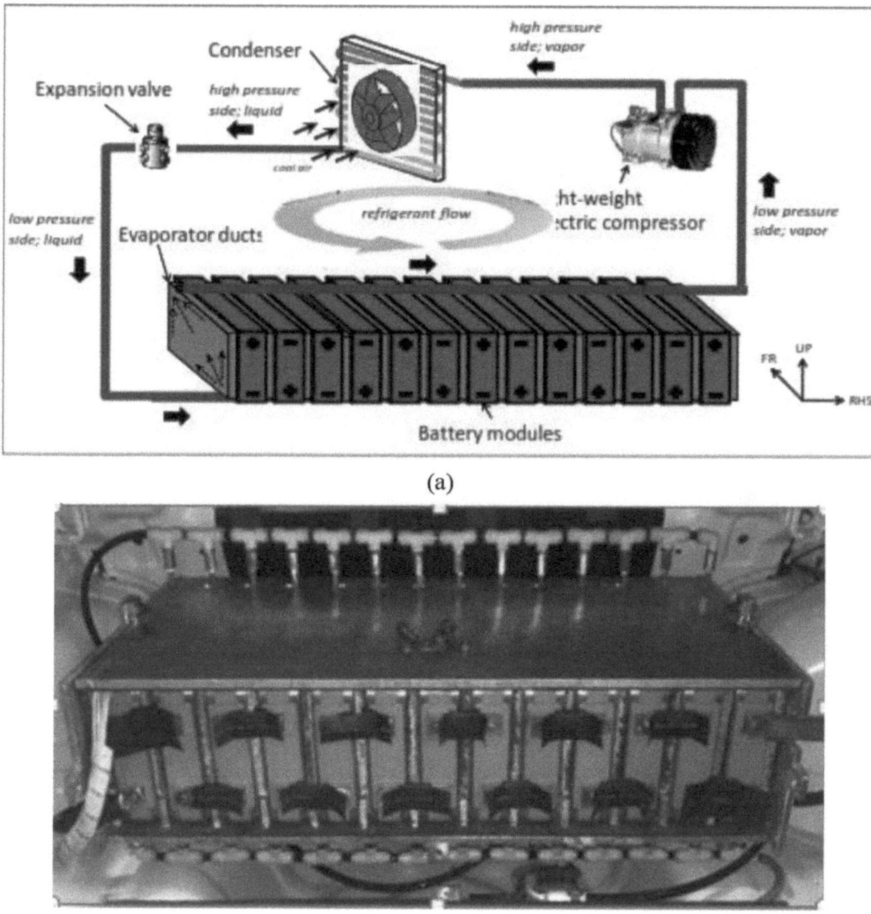

(a)

(b)

Fig. 2.9 EC-BThMS [12]

Cooling has been accomplished using the R-134a refrigerant. The EC-BThMS is an energy-intensive system since it requires 1.6 kW of power to run the electro compressor continuously, which keeps the battery temperature between 25 and 40 °C.

Battery Monitoring and Recycling

AI-enabled battery analytics and monitoring for electric cars and fixed storage allow for usage optimization and predictive maintenance, which can prolong battery life and lessen supply chain strain and the need for new batteries. Lithium-Ion battery recycling can be made easier and more efficient by using the data to inform better decisions about when to recycle or repurpose batteries and to identify broken individual cells rather than recycling the full battery pack.

2.7 Conclusion

EV sustainability is viewed favorably by the general people. According to a survey, 13.2% of participants strongly agree and 28.8% of respondents believe that EVs are more environmentally friendly vehicles than traditional modes of transportation. On the other hand, it is totally dependent on energy security, battery affordability, and sustainability according to its state-of-health (SoH).

Acknowledgements For funding this work, the authors are thankful to the Malaysian Ministry of Higher Education and the Research Management Centre (RMC) of the International Islamic University of Malaysia. The authors would like to express their gratitude to Eastern Michigan University for their ongoing cooperation in implementing this study.

References

1. Electric Vehicle Charging Station (2022). https://www.alliedmarketresearch.com/electric-vehicle-charging-station-market-A17391, Retrieved date: October 2022.
2. Rahman, A., Afroz, R., & Alam, Z. (2014). Development of electric vehicle: Public perception and attitude, the Malaysian approach'. *World Review of Intermodal Transportation Research, 5*(2), 149–167.
3. Rafia, A., Rahman, A. & Jarita, D. (2013). Malaysian perception and attitude towards electric vehicle. In *2nd Int Conf Mech Automot Aerosp Eng (ICMAAE 2013).*
4. Kline, R. B. (2010). *Principles and practice of structural equation modeling* (3rd ed.). The Guilford Press.
5. Hair, J. H. B. C. W., Babin, B. J. & Anderson, R. E. (2010). *Multivariate Data Analysis: A Global Perspective.* Upper Saddle River, NJ: Pearson.
6. Byrne, B. M. (2010). *Structural equation modeling with AMOS: Basic concepts, applications, and programming* (2nd ed.). Routledge Academy.
7. Fornell, C., & Larcker, D. (1981). Evaluating structural equation models with unobservable variables and measurement error. *Journal of Marketing Research, 1*, 39–50.
8. Chiu, C. M., & Wang, E. T. G. (2008). Understanding Web-based 242 learning continuance the role of subjective task value. *Information Management, 45*(194–201), 25.
9. Evgo_Rivian (2023). https://chargedevs.com/newswire/evgo-adds-rivian-to-its-autocharge-seamless-charging-system/. Retrieved date: October, 2023.
10. Rahman, A., Sany, I., & Waqar, A. (2023). *Fast charging using the renewable energy for EVs: A flagship project of international Islamic University Malaysia.* Office of Strategy Institutional Change (OSIC), IIUM.
11. California Building Energy Standard (2022). https://www.energy.ca.gov/programs-and-topics/programs/building-energy-efficiency-standards.
12. Rahman, A., Farhana, S., & Hawlader, M. N. A. (2015). Two phase fuzzy controlled evaporative thermal management system for EVs/HEVs. *International Journal of Automotive Technology, 28*(4), 297–304.

Chapter 3
Fuel Cell Hybrid AI-Driveline

Abstract Hybrid electric fuel cell (FC) all-terrain vehicle prototype has been created for peat soft terrain with usual ground pressures between 7 and 8 kN/m2. For the best fuel efficiency and fuel cell longevity, the vehicle's power system uses a 60 V, 3.6 kW FC with 500 g of hydrogen and a 60 V, 5.2kWh LiFe3PO4 battery coupled in parallel with the DC/DC converter. Power is switched from FC to battery for traction and from battery to FC to maintain high vehicle speed using an adaptive neuro-fuzzy intelligence system (ANFIS). This vehicle can go 31 km with its FC hybrid driveline.

Keywords Fuel cell hybrid power system · ANFIS · All terrain EV · Vehicle mobility

3.1 Introduction

Globally, transportation is responsible for around 40% of all greenhouse gas (GHG) emissions. Making the switch to cleaner Fuel Cell Electric Vehicles (FCEVs) is a long-term, low-carbon mobility solution that aims for a zero-carbon future and high capacity. In terms of climate change and air quality, FCEVs benefit society. When compared to conventional gasoline-powered vehicles, these hydrogen-powered vehicles emit 30–60% fewer greenhouse gases and no harmful air pollutants. For zero-carbon university campus transportation, an electric shuttle bus (ESB) powered by a fuel cell hybrid is one of the most promising options. This study will look into several topics, including stakeholder engagement procedures, the design of an autonomous fuel cell hybrid driveline for the university ESB, infrastructure requirements for fast battery charging and hydrogen tank refueling, and mechanisms for monitoring and assessing the university's carbon footprint and community health.

Mobility and transportation were unanimously acknowledged as being essential to sustainable development at the United Nations (UN) Conference on Sustainable Development (Rio + 20). Accessibility and economic growth can both be improved by sustainable mobility, which also improves economic integration while protecting the environment, enhancing health, lowering greenhouse gas emissions,

A. Rahman et al., *Autonomous Driveline System for Electric Vehicle*, Green Energy and Technology, https://doi.org/10.1007/978-981-96-8418-2_3

and promoting social justice. In 2016, Malaysia produced 6.93 tons of CO_2 per person, out of 216.8 million tons overall [1]. Compared to the global average of 24%, the transportation sector's total weight is 62.8 million tons, or 29.05 percent. According to the Malaysian government, a 20% renewable energy target will be met by 2025. Malaysia will see a notable increase in renewable energy starting in 2025 [2], with hydrogen as an alternative and a 5.6% CAGR.

Action against climate change and its effects is therefore urgent. Vehicles with zero emissions, as batteries-EV and H2-FCEV, are crucial in reaching the goal [3]. In order to fight climate change and establish Malaysia as a hub for advanced automotive technology by 2030, the National Automotive Policy (Malaysia) 2020 calls for the introduction of next-generation vehicles, such as battery-powered and hydrogen fuel cell vehicles.

The hydrogen fuel cell is a crucial technique for the development of a hydrogen economy and sustainable mobility, and it is the greatest choice for a sustainable energy conversion system. Recently, hydrogen fuel cell cars (FCVs) have become a viable alternative to battery electric vehicles (EVs) with zero tailpipe emissions. Similar to battery vehicles, FCVs have no emissions from their tailpipes (except from water vapor), and when combined with renewable energy sources, they can potentially have almost zero emissions throughout their entire fuel cycle. Being the lightest element in the universe, hydrogen has a number of inherent qualities that make it a desirable energy carrier for transportation. It may be made in huge quantities from a variety of main energy sources and has a high energy density by weight [4].

Moreover, hydrogen may be refueled at rates similar to those of gasoline, unlike battery recharging. Governments and the automotive industry are very interested in FCVs because of these benefits. Low-temperature fuel cells frequently use platinum (Pt) as a cathode and anode catalyst. Significant obstacles to the broad adoption of Pt-based fuel cells are the price of Pt and its limited global supply. Furthermore, because of the decreased activity of the oxygen reduction reaction (ORR), platinum employed as an anode material is easily poisoned, fermented (producing alcohol), and releases carbon monoxide (CO) [5–7]. As a result, the average efficiency of hydrogen fuel cells is 30%.

According to the 2018 Global Automotive Executive Survey, which surveyed over 2,100 global consumers and nearly 1,000 senior executives from the top automakers, 78% of executives believe that fuel cell hydrogen vehicles will be the power source for electric and driverless vehicles in the future.

3.2 Fuel Cell Hybrid EV

According to the National Automotive Policy (NAP) 2019 and Industry Revolution (IR) 4.0, the project's objective is to develop a comprehensive electric vehicle technology with an emphasis on the entire energy storage system needed for an FCHEV to go 500 km. The emphasis should be on a technological innovation that optimizes the overall energy storage system design with a capacity of roughly 80 kWh based on IEA guidelines, enabling FCHEV to reach a range of up to 600 km [8]. The 33 kWh LiFePO4 battery pack and the 47 kWh fuel cell will make up the FCHEV's hybrid energy system.

Designing a power assist autonomous programmable controller for the car will enable power assistance for the FC during both cruising and acceleration. The FC pack will have a 53 kW capacity to push the vehicle around 400 km while maintaining the mean power consumption, while the battery pack is designed to provide the propulsion power for the FCHEVs for roughly 220 km without the assistance of FC. Interestingly, FCHEV's power system would also be capable of supplying electricity to the vehicle in three distinct modes: BEV, FCEV, and FCHEV. It is designed such that the other system will take over the vehicle's propulsion in the event that one of the systems fails.

By 2022, the IIUM Research Group hopes to have a modified FCHEV prototype with a 600 km range for Malaysia's future mobility as a next-generation fuel thanks to financial support from Green Tech Malaysia. In fact, IIUM has adapted two EVs: the IIUM electric coaster from the Mercedes Benz D140 in 2019 and the IIUM EV from the PROTON saga FXL in 2012. In the Proton Green Mobility Challenges 2012, Team IIUM won the overall title and took first place in two events: the two fastest laps (11.4 km) and the farthest distance (69 km) on the F1 Sepang circuit using an 8.6 kwh battery. The IIUM EV was modified for the competition. For IIUM's flagship project, IIUM Electric Coaster Innovation, a retrofitted electric coaster was created. The AC induction motor, Curtis controller, battery pack with R143a cooling system, and 2-speed gearbox with stainless steel motor attached adapter make up the highly energy-efficient power train for both projects, which may be regarded as IIUM's contribution to Green Tech Malaysia.

3.3 FC Hybrid EV Power Characterization

The hydrogen FC hybrid air-cushion tracked vehicle's powertrain configuration is depicted in Fig. 3.1. The powertrain's main parts include a 3.6 kW continuous power fuel cell with a 200 V terminal voltage, a lithium-ion battery with a 200 V terminal voltage and 220 Ah capacity, a power switching microcontroller, a DC to AC converter, and two parallel 470 micro-Farad capacitors. When the vehicle needs to be accelerated, the battery provides current to the traction motor at a steady rate of 70 A and a maximum of 440 A. In contrast, the fuel cell provides current to the traction motor at a maximum of 110 A as long hydrogen supply for the vehicle to operate in the speed range of 7–15 km/h. To regulate the FC hybrid power switching mechanism, an ANFIS controller is employed. If the car needs to satisfy its power requirement, the ANFIS controller switches the power switching from FC to battery.

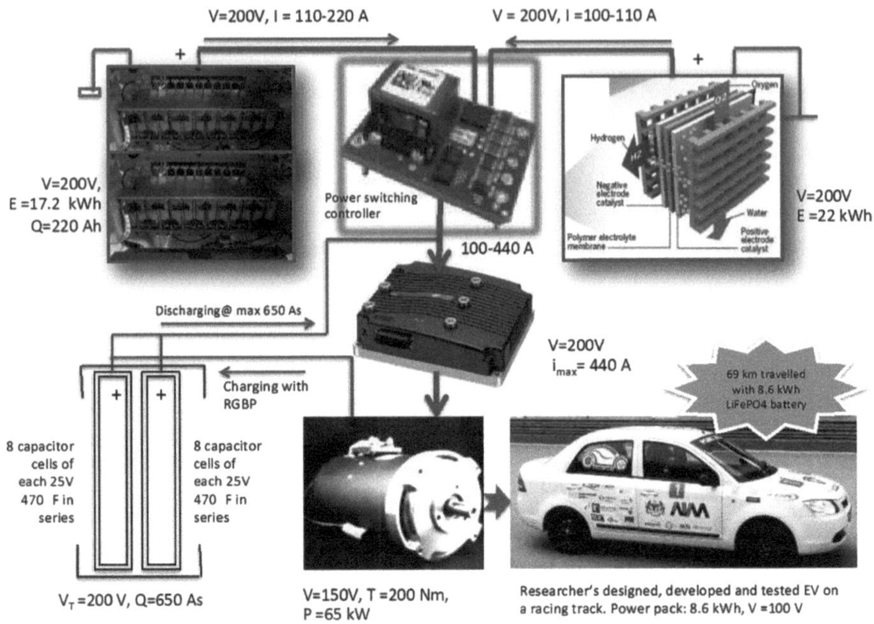

Fig. 3.1 FC hybrid power system for EV

Power Assist Switching Control Strategy

Two 60L hydrogen fuel cells and a 33 kWh lithium ion battery make up the EV storage system. The voltage of 200 V is the same for both. In order to satisfy the specified load demand, the vehicle's primary energy source, the fuel cell, is connected to a unidirectional DC/DC converter and transfers energy through the DC bus. In the meantime, the battery acts as an additional energy source to help the system satisfy peak power demands (such as during acceleration) and power demands when the hydrogen pressure is less than 1 atm. Furthermore, the rechargeable battery is linked to a bidirectional DC/DC converter in order to recover energy through regenerative braking.

As seen in Fig. 3.2, an intelligent switching controller with an optimization method is used to regulate the power supply and guarantee that the output voltage demand is met with the necessary current quantity. The switch controller tracks a number of variables, such as the fuel cell's hydrogen pressure, the lithium ion battery's state of charge, and driving conditions including the vehicle's speed and the slope of the road. There are two modes in the switching network:

- Mode 1: When the hydrogen pressure is greater than 1 atm and the fuel cell's voltage is 150 V, the fuel cell is utilized to meet the load requirement.
- Mode 2: When the hydrogen pressure is less than 1 atm and the battery state of charge (SOC) is greater than 55%, the battery is used to provide the load demand.

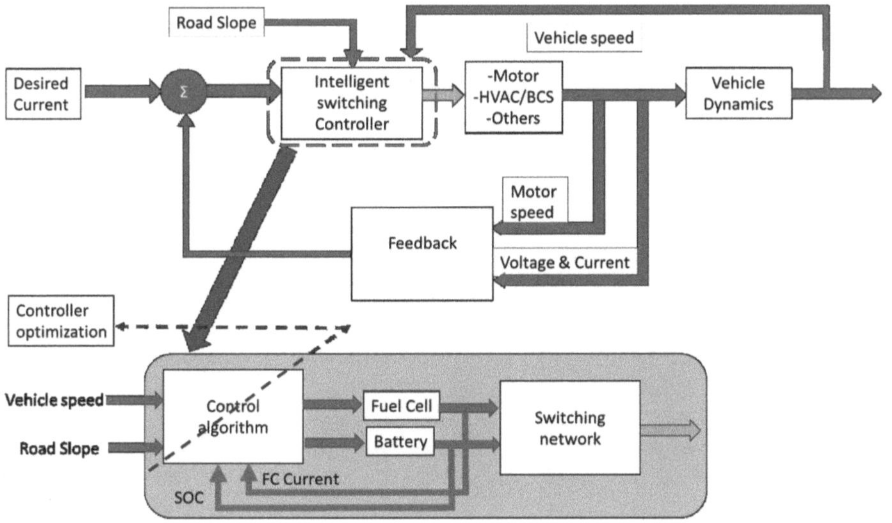

Fig. 3.2 FC and battery power switching system

- Mode 3: When the vehicle's speed and road slope suggest peak power need, such accelerating, both the fuel cell and the battery are employed.

Chemical Reaction at FC: Generate Electricity

The PdLaCoO3_CFC electrode, the mass flow rate of hydrogen and oxygen, and the thickness and size of the polymer PEM (electrolyte film) will all be used in the fuel cell modeling process, as seen in Fig. 3.3. The relationship between open-circuit voltage (Voc) and current density (Jsc) from simulations using the fundamental phenomenological equations will be used to characterize the steady-state electrical performance of FC. The Stefan–Maxwell equation is used to transfer the hydrogen and oxygen phases, while the Nernst–Planck equation describes the transfer of substances.

3.4 Mathematical Modeling for FC Hybrid EV

The FC's electrical activity is focused on the transformation of chemical energy contained in reactants to electrical energy. The amount of energy released as a result of the reaction is

$$2H_2 + O_2 \rightarrow 2H_20 + Heat \tag{3.1}$$

The available energy of a chemical reaction that can be used to do work is called Gibb's free energy (ΔG). Changes in Gibbs free energy can also be expressed this

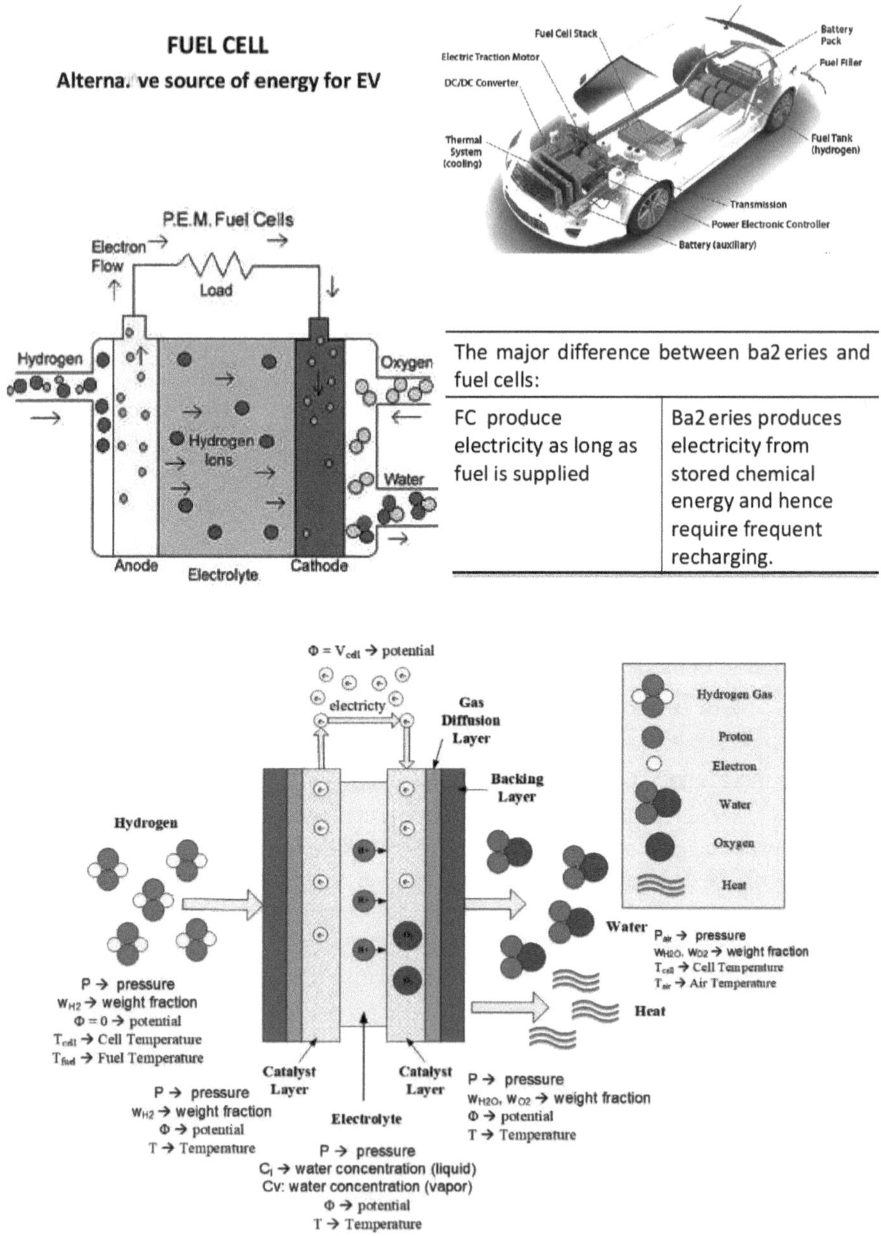

Fig. 3.3 Model of PEM_FC

form:

$$\Delta G = G_{Products} - G_{Reactants} \tag{3.2}$$

$$\Delta G = \Delta H - \Delta TS \tag{3.3}$$

where G is the Gibbs free energy in J, S entropy J/K, H enthalpy ($\Delta H = mc_p \Delta T$, m is mass, g, c_p specific heat, J/g*C), and T temperature °C, J (At ambient conditions (Pressure 1 bar), water boils at 100 °C, and the specific enthalpy of water is 418 kJ/Kg, and T temperature K. The enthalpy of water at 100 °C is 40.7 kJ/mol and entropy 109.1 J/(mol*K) (entropy, $\Delta S = \frac{\Delta T}{T}$). $Entropy\ of\ hydrogen, S_{1\,bar} = 130.68^{J}/_{mol\,*\,K}$

By considering the independent activity (*a*) of hydrogen (H_2), oxygen ($O^{1/2}$) and water (H_2O), The changes of Gibbs free energy per mole can be estimated as.

$$\Delta G = \Delta G^0 - RT\ ln\left[\frac{a_{H2}a_{O2}^{\frac{1}{2}}}{a_{H_2O}}\right] \tag{3.4}$$

$with\ \alpha = \frac{P_I}{P_O}$, where, G° is the Gibbs free energy reference values 8.6 kJ/mol at 289 K and 1 atm, R gas constant (R = 8.3145 J_{SC}/mol*K), *a* is the activity, P_i partial pressure in P_a, P_O standard pressure P_a (p_O = 1 atm = 101.3 kPa), $^a{}_{H2}$, $^a{}_{O2}$, and $^a{}_{H2O}$ are the activity of H_2, O_2, and H_2O, respectively.

$$E_{Nernst} = E^0 - \frac{RT}{nF} ln\left[\frac{a_{H2}a_{O_2}^{\frac{1}{2}}}{a_{H_2O}}\right] \tag{3.5}$$

$$with\ E = E_0 - (T - T_0)\left(\frac{\Delta S}{nF}\right)$$

where n is the number of electrons moving (n = 2), *F* is the Faraday's contact (96,485 C/mol), E° is the Nernst reference potential, V, S is the entropy reference (known values) J/K, E_0 reference potential of FC (1.22 V) at steady state condition V, and T_0 temperature at room temperature (T_0 = 298.15 K).

Nernst Voltage

Nernst voltage can be estimated with considering the effect of temperature and pressure of the hydrogen, oxygen, and water, using the following equation as

$$E_{nernst} = \begin{vmatrix} E0 + (T - 298)\frac{-44.43}{nF} + \frac{RT}{nF}ln\left(P_{H_2O}P_{O_2}^{1/2}\right) & when\,T \leq 100\,°\mathrm{C} \\ 1.229 + (T - 298)\frac{-44.43}{nF} + \frac{RT}{nF}ln\left(P_{H_2O}P_{O_2}^{1/2}\right) & when\,T > 100\,°\mathrm{C} \end{vmatrix} \tag{3.6}$$

where P_{H2O} and P_{O2} are the partial pressure of water vapor inside the stack (atm), ω percentage of water vapor in the oxidant (%).

Overvoltage Activation

The instantaneous current density of the FC can be estimated using the following equation:

$$J_{sc} = J_0\left[e^{\frac{-\alpha nF\eta_{act}}{RT}} - e^{\frac{(1-\alpha)nF\eta_{act}}{RT}}\right] \tag{3.7}$$

with

$$V_{act} = -\frac{RT}{\alpha nF}ln\left[\frac{J}{J_0}\right]$$

where J_{sc} is density of current, A/cm^2, J_0 is exchange current density A/cm^2 at no load, h overvoltage, and α transfer coefficient.

Overvoltage Resistive

$$V_{Ohm} = -J_0\left[e^{\frac{-\alpha nF\eta}{RT}} - e^{\frac{(1-\alpha)nF\eta}{RT}}\right]R_a \tag{3.8}$$

where V_{ohm} is the overvoltage resistive, V, R_0 specific resistance per area, ohm/cm^2,

Overvoltage Concentration

$$V_{con} = -\frac{RT}{nF}ln\left(1 - \frac{J}{J_0}\right) \tag{3.9}$$

Mass flow rate of hydrogen required to get the 70 A current with 100 cells can be estimated with considering the heating effect of hydrogen/min as

$$N_{cell} = \frac{nF * \dot{m}_{h20}}{\Delta fH^0(H_2O\_gas) * \eta_{nom}}, \quad Hydrogen\ conversion\ efficiency,\ \eta_{nom} = 55\% \tag{3.10}$$

where ΔfH^0 $(H_2O(gas))$ is the change of enthalpy of water gas = $241.8^3 \times 10^3$ J/mol.

3.5 FC Performance: Theoritically

Table 3.1 shows the sample calculation of FC performance inline with the vehicle power requirement.

The PdLaCo2O3 catalyst electrode model for PEM_FC STACK, which can produce 1.0 A and a nominal voltage of 1.0 V, is depicted in Fig. 3.4. To evaluate the effectiveness of PdLaCoO3 catalyst electrodes with PEM, a basic structural PdLaCo2O3 PEM_FC STACK will be created.

Table 3.1 Sample calculation for the simulation

Pd/Co fuel cell design parameters		
Mass flow rate of hydrogen	Model Equations: Estimation of hydrogen flow rate, g/s	
	$\dot{m}_{H2} = \frac{1}{nF} * N_{cell} * M_{H2}$ Values considered: Faraday's constant, F = 96,500 [C/mol] N_{cell} = Number of cells = 100 cells M_{H2} = hydrogen molar weight = 2 [g/mol] *x* of opening the hydrogen tank valve: %	If x = 20% Mass flow rate of hydrogen will be 0.021 g/s
Mass flow rate of hydrogen	$\dot{m}_{O2} = \frac{y}{nF} * N_{cell} * M_{O2}$ Value considered: M oxygen molar weight = 32 [g/mol] y = opening of oxygen compressor valve	If y = 20%, Mass flow rate of oxygen will be 0.06 g/s
Vehicle power required for the operational speed of 50–60 km/h		
Traction torque	$T_w = T_m * N_g * N_d * \eta_m$ Values considered: T_m = motor torque = 60 Nm N_g = gear ratio = 1.0 N_d = differential ratio = 3.5 η_L: motor efficiency = 85%	T_W = 178.58 N.m
Traction power required for vehicle speed 60 km/h	$P = \frac{p_w}{\eta_m} = \frac{T_w * \omega}{\eta_m} = (VI)_{FC}$ Value considered: $V_{FC\,(max)}$ = 100 V and $I_{FC(max)}$ = 72 A Wheel radius, r_w = 0.35 m V_{Speed} = 90 km/h or $\omega = \frac{V_{Speed}}{r_w} = \frac{90}{3.6 * 0.35} = 71.4\ rad/s$	$P_w = 15.293\ kW$
FC current required to meet vehicle load demand	$I_{FC} = \begin{cases} 82.66A & if\ V_{FC} = 100V;\ 0.15 > \dot{m}_{H2} \leq 0.18\ and\ T_{FC} \cong 65\,°C \\ 68.88 & if\ V_{fFC} = 120\,V;\ 0.12 > \dot{m}_{H2} \leq 0.13\ and\ T_{FC} = 65\,°C \end{cases}$	

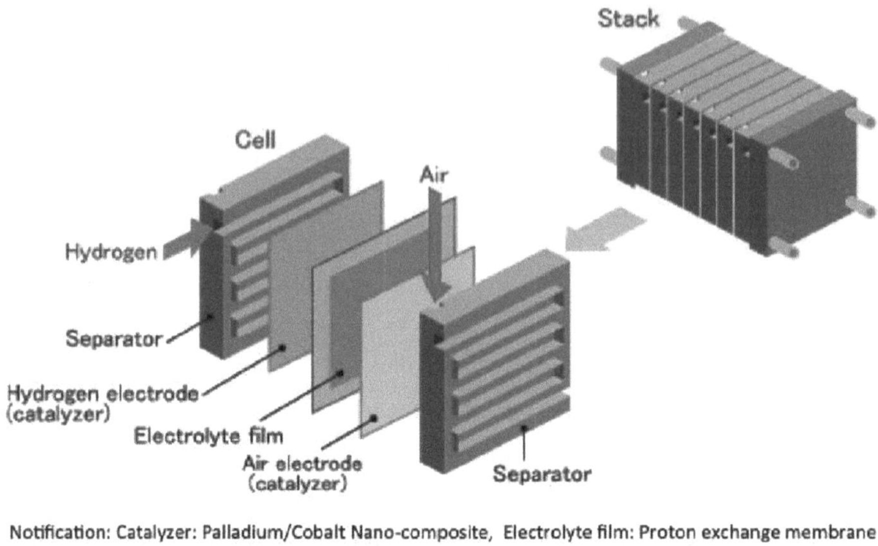

Fig. 3.4 A simple $PdLaCoO_3$ PEM_FC STACK model

Figure 3.5 shows the battery pack of 100 V, 130 Ah, and 13 kWh, design for a test EV with a lithium-ion battery cell of 3.75 V and capacity of 43 Ah.

3.6 Battery Characterization

Battery pack has been characterized based on the power requirement of the vehicle, e.g., for vehicle traction in hilly steep slope and/or start from the rest [10].

Battery Discharge Rate

- Discharge rate, Q/t = C/t
- C = total charge, Ah, t = time, h
- Rated capacity of lead acid battery are usually specified as: 8, 10 or 20 h rate (C/8, C/10 and C/20)
  - Find the discharge rate for C/2, C/75 min, 1C, 2C, 4C, C/16.
  - If C = 86 Ah, supply current will be for C/2, i = 86/2 = 43 A.
  - For 1C, i = 86 A for 1 h, for 2C = C/0.5 = 172 A for 30 min, 4C = C/0.25 = 86*4 = 342 A for 15 min.
  - If the battery capacity is 8 kWh battery of voltage 100 V, the battery can deliver current, i = 8*1000/100 = 80 A for 1 h continuously

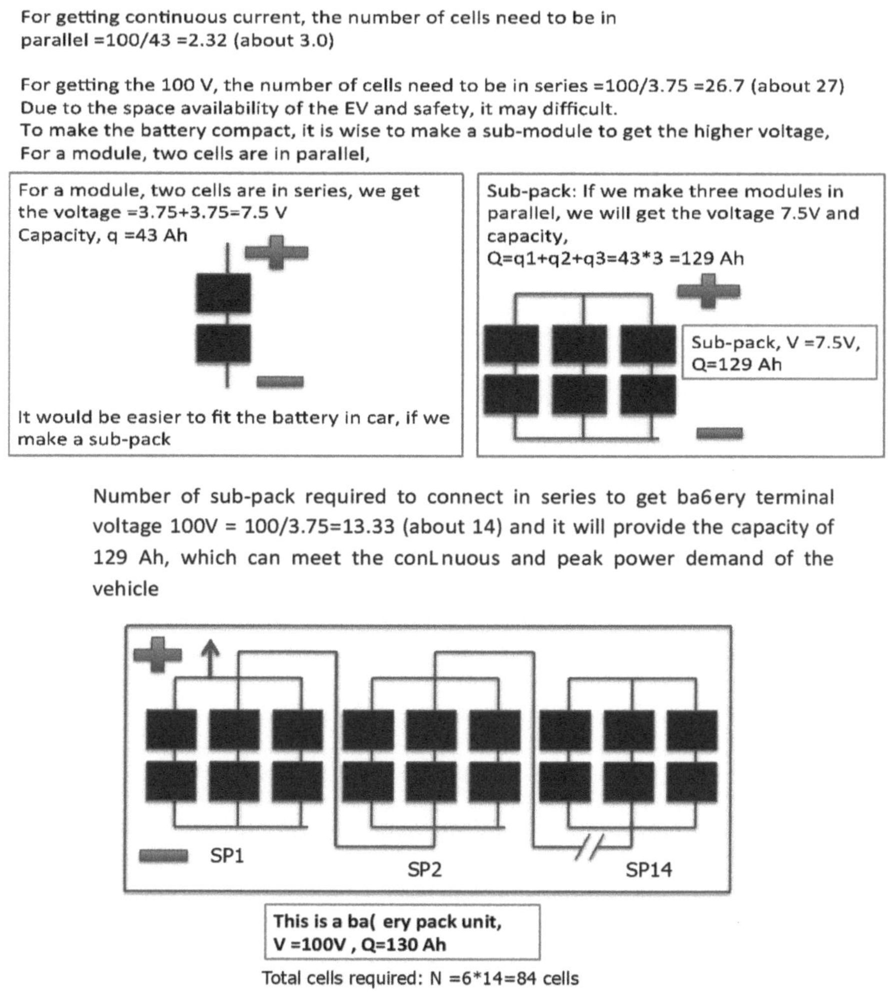

Fig. 3.5 Battery modeling for fuel cell hybrid vehicle

3.7 Integrated Cooling System of FC Hybrid EV

Batteries and fuel cell energy packs will use the intelligent integrated thermal management system (ThMS) (Fig. 3.6). It will, however, be built on the IIUM direct refrigerant-based two-phase evaporative cooling system, which operates on its own using a programmed controller and temperature sensor signals. The battery can be kept between 20 and 35 °C via the IIUM IthmS. As the LiFePO4 rises to 40 and above, battery performance drops by 20% and battery life is shortened by 15%.

However, as the temperature and pressure increase, the proton exchange membrane fuel cell performs better. A proton exchange membrane fuel cell's total

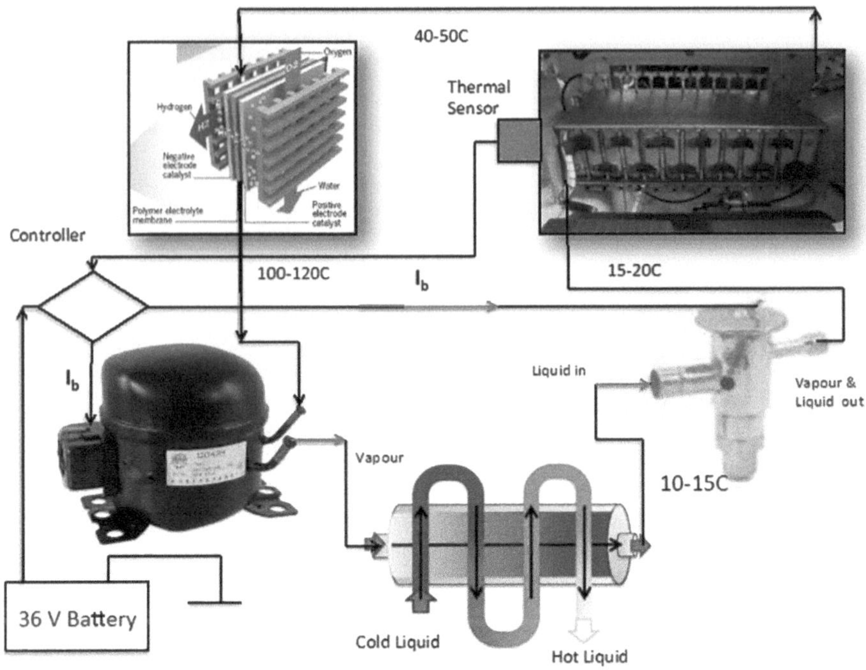

Fig. 3.6 Proposed FCHEV's FC and batteries cooling strategy [10]

performance, including its current, current density, voltage, and electricity output, improves as its self-generated thermal energy increases. When the operating temperature in a PEMFC increases from 50 to 80 °C, the power generation density increases by 16%. Higher operating temperatures, however, have the potential to limit the fuel cell's lifespan by causing membrane drying, an increase in the rate of hydrogen crossing, and the deterioration of parts including electrocatalysts, gasket materials, and bipolar plates. For optimal performance, the ideal temperature is therefore necessary.

The heat produced by the cell must be cooled, either by air or a fluid, for increased efficiency and steady production. The FC hybridized battery's iThMS is displayed in Fig. 3.6. The energy-efficient system can regulate the FCHEV energy system's temperature, as the figure illustrates [2, 9, 10].

3.8 Conclusion

A hybrid fuel cell (FC) EV prototype with a normal ground pressure of 7–8 kN/m2 would be a viable option for mobility over peaty soft terrain. Using a 60 V 3.6 kW FC with 500 g of hydrogen and a 60 V 5.2kWh LiFe3PO4 battery linked in parallel with the DC/DC converter, the ideal FC hybrid power system can meet the vehicle's power needs. To ensure high speed, an adaptive neuro-fuzzy intelligence system (ANFIS) shifted the vehicle's power from FC to battery and from battery to FC for traction. This car can go 31 km on 500 g of hydrogen and a fully charged battery.

Acknowledgements For funding this work, the authors are thankful to the Malaysian Ministry of Higher Education and the Research Management Centre (RMC) of the International Islamic University of Malaysia. The authors would like to express their gratitude to Eastern Michigan University for their ongoing cooperation in implementing this study.

References

1. Ataur, R., Rafia, A., & Abdul, H. J (2020). *Green transportation system.* IIUM Press.
2. Rahman, A., Rahman, M., Sany, I. I., & Ahmad, F. I. (2023). Fuel cell hybrid electric air-cushion tracked vehicle for peat swamp. *International Journal of Heavy Vehicle System, 2023.*
3. Indati, M. S., Ghate, A. T., & Leong, Y. P. (2013). Towards greener environment: Energy efficient pathways for the transportation sector in Malaysia. *IOP Conference Series: Earth and Environmental Science, 16*, 121–122.
4. Coutanceau, C., & Baranton, S. (2016). Electrochemical conversion of alcohols for hydrogen production: A short overview. *Wiley Interdisciplinary Reviews: Energy and Environment, 5*(2016), 388–400.
5. LaHood, R. (2010). U.S Department of Transportation. In *Transportation's role in reducing U.S. greenhouse gas emissions, vol: 1. Synthesis Report to Congress.*
6. Barlow, T. J., Latham, S., McCrac, I. S., & Boulter, P. G. (2013). *European driving cycle.* Accessed May 20, 2013.
7. Karrer, T. (2010). *General motor.* https://www.evdriven.com/2010/ev1/general-motors/. Accessed January 1, 2012.
8. https://www.iea.org/reports/world-energy-outlook-2007. Retrived date: 27 January 2022.
9. Qussem, A. A (2017). *Decarbonization through the introduction of hydrogen fuel cell vehicles in Malaysia: Prospect and challenges.* MIT-UTM Malaysia Sustainable Cities Program, Massachusetts Institute of Technology.
10. Rahman, A., Rahman, M., Sany, I. I., & Ahmad, F. I. (2018). Power optimization of electric coaster. *International Journal of Electric and Hybrid Vehicles, Inderscience Publisher, 10*(2), 272–278.

Chapter 4
Prospect and Challenges of Electric Vehicle Adaptability: An Energy Review Malaysia

Abstract The absence of electrification has prevented the development of commercialized electric cars (EVs) from progressing more quickly. When it comes to lowering emissions in passenger road transportation, electric vehicles that are powered by low-emission electricity are a crucial choice. The advent of electric vehicles will unavoidably increase the interplay between power generation and mobility. The demand for electricity from electric automobiles becomes a significant issue in the energy system and affects the operation of power plants and grid infrastructure as the prevalence of electric cars increases. The selected objective of electric car demand management may involve the following smart charging strategies: lowering grid limitations, optimizing the use of renewable energy production to lower CO_2 emissions, or decreasing emissions by switching electricity generating from coal and gasoline. The environmentally friendly smart electric car body panel that is presented in this study is thought of as an energy generator that uses solar power and vehicle deceleration and storage before being released to satisfy the demand of the auxiliary power system. In addition to reducing the vehicle's speed from to 25 km/h and producing 1.43 kW of renewable energy from solar power, the smart body panel is anticipated to save 10% of the vehicle's propulsive force by reducing the weight of the vehicle by 15%.

Keywords Electric vehicle · Installed electric capacity · Renewable energy · Body panel functions as energy generator and storage

4.1 Introduction

This study introduces an environmentally friendly smart electric car body panel that can be used as a solar-powered energy generator. The vehicle's storage and deceleration units can then be released to satisfy the demands of the auxiliary power system. The smart body panel anticipates producing 1.43 kW of renewable energy from solar power and reducing the vehicle's speed from 125 to 25 km/h. It also

A. Rahman et al., *Autonomous Driveline System for Electric Vehicle*, Green Energy and Technology, https://doi.org/10.1007/978-981-96-8418-2_4

anticipates saving 10% on vehicle pushing power by reducing the car's weight by 15%.

In order to meet the requirement of the auxiliary power system, the eco-friendly smart electric vehicle body panel is presented in this study. It is seen of as an energy generator that draws electricity from solar power and the car's deceleration and storage unit. It could lessen grid restrictions and increase the capacity of power plants. It is anticipated that the smart body panel will produce 3.66 kW of renewable energy from solar power, reduce vehicle speed from 125 to 25 km/h, and save 10% of vehicle propulsive force by reducing vehicle weight by 15% [1].

The primary energy source for the transportation industry is fossil fuels. Over the past few decades, greenhouse gas (GHG) emissions have increased, primarily due to global transportation. The use of low-carbon fuels, such as synthetic and renewable biofuels, can aid in lowering greenhouse gas emissions. But it wouldn't address other issues like traffic, noise, road safety, human food consumption (corn, sugar, etc.), and land use. Traffic avoidance and a transition to more efficient, non-motorized modes of transportation are two ways to lessen adverse consequences from the transportation sector. Electric vehicles (EVs), an alternative advanced vehicle drive technology, can directly use renewable electricity as its energy source.

Electric cars are therefore a viable means of reaching the long-term GHG target of a cross-sectoral emission reduction of roughly 80–95% by 2050 as compared to 1990. Additionally, electric cars lessen reliance on crude oil. The transportation industry uses a variety of energy sources. Additionally, electric vehicles' low exhaust emissions (GHG and air pollutants) lessen urban air pollution.

Higher electricity consumption are a result of the growing number of electric vehicles. Coordination of battery charging with wind and solar power generation availability will become increasingly crucial as the proportion of renewable energy sources increases. Therefore, in order to meet the goal of reducing GHG emissions overall, the energy sector's layout needs to be adjusted to the new electricity consumer. With an 80% proportion of electric vehicles by 2050, the European Union will need to generate more electricity to fulfill the increased energy demand. The proportion of electricity used by electric vehicles in Europe is expected to rise from about 0.03% in 2014 to about 4–5% by 2030 and 9.5% by 2050 [2].

The widespread use of electric vehicles creates a variety of integration challenges for grid infrastructure and power generating combinations in different nations and areas. In places where the availability of renewable energy fluctuates a lot, it will be very difficult to balance the demand for electric vehicles with the supply. Countries that prioritize solar energy (preferred charging peak during the day) should, however, use different charging techniques than those that solely concentrate on wind power (preferred charging peak owing to wind power is more variable). The desire for electric vehicles could result in significantly higher CO_2 emissions in nations like Poland and the Czech Republic that rely heavily on fossil fuel plants to provide their electricity. However, by using smart charging techniques, it may be possible to gather renewable energy and lessen the need for more capacity for some technologies like coal, gasoline, and nuclear electric power producing plants.

4.2 Electric Vehicle Embark in Malaysia

Malaysia is not an anomaly in the global shift toward electric vehicles. As part of the Electric Mobility Blueprint (EMB) and the National Green Technology Master Plan, Malaysia intends to import 100,000 electric cars by 2030 [3, 4]. As a result, there must be sufficient electricity in the nation to charge the electric automobiles on a regular basis. Therefore, a detailed mapping of Malaysia's power generation resources is required.

Malaysia's history of producing high-voltage electricity started slowly in 1964. That occurred when the Connaught Bridge Power Station was connected to the Bangsar Power Station. In 1965, a plan was created to distribute the production of power throughout the nation. Paka in Terengganu, Temengor, Kenering, Bersia and Batang Padang in Perak, Connaught Bridge, Kapar and Serdang in Selangor, Cameron Highlands in Pahang, Perai in Penang, Port Dickson in Negeri Sembilan, Pergau in Kelantan, Pasir Gudang in Johor, and Melaka are the locations where the plants that were previously located will be connected. When Kota Bharu was added to the grid in the late 1980s, the loop was finally completed.

Malaysia intends to become a marketing hub for electric vehicles, according to Datuk Seri Panglima Dr. Maximus Johnity Ongkili, Minister of Energy, Green Technology, and Water, as published by the Sundaily News on April 11, 2017, at 7:13 p.m. According to him, by 2030, Malaysia wants to have 125,000 charging stations, 100,000 electric automobiles, 2,000 electric buses, and 100 electric motorcycles [5]. This will lessen reliance on fossil fuels. All forms of transportation—air, sea, and land—rely heavily on fossil fuels, as we are all aware. Thus, the electric car is an attempt by the government to lower the emissions from conventional internal combustion engines that consume diesel and petroleum. However, since electric vehicle batteries must be charged every day, we must ensure that the nation's electrical supply is enough before deploying electric automobiles. Therefore, mapping Malaysia's electricity generation sources is crucial.

This study examines the mix of power generation and overall consumption, with a focus on the flexibility of introducing electric vehicles. With the help of the national automaker, Malaysia intends to launch 100,000 electric cars and 2000 electric buses by 2030 [6]. Reducing reliance on fossil fuels for transportation and lowering greenhouse gas emissions are the main goals of the EV introduction. Using a 33kWh battery pack, the electricity required to charge 100,000 EVs is 3,300 MW for every 160 km of travel. As a result, the nation must expand the plant's installed capacity to produce roughly 3,300 MWh (3.3 GWh), or 1% of the power plant's existing installed capacity.

The coal and gasoline plants are responsible for 1.2 and 1.4 kg/kWh of emissions, respectively [7]. Malaysia must raise the installed capacity of the hydro project by roughly 157% of its 2.105 GW installed capacity in order to potentially minimize emissions from electricity generation. By 2025, Malaysia intends to use a hydroelectric project to produce over 7,400GW of power. Furthermore, an electric car battery's charging time might range from 1.30 to 12 h, which may exceed the waiting period.

Consequently, it is imperative to develop a partially energy-independent electric vehicle (EV) that can increase the vehicle's range by roughly 200–220 km [8].

If electric vehicles continue to advance quickly, by 2030, they will account for 4–5% of all electricity use in a number of European nations. By 2050, the EU-28 countries' electric vehicle demand share of total electricity demand ranges from 3 to 25%, according to the EV-high scenario. In many nations, it will rise to over 10%. By 2050, it will account for 9.5% of the EU on average, with 80% of cars being electric. By 2050, electricity would account for around 50% (450 TWh/a) of the fuel mix for passenger cars, making it a viable energy source for transportation.

As the number of electric vehicles rises, the transportation sector's emissions of CO_2 and other local air pollutants decrease. These beneficial benefits will be (partially) countered by additional emissions in the energy sector brought on by the increased demand for electric passenger cars, depending on how the power sector is set up (assuming the same generation mix as in the Reference Scenario 2013 for emission balancing). Conversely, using less liquid fuel lowers the WTT emissions associated with fuel manufacturing. However, high electric car penetration share (EV-high scenario) does have a substantial positive CO_2 impact compared to other scenarios, with a net reduction of 255 Mt CO_2/a for the year 2050 (avoided emissions in the transport sector and WTT emissions from power and fuel production).

The overall emission decrease (including WTT emissions) for 2050 suggests an 84% reduction in CO_2 emissions from passenger road transportation when compared to 2010 (−395 Mt CO_2/a). This includes energy efficiency enhancements and the incorporation of electric cars, both of which are included in the Reference Scenario 2013 [9].

4.3 Electricity Generation Capacity: Malaysia

Malaysia's electricity generation is heavily reliant on fossil fuels, including coal and natural gas [4, 5]. Fossil fuel combustion may cause climate change and environmental deterioration. It also results in the depletion of fuel reserves. In order to achieve sustainable development, Malaysia is working to become a developed nation. A few energy policies have been initiated by the nation to increase energy efficiency, safeguard the environment, and improve the quality of services provided to its population. There are numerous studies on the energy sector in Malaysia, including its present production capacities, energy resource availability, prospects, potential, and constraints [10].

Malaysia uses thermal and hydroelectric power facilities to generate electricity. A thermal power plant produces electricity by burning coal, natural gas, biomass, and fuel. In contrast, hydropower facilities use water turbines to convert water heads into energy. About 16% of Malaysia's electricity is generated hydroelectrically, with the remaining 84% coming from thermal sources. The hydro plant's installed capacity is approximately 2.105 GW, as shown in Table 4.1. In 2015, Malaysia's total electric energy development capacity was roughly 138,000 GWh. 105,861 GWh of electricity were used in 2013. About 81 and 88% of Malaysia's installed power generation capacity in 2014 and 2015, respectively, came from fossil fuels like coal and natural gas [9].

The majority of Malaysia's electricity was produced by thermal stations. The energy reserves produced 31,000 TWh of natural gas, 7700 TWh of crude oil, and 15,800 TWh of coal. Crude oil output reached 73.5 thousand cubic meters per day and natural gas production reached 190.6 million cubic meters per day in 2013. If the pace of use of natural gas is maintained, it is anticipated that it will last 41 years, whereas crude oil will last 29 years [11–13]. As a result, after 29 years, Malaysia will no longer rely on natural gas and crude oil to generate electricity. Coal is Malaysia's second-largest energy source.

Coal is widely accessible, has a low price, and is a necessary component of the energy mix. In actuality, coal has been the primary energy source in both China and the United States. Malaysia, where coal is the primary energy source, is no exception. Coal accounted for 43.7% of the nation's energy mix in 2013 and 48% in 2015 [14]. In 1993, the total power generating mix, including alternative (renewable energy) sources, was 33,323 GWh; in 2013, it was 141,266 GWh. This figure shows Malaysia's electricity consumption between 2014 and 2016. Approximately 136,000 GWh of electricity were used in Malaysia in 2016, a 9% increase from 2015 (Fig. 4.1). But the average yearly growth rate However, the average annual growth rate of electricity consumption (Table 4.1) is considered to be 5.6% [15].

Electricity Consumption in Malaysia

Table 4.2 shows the total amount of energy used in Malaysia from 2010 to 2015 across all sectors, including residential, commercial, transportation, industry, and agriculture. According to energy balance, Malaysia has 12,274 GWh of excess energy in 2015 [2]. However, if 2,000 electric buses and 100,000 EVs are introduced in 2030, and each EV has a power of 33 kWh, and the electric bus has a power of 70 kWh, then 3.44 GWh of electric power will be needed to charge the 2,000 buses and 100,000 EVs, which is sufficient for the nation's electrical capacity. However, the existing capacity for electric energy generation is insufficient if the average annual growth in electric energy demand is taken into account, which is 5.6% until 2025.

Table 4.1 Electricity generation [16]

Year	Electricity generation mix (GWh)						
	Hydro	Gas	Coal	Oil	Diesel	Others	Total
2010	6361	81,342	495,401	933	726	170	584,933
2011	8056	55,732	54,302	4295	5108	1576	129,069
2012	9251	60,992	55,615	2279	4344	1596	134,077
2013	11,799	71,174	53,663	1571	1741	1318	141,266
2014	13,540	74,466	52,693	376	756	995	142,826
2015	15,524	66,919	59,335	45	1516	1226	144,565

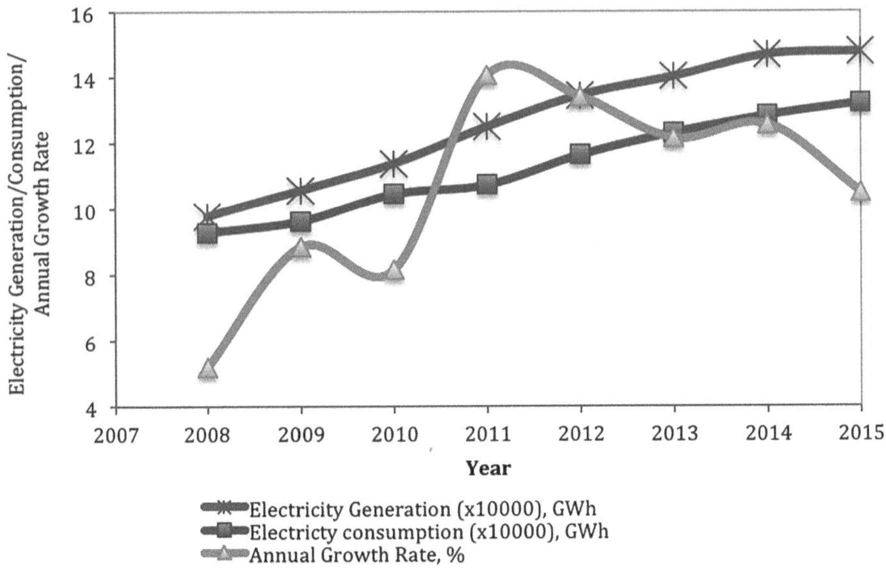

Fig. 4.1 Electricity status in Malaysia [2]

Table 4.2 Malaysia energy consumption [16]

Year	Electricity consumption (GWh)					
	Agriculture	Commercial	Transport	Industrial	Residential	Total
2010	279.12	35,122.6	209.34	46,450.22	22,527.31	104,588.59
2011	302.38	36,890.36	209.34	47,043.35	22,957.62	107,403.05
2012	348.9	38,669.75	244.23	52,439.67	24,725.34	116,427.89
2013	372.16	40,309.58	244.23	55,928.67	26,307.06	123,161.7
2014	418.68	41,472.58	255.86	58,987.36	27,283.36	128,417.84
2015	465.2	42,554.17	267.49	60,685.34	28,319.05	132,291.25

The country needs to increase the electric energy generation capacity to meet the country's electric energy peak demand and electric vehicle and buses charging energy. The reserve margin of electricity can be estimated as

$$Reserve\ margin = \frac{Installed\ capacity - Peak\ Demand}{Peak\ Demand} \tag{4.1}$$

where peak demand is the power demand registered by the customer in a stated period of time and installed capacity is the maximum capacity that can be provided by the plant.

The Feed-in Tariff (FiT) is intended to attain 985 and 2080 MW by 2015 and 2020, respectively, from a variety of renewable energy (RE) resources under the Tenth

Malaysia Plan (2011–2015) [7]. To keep up with the growing demand, Malaysia also intends to install 6 GW of new generating capacity between 2015 and 2020. Between 2009 and 2010, the gross production of electricity increased by 2.3%, from 105,706 to 108,175 GWh [8]. The demand for electricity has grown from 96,302 to 104,519 GWh, an increase of 8.53%. Natural gas was 46% of the commercial primary energy mix in 2010, followed by coal at 19%, hydropower at 2%, and crude oil and petroleum at 33% [3] (Economic Planning Unit, 2012). Renewable energy represents a very small percentage.

However, according to the Economic Planning Unit's prediction, Malaysia's electricity growth increased by 3.52% in 2012 as opposed to 3.48% in 2011. Strong demand growth from the residential and commercial sectors has been the main driver of this expansion. In 2007 and 2012, respectively, the proportion of electricity use to total energy consumption rose from 17.4% to 21.7%. Gas continues to be the primary fuel source, accounting for 52.7% of the overall power generation fuel mix in 2012, with coal coming in at 38.9%, hydro at 7.3%, oil at 1%, and others at 0.2%. The total electricity production was recorded as 122120 GWh.

Emissions from Power Generation

Coal, natural gas, and fossil fuels are the three main sources of electricity generation in Malaysia. Greenhouse Gases (GHGs), which are released when fossil fuels are burned, pollute the environment and alter the climate. The earth's temperature rises due to GHG emissions, causing natural calamities like storms, soil erosion, floods, and changes in the landscape. Carbon dioxide, methane, nitrous oxide, and fluorinated gases are the greenhouse gases (GHGs) that are created when fossil fuels are burned. Carbon dioxide is released into the atmosphere when fossil fuels like coal, natural gas, and oil are burned.

Allowing plants to absorb carbon dioxide through the biological carbon cycle is the only method to get rid of it. Methane and nitrous oxide are additional emission gases that were released as a result of the extraction and transportation of coal, natural gas, and oil. Additionally, it is released by cattle, agricultural operations, and the decomposition of organic waste. Finally, when fossil fuels burned, fluorinated gases were also released. Strong greenhouse gases produced by numerous industrial processes include nitrogen trifluoride, sulfur hexafluoride, hydrofluorocarbons, and perfluorocarbons. They are also known as gases with a high potential for global warming. The proportion of energy produced using different fuels and the factors that contribute to CO_2, SO2, NO2, and CO emissions from fossil fuels for a unit of electricity generation as shown in Tables 4.3 and 4.4.

According to predictions, Malaysia's energy production is expected to grow at an annual growth rate of 5.3%, from 87,306 GWh in 2005 to 314,984 GWh in 2030. Although Malaysia has abundant hydroelectric resources, building a hydroelectric power (HEP) plant is expensive and extremely complicated because it involves not only the design, building, and operation of the dams but also significant environmental, social, and political factors. However, the benefits are many since hydroelectric power is renewable and less susceptible to changes in fuel prices. Significant hydropower resources are available in Malaysia.

Table 4.3 Emission for a kWh power generation

Fuel	Emission (Kg/kWh)			
	CO_2	SO_2	NO_x	CO
Coal	1.18	0.014	0.01	0.0002
Petroleum	0.85	0.017	0.003	0.0002
Gas	0.53	0.001	0.001	0.001
Hydro	0.00	0.00	0.00	0.00

Table 4.4 Power plant contributes to emission generation

Year	CO_2 (Ton) by plant			SO2 (Ton) by plant			NOx (Ton) by plant		
	Coal	Gas	Petroleum	Coal	Gas	Petroleum	Coal	Gas	Petroleum
2010	72,383	72,384	1958	859	859	232	319	319	8
2011	64,076	65,764	11,096	760	780	132	282	290	49
2012	65,626	71,971	7815	779	854	93	289	317	34
2013	63,325	83,985	3908	751	996	46	279	370	17
2014	62,178	87,870	1336	7378	1043	16	274	387	6
2015	70,015	78,964	1896	831	937	23	309	348	8

The hydropower potential is estimated at 29,000 MW, of which only 2,000 MW is currently utilized. The mathematical equation to estimate the energy required for the electric vehicle to propel the can be modeled as

$$E = \frac{(v * t)}{1000}\left| \begin{array}{ll} (\mu mg) & Acceleration\,speed,\ 0-10km/h \\ (mg)(f_r + sin\theta) & Acceleraion\,speed\,at\,10\%\,grade,\ 40\,km/h \\ 0.5\rho_a C_d A_f v^2 + f_r mg & Cruising\,speed,\ 90km/h \end{array} \right| \tag{4.2}$$

where m is the vehicle mass, kg, μ is the road friction coefficient, f_r rolling motion resistance coefficient, sinθ ≅ G, G is the grade, ρ is the air density, kg/m^3, C_d drag coefficient, A_f frontal area, m^2 v vehicle speed, m/s, and t time, h.

G is 10% for the urban mode, ρ is 1.16 kg/m3, Cd is 0.2, Af is 1.67 m2, t is two hours for cruising, and m is 1000 kg, μ is 0.4, and fr is 0.012 in θ ≅ G. The simulation results for the energy needed by the vehicle in each mode are displayed in Table 4.5. The cruising speed is the primary basis for calculating energy. The vehicle can meet the traction power requirements of the other modes, such as beginning and hill climbing, if it has a 55.5 kWh power pack. Consequently, a 55.5 kWh battery pack may be used to enable the car to go 200 km between charges. On the other hand, if the battery pack is 33 kWh, Table 4.5 shows the total power needed to charge the battery pack. The ratio of electric car power demand to overall electricity demand is a crucial metric for evaluating the effect that electric car charging has on electrical systems. In order to produce the electricity needed to charge the battery pack, the plant's emission level took into account the 1.19 kg/kWh of coal, 0.85 kg/kWh of petroleum, and 0.53 kg/kWh of gas emissions (mostly CO_2).

According to Table 4.6, the country will avoid 3072 tons of emissions from burning 1.3 × 106 L of gasoline if 100,000 EVs are adopted. However, the plant will produce 2805 tons of emissions from coal plants, 1628 tons from gas plants, and 1921 tons from petroleum plants in order to meet the energy need of charging the 100,000 EVs. The gas plant can provide extra energy to satisfy the EV charging energy demand, depending on the emission level. However, much as with petroleum, its appropriateness is dependent on the nation's gas supply.

In order to fulfill the higher electricity demand of 80% car stock penetration, alternative renewable energy plants can be taken into consideration. In 2030, a large amount of additional generation capacity might be needed to meet the battery charging requirement for 100,000 EVs. Electric vehicles contribute up to 150 GW of generation capacity to the energy system (130 GW). With the applicable first-order

Table 4.5 Energy required for an EV of weight 10.8 kN for traveling 200 km

Traction mode	Speed (km/h)	Traction force (kN)	96 V rating motor required to propel 10.8 kN Electric Vehicle		
			Power (kW)	Current (A)	Energy (kWh)
Initial starting (Reduction)	0–10	5.88	16.33	170	10
Urban mode at 10% gradian	40	1.26	19.1	145	10
Cruising speed	90	1.15	28.75	290	55.5

Table 4.6 Power and emission level for the EVs introduction

Units of EV	Fuel save (liter)	Emission save (ton)	Power required (GWh)	CO_2 Emission by plant (ton)		
				Coal	Gas	Petroleum
10,000	133,300	307	362,562	281	163	192
20,000	266,600	614	725,125	561	326	384
30,000	399,900	922	1,087,688	842	489	576
40,000	533,200	1229	1,450,250	1122	651	769
50,000	666,500	1536	1,812,813	1403	814	961
60,000	799,800	1844	2,175,376	1683	977	1153
70,000	933,100	2151	2,537,938	1964	1140	1345
80,000	1,066,400	2458	2,900,501	2244	1303	1537
90,000	1,199,700	2765	3,263,064	2525	1466	1729
100,000	1,333,000	3073	3,625,626	2805	1628	1921
300,000	3,999,000	9218	10,876,880	8415	4885	5765

estimate, this comprises up to 47 GW wind, 25 GW solar, 41 GW fossil, and 11 GW nuclear capacity, assuming constant generation mixes to the Reference Scenario 2013. These capabilities increase the amount of land used for electricity generation and necessitate large extra investments, especially for wind and solar power.

Additional nuclear and coal power facilities should be avoided since they have major negative environmental effects and do not fit into the future energy sector, which needs to supplement the erratic renewable energy supply. Rather, the necessary additional coal and nuclear capacity should be replaced with more renewable capacity. This would comprise 13 GW of biomass, 24 GW of hydropower, 45 GW of solar power, and 87 GW of wind power. The energy-efficient car body panel, which may be utilized as both an energy generator and a store of energy, is an additional option.

4.4 Potential EV Body Panel: Enery Renewable

In order to create a laboratory-scale automobile body, the author sandwiched ZnO filler epoxy resin with carbon fiber and lithium foil and CuO filler epoxy resin with carbon fiber (CF) and lithium foil. The exceptional physical and chemical features of nanostructured materials have generated significant attention in both basic and practical research fields. As the battery's energy density and power density have increased recently, the use of nanostructured materials as electrodes in lithium ion batteries (LIBs) has shown promise.

A laboratory-scale EV body that stores electrical energy has been experimentally studied by the author [14] using the following criteria: ionization properties of Li thin plates, electrical energy charge and discharge, tensile and impact testing of CF, microscopic testing of CuO filler ER, and thermal and electric conductivity of CuO filler ER (Figs. 4.2 and 4.3). According to the author, as more CuO nanoparticles are added to the polymer film, the ionic conductivity of the polymer electrolyte falls. At low filler compositions, the polymer electrolyte system can be thought of as a conducting medium with uniformly distributed, randomly distributed filler grains.

The existence of the filler grains may result in the development of more advantageous conducting routes close to the grain surface. These surface interactions may be responsible for the observed rise in conductivity with increasing filler concentration at lower filler compositions where the grains are still well separated. However, with larger filler concentrations, the grains get excessively near to one another, increasing the geometrical constrictions or blocking effect caused by the more prevalent alumina grains. Lower conductivity results from the lengthy polymer chains becoming more "immobilized." This results in a peak in conductivity and a subsequent decrease.

The higher the filler concentration, the closer the filler grains become to one another, the more intertwined the high conducting zones are around the grain surfaces. The increase in conductivity results from the migratory ionic species' ability to move along and between these interconnected high conducting channels. According to Tables 4.7–4.9, the author also said that if the car had a CuO filler ER solid

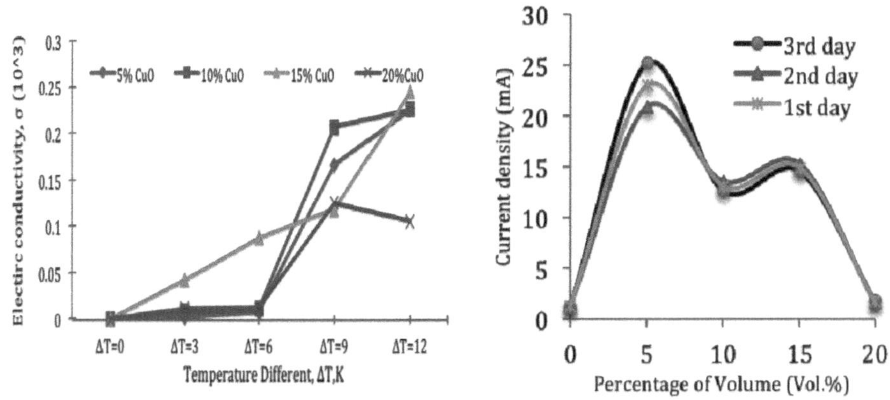

Fig. 4.2 Thermal conductivity of CuO filler ER [14]

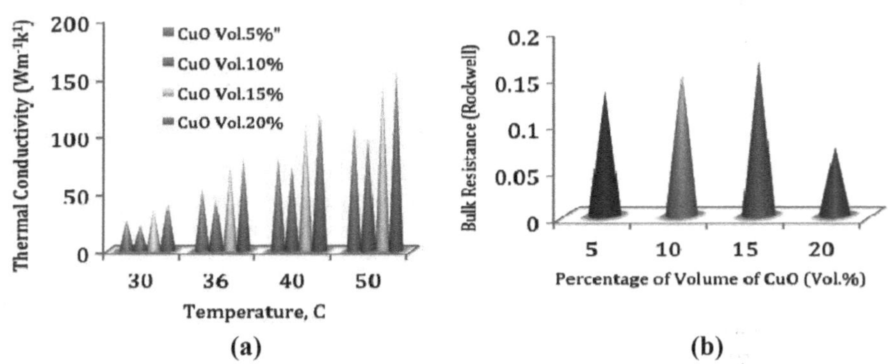

Fig. 4.3 Mechanical properties of CuO filler ER **a** thermal conductivity, **b** hardness [14]

electrolyte battery installed as the roof panel, the panel might save power since it would weigh less at 264, 304, 355, and 289W for filler concentrations of 5%, 10%, 15%, and 20%, respectively.

4.5 Impact of Organic EV Body on Environment

Carbon composites have only been used in race cars, high-end performance automobiles, and few high-end luxury cars in the automotive industry. The performance benefits of these materials make their use in race cars more warranted, while their higher cost is more readily justified in luxury automobiles. Numerous design changes have the potential to increase engine efficiency, reduce structural weight, and increase fuel economy. As a result, this lowers engine weight in addition to fuel consumption, which further improves fuel economy and lowers emissions.

Table 4.7 Application of the specimen as vehicle roof panel

Composition of CuO (wt%)	Vehicle weight saving (kg)	Fuel consumption (kg/s)	Power charge/ discharge		*Power saving (kW)
			A_P (m^2)	A_v (m^2)	
5	21	5.46×10^{-3}	105.6	264	2.38
10	20	5.58×10^{-3}	121.6	304	2.43
15	18	5.46×10^{-3}	142.1	355	2.48
20	17	5.47×10^{-3}	115.6	289	2.41

Notification: A_p = Prototype size = $0.02 m^2$, A_v = Vehicle roof size = $3.0 m^2$
*Power saving = power from weight saving + 1.1 kW from heat insulation + 1.02 kW from mobility due to diffusion.

This study considers using a nano filler conductive polymer composite body for the vehicle's body side and roof. According to the figure, the combined weight of the steel body, roof, and body side is 145 kg, whereas the carbon fiber body would be 86 kg. The car can accelerate from 0 to 30 km/h with a weight reduction of 59 kg and a traction power reduction of 4.82 kW. Table 4.8 lists the emission savings for the 4.82 kWh of energy generated by the gas, coal, and petroleum plants. Additionally, the battery weight is reduced by roughly 60 kg (0.56 kN) thanks to the nano filler conductive polymer-carbon fiber composite body, which may contribute on the saving of traction power for the acceleration speed of 0–30 km/h about 3.27 kWh and save the environment from emission about 5.06 kg for coal, 3.72 kg for petroleum and 2.31 kg for gas plant to stop producing the electric energy 4.98 kWh. The emission emitted by the plant is estimated by considering 1.18 kg/kWh for a coal plant, 0.85 kg/kWh for petroleum plant and 0.54 kg/kWh for the gas plant. Table 4.9 shows the vehicle mileage with the power pack of 33 kWh lithium-ion (LiFePO4) battery.

Table 4.8 Emission for the body weight

Electric vehicle body weight (kN)			Weight saving by CFC body (kN)	Battery energy save (kW)			Emission (kg)		
				CFC	RGBs	INS	Coal	Petroleum	Gas
Steel body	Side	0.18	–	–	–	–	1.6	8.6	0,31
	Roof	0.426	–	–	–	–	4.18	3.017	1.92
CFC body	Side	0.84	0.34	0.83	0.7	0.00	0.26	5.88	0.78
	Roof	0.25	0.18	0.36	0.00	0.1	0.46	1.78	0.13

Notification: CFC = Carbon fiber composite, RGBs = Regenerative braking system, INS = Insulation.

Table 4.9 Vehicle mileage and battery life span

Material for vehicle (Body side + Roof)	Power pack (Lithium-ion battery) (kWh)	Vehicle Travelling distance (kM)	Energy efficient (5)	Battery life span (%)
Steel	33	160	–	–
Nano 15% CuO filler composite body sandwiched by CF and Lithium thin plate	33	184.145	15%	10%

4.6 Conclusion

The widespread use of electric vehicles causes a variety of integration problems with grid infrastructure and electricity generation mixes in different nations and areas. The sharing of renewable technologies and production capacity can be used to generate electrical power to fulfill charging power demands, which can result in considerable reductions in CO_2 and most air pollutant emissions from electric mobility. The development of body panels with nano-conductive filler epoxy resin can also lower the charging power demand by reducing the body weight by approximately 15% and reducing the battery size by using the body as a capacitor/energy storage, in addition to increasing the mix electric generation capacities and highlighting the generation of renewable energy (such as solar and wind energy).

Acknowledgements The Ministry of Higher Education, Malaysia, the Research Management Centre (RMC), International Islamic University, Malaysia, and Eastern Michigan University for funding this study and permitting it to be continued for the portable power generation of electric vehicles are acknowledged by the authors.

References

1. Peter, K., Joß, B., Markus, H., & Joko, P. (2016). Electric mobility in Europe—future impact on the emissions and the energy systems. Oko Institute e.V, Berlin, Germany.
2. Malaysia Energy Commission (2016). *Peninsular Malaysia Electricity Supply Industry Outlook 2016*.
3. National Energy Balance (2010). *Energy commission Malaysia*. ISSN No: -0128–6323.
4. Mustapha, I. (2008). Renewable energy outlook and development strategies for Malaysia. In *Paper presented at National PV conference*, August 14, KL.
5. National Energy Balance. (2020). *Malaysia Energy Statistics Handbook*. Putrajaya.
6. Ong, M. C. (2011). A review on energy scenario and sustainable energy in Malaysia. *Renewable and Sustainable Energy Review, 15*, 639–47.
7. Tan, C. S. (2013). Electricity energy outlook in Malaysia. In *IOP Conf. Series: Earth and Environmental Science* 16, 012126.

8. Suruhanjaya Tenaga: Energy Commission (2020). Malaysia Energy Statistics Retrieved date 7/3/2018. https://www.st.gov.my/en/contents/files/download/116/Malaysia_Energy_Statistics_Handbook_20201.pdf.
9. Harris, A. H. (2010). Introduction and the Malaysian feed-in tariff scenario. In *Paper presented at the Malakoff Community Partnership: Energy Expert Series*, 2010 April 12, Kuala Lumpur.
10. Martin, E., Shaheen, S., & Lipman, T. (2009). Behavioral response to hydrogen fuel cell vehicles and refuelling. *International Journal of Hydrogen Energy, 34*, 8670–8680.
11. Singh, S., Jain, S., & Tiwari, A. K. (2015). Hydrogen: A sustainable fuel for future of the transport sector. *Renewable and Sustainable Energy Reveiws, 51*, 207–214.
12. Oscar, L. J., Rhian, S., & Scott, K. (2019). Content analysis of interviews with hydrogen fuel cell vehicle drivers in Los Angeles. *Journal of Transportattion Research Board, 2673*(9), 2019.
13. Shaheer, K.A., Rahman, M.A., Ibrahim, F. B., & Dafa-Alla. (2021). The impact of film thickness on the properties of ZnO/PVA nanocomposite film. *Materials Research Express, 8*(7), 1–12.
14. Rahman, M. A., & Golam, K. (2021). Al-ZnO/CuO_Cfrp organic structural auxiliary energy storage system for electric vehicle. *Journal of Solid State Technology, 28*(1 & 2), 103–116.
15. Frank, M., Arvids, S. (2010). *Elsevier Journal of Physics and Chemistry of Solids, 71*(2010) 784–787 (2010).
16. Nadzirah, A. (2016). Development of Nano CuO filler epoxy resin solid electrolyte for energy technology. In *Undergraduate final year project thesis, faculty of enginering*. International Islamic Uniersity Malaysia.

Part II
AI Control Energy System for EV

Chapter 5
Impactful of Organic Ultra-Capacior in Energy Harvesting

Abstract Climate change is regarded as one of the biggest environmental risks that the world is now experiencing. Self-powered electric vehicles are a great advancement in transportation technology that extends their driving range. Structural organic solar supercapacitors (SOSCs) are one of the technological alternatives for EVs that increase their energy efficiency. With a power of roughly 800 W, which may make up around 10% of the battery power, an EV fitted with a CuO-filled solid electrolyte supercapacitor ER as a roof panel can help the EV become more energy efficient.

Keywords Climate change · Power generating structural capacitor · EV energy efficient · CuO/ZnO_RP solar dependent capacitor

5.1 Introduction

The absence of electrification has prevented the development of commercialized electric vehicles from progressing more quickly. Integrating various renewable energy sources through advancements in storage technology and power system cost reduction could help address the EV energy shortage. The new technology that energy independent cars are advocating could hasten the deployment of EVs. Due to their less reliance on electrical infrastructure, they are more scalable. Electric vehicles are seen as the green technology of the future since they use renewable energy sources in the power system and minimize greenhouse gas emissions, costs, and weight.

These days, one of the most important components of an electric car is the battery pack. If the EV becomes energy independent, it might lose significance. If multimode energy harvesting technologies, an efficient power train, and lightweight electric cars are introduced, it may be possible to eliminate the need for batteries entirely. Regenerative deceleration mode (RDM), radio wave generators, smart nano-filler epoxy composite body panels, and electromagnetic shock absorber generators are examples of modern multimode renewable energy technology.

The composite body might be used as a solar energy harvester and structural battery or capacitor, contributing up to 30% of the energy used by EVs. However,

A. Rahman et al., *Autonomous Driveline System for Electric Vehicle*, Green Energy and Technology, https://doi.org/10.1007/978-981-96-8418-2_5

there is a lack of laboratory-based research on the smart body panel. The creation of smart body panels for EVs has been the main focus of research by the International Islamic University Malaysia EV/Energy team, VOLVO, and the James Dyson Foundation. Additionally, it is very desirable to be able to collect both solar and electrical energy from a single hybrid system, which is a new trend in all-in-one multiple energy harvesting technology.

An excellent transportation technology that increases the EV's range is the energy-independent EV. The development of a panel-type composite body with high conductive metal oxides (such as graphene oxide, copper oxide, zinc oxide, etc.) sandwiched by carbon fiber and lithium thin plates, suspension generator, and RDM system should enable EV independence. By lowering weight and acting as a battery or capacitor that can store the energy of the RDM system from the deceleration of the vehicle's motor speed range of 3000–2500 rpm, the nano-micro filler epoxy resin may play a significant role in clean energy technology. However, by converting the sprung mass movement, the suspension generator may provide 250–300 W.

The James Dyson Foundation, and Rahman et al. [1–3] have all suggested panel-style car bodies for the structural battery or capacitor described above. The multipurpose panel type battery can store and supply energy on demand, save weight, and act as an insulator to shield the vehicle's body from heat from the outside. Based on their research, it can be concluded that adopting a lightweight and effective composite body panel can reduce the size of the battery pack and replace the capacitor.

Importance of Organic Super-Capacitor

Currently, one of the biggest environmental concerns to the planet is thought to be climate change. Governments have backed the commercialization of electric vehicles (EVs) in recent years because they can lessen the impact of IC-powered vehicles, which are responsible for about 90% of the country's greenhouse gas (GHG) emissions. In the future, electricity could replace gasoline and greenhouse gas emissions in automobiles as a sustainable energy source. A major obstacle is the scarcity of EV rapid charging facilities. We are still in the early stages of developing electric charging stations. Additionally, EVs have speed and range restrictions. The majority of EVs have a range of 50–100 km. EV batteries take longer to charge.

EVs require between 4 and 6 h to fully charge, but internal combustion engine (ICE) vehicles can fill up their fuel tanks in a matter of minutes. All EV batteries need to be changed every three to ten years, depending on the battery's size, type, and usage. Additionally, the electrical system, radios, air conditioners, and automotive accessories overwhelm the EV's primary traction batteries. These uses shorten the EV's range and accelerate battery depletion. Furthermore, EV battery packs are rather hefty. An average electric vehicle's battery pack weights roughly 1000 pounds, or 450 kg. Because of their large weight, batteries are under more strain and lose energy more quickly.

Since the inception of electric and hybrid vehicles, researchers have been investigating new technologies to get beyond the limitations of their use, like battery charge-drain and mileage. Structural organic solar supercapacitors (SOSC) are one possible technological option for the body panel of an electric car. In EV and hybrid vehicles, supercapacitors have already been used for regenerative braking, powering start-stop systems, and other purposes. Traditional supercapacitors have two drawbacks: they are costly per Watt-hour and have a low specific energy (Wh/kg), which is a small portion of that of a typical battery.

Energy density and rapid discharge can be enhanced by using SOSC as the EV's body panel. In order to produce electricity by trapping solar heat, this research aims to construct a structural supercapacitor as an EV body panel using carbon-dopped zinc oxide (C-ZnO) as the n-type and copper oxide (CuO) as the p-type. In order to produce electrical energy by trapping solar heat, this study aims to develop a structural supercapacitor as an EV body panel employing carbon-doped zinc oxide (C-ZnO) as the n-type and CuO as the p-type. An EV's SSC body panel has a good probability of becoming a real body structure for a commercialized EV since it is more durable, lighter, and multipurpose than an EV's body.

5.2 Overall Goal of SOSC

The overall objective of the structural organic supercapacitor (SOSC) is to create a solar super-capacitor with an organic structure for charging an electric vehicle's auxiliary battery in order to reduce the strain on the traction batteries. The specific objectives of this studies are:

1. To characterize and modeling of organic structural solar supercapacitor made with ZnO/CuO_ER with CF reinforced for the structural super-capacitor of EV.
2. To design power management system with PI and Fuzzy Logic controllers in Simulink for the organic solar supercapacitor.
3. To investigate the performance of solar structural supercapacitor system.
4. To validate the performance of the ZnO/CuO_ER-CF reinforced SSC with a published result and ANFIS controller.

5.3 Overview of SOSC

The large capacitance of a supercapacitor, sometimes referred to as an ultra-capacitor or double-layer capacitor, sets it apart from a typical capacitor. Rather than storing energy through an electrochemical reaction, a capacitor can store energy as a static charge. A voltage difference is applied to the positive and negative plates of the capacitor to charge it. Because a supercapacitor includes two electrodes immersed in an electrolyte with a separator between them, it is essentially built similarly to a battery. The electrodes are made of a porous substance with a large surface area and pore sizes in the nanometer (nm) range. An electrochemical capacitor's electrode materials have a surface area of 500–2000 m2/g, which is significantly larger than that of battery electrodes [4–6].

Supercapacitors have ratings that are thousands of times higher than those of regular capacitors and are typically expressed in farads (F). In order to store energy, the supercapacitor can go through repeated cycles of charging and discharging at high current and brief duration [7]. Double-layer electrodes significantly improve energy storage. With the use of specialized electrodes and electrolyte, the supercapacitor has advanced to the point where it is comparable to battery technology. Voltage limitations apply to all capacitors. The supercapacitor can only handle 2.5–2.7 V, whereas the electrostatic capacitor can tolerate higher voltages. A shorter service life can enable the use of supercapacitors with greater voltages.

Compared to Li-ion batteries, the supercapacitor's specific energy range is 10–50 times lower, ranging from 1 to 30 Wh/kg. As seen in Fig. 5.1, constant current charging (CCV) causes the voltage to rise linearly. The current automatically decreases when the capacitor is fully charged. As seen in Fig. 5.2, the voltage decreases linearly during discharging. All of the ions are dispersed at random throughout the cell when it is discharged. Every positive ion moves to the negative terminal while in a charged state, and vice versa.

Supercapacitors improve both energy and capacitance because of their larger electrode surface areas, thinner dielectrics, and closer spacing between electrodes. Furthermore, supercapacitors can attain power densities that are comparable to Li-ion batteries while maintaining a low ESR characteristic. Supercapacitors outperform electrochemical batteries and fuel cells in a number of areas, including higher power density, faster charging times, and longer cycle and shelf lives [8–10].

The three main categories of supercapacitors that are currently in use are hybrid capacitors, electrochemical double-layer capacitors (EDLC), and pseudocapacitors. A unique technique for storing charge governs each group. These include Faradaic, non-Faradaic, and a mix of the two. Oxidation–reduction reactions and other Faradaic processes refer to the transfer of charge between the electrode and the electrolyte. A chemical reaction is not a part of a non-Faradaic process. Charges are dispersed

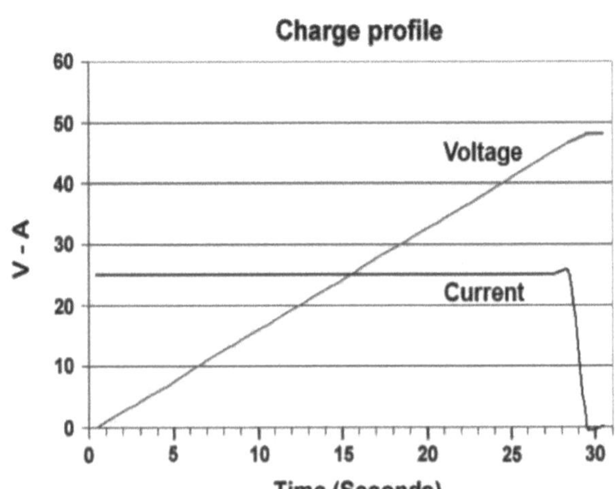

Fig. 5.1 Charge profile of supercapacitor

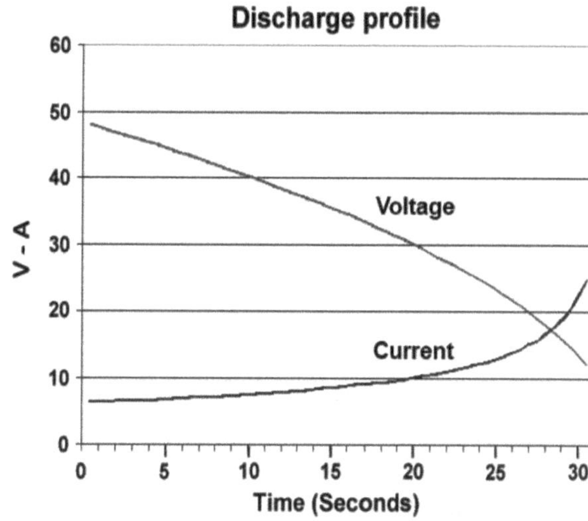

Fig. 5.2 Discharge profile of supercapacitor

over surfaces by physical processes that do not entail the formation or disruption of chemical bonds.

An EDLC's performance characteristics can vary depending on the type of electrolyte used. For EDLC, an organic or aqueous electrolyte can be utilized. Compared to organic electrolytes like acetonitrile, aqueous electrolytes like H2SO4 and KOH often have lower ESRs and lower minimum pore size requirements [10]. Ultracapacitors store charge faradaically by transferring the charge between the electrode and the electrolyte, as opposed to electrostatically, as EDLCs do. These faradaic procedures could make it possible to In comparison to EDLCs, faradaic techniques enable pseudocapacitors to reach higher energy densities and capacitances [11–13]. Conducting polymers and metal oxides are two typical electrode materials used to store charge for ultra-capacitors.

In order to improve performance characteristics, hybrid capacitors take use of the relative benefits and reduce the relative drawbacks of EDLCs and ultracapacitors. Because hybrid capacitors use both Faradaic and non-Faradaic techniques to hold charge without compromising price and cycling stability, they can reach higher energy and power densities than EDLCs. Hybrid capacitors combine several novel performance features. A hybrid supercapacitor combines the best aspects of pseudocapacitors and EDLCs [14, 15].

5.4 Electrode Materials for Organic Supercapacitor

The most common material used in commercial supercapacitor electrodes is carbon. Interest in carbon has grown significantly. Because it is easily accessible, comes in a variety of allotropies, has a large surface area, is inexpensive, has excellent corrosion inhibition, is highly conductible, stable at high temperatures, and is environmentally friendly. Graphene, carbon fibers, carbon nanotubes (CNTs), activated carbon, and carbon aerogels are all known types of carbon that are used as supercapacitor electrodes [16].

Carbon aerogels are among the most promising materials for supercapacitor electrodes. For carbon aerogel, the maximum specific capacitance (SC) value is 180 F/g [17]. Carbon aerogels come in a variety of forms, including monoliths, composites, thin films, powders, and microspheres. Additionally, carbon aerogels' surface area can be significantly increased through surface activation.

CNTs, which come in two varieties single wall and multiwall, are among the significant forms of carbon. CNTs' excellent mechanical and electrical qualities make them a promising subject for further study. Additionally, compared to copper wire, it can generate and transport an electric current 1000 times greater [18]. Capacitive behavior is a natural property of pure carbon nanotubes. However, because CNT manufacture is expensive, its potential use in a variety of applications is constrained.

Additionally, carbon fiber is a desirable material for the electrodes of supercapacitors. Because it is one dimensional, it has excellent charge transportation properties. It has been shown to be the most promising option for EDLC electrodes because of the fiber's many pores [19]. Using carbon fiber composites reduces weight significantly as compared to other materials [20].

Conducting Polymer

Organic compounds called polymers are composed of lengthy, repeating chains of molecules. These molecules may have special characteristics depending on the kind of bonding and the bonding procedure. Its mechanical characteristics can drastically shift, exhibiting brittle, glass-like behavior at low temperatures and rubber-like behavior at high ones. Moreover, polymers are highly susceptible to the strain rate, or rate of deformation. The stress–strain behavior of polymers is comparable to that of metals, as illustrated in Fig. 5.3, and decreasing the rate of deformation has the same impact as raising.

A novel option for supercapacitor electrodes is conducting polymer. Conducting polymers have several benefits over metal and carbon oxides, including strong conductivity, substrate independence, low cost, vast scale, and quick redox reactions [21–23]. For conducting polymers, the primary drawback is their poor mechanical stability. Performance is gradually deteriorated by a number of physical changes that occur throughout recycling, including swelling, shrinking, doping, and undoping. Thus, the suggested method to enhance the mechanical property is the production of metal oxides like ZnO/Cuo and polymers. For the development of enhanced pharmaceuticals and composite materials like carbon polymers for the automobile industry, researchers are experimenting with a variety of polymer kinds.

Nano Metal Oxides: Nano ZnO/CuO

Superior pseudo-capacitive behavior of metal oxide makes it a potential material for supercapacitor. Extensive researches have been conducted for supercapacitor applications on metal oxides such as ruthenium oxide, nickel oxide, manganese oxide, and

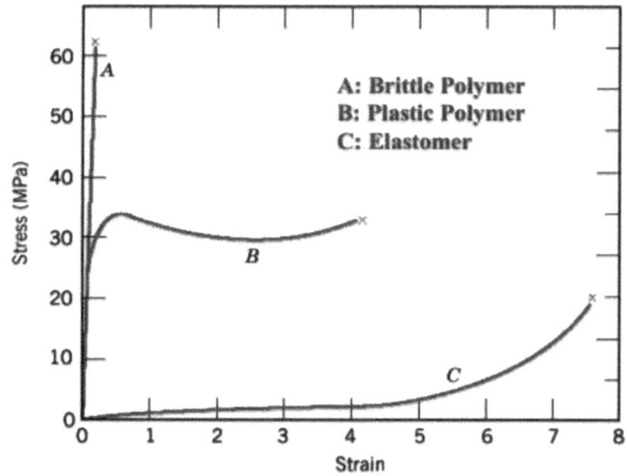

Fig. 5.3 Stress–strain behavior of Polymer (MSE 2009)

cobalt oxide etc. [24–26]. Due to their desirable properties, Nano-sized particles of semiconductor materials are being utilized for applications in different areas such as sensors, catalysts and photoelectron devices. Novel electronic, structural and thermal properties of nano-materials are of great scientific interests in basic and advanced fields. ZnO and CuO have been proposed as supercapacitor electrodes materials due to super conductivity and low cost.

One metal oxide semiconductor with a narrow bandgap of about 1.2 eV in bulk is copper (II) oxide (CuO). It is a p-type semiconductor by nature. CuO attracts study interest due to the low cost and availability of the beginning growth material as well as the low cost of the preparation techniques [27]. CuO nanostructures have sparked extensive research because of their high surface area to volume ratio. CuO NSs have outstanding electrochemical activity and the capacity to facilitate electron transfer at low potentials, making them an ideal material for sensors [28].

CuO NSs are also appropriate materials for the creation of solar cells because of their photoconductive and photochemical qualities [29, 30]. For research reasons, CuO-based materials' high temperature super conductivity is widely recognized. Thus, it is crucial to synthesize new sizes and forms of CuO NSs so that their use as a catalyst can be enhanced. Table 5.1 provides a summary of some of CuO's physical characteristics.

The concentration of filler material has a significant impact on the synthesized composite material's performance. Higher filler concentrations for CuO cause the filler particles to compress, which leads to the interconnection of high conducting zones along the grain surfaces. As seen in Fig. 5.4, the rise in conductivity is the result of the roaming ionic species' ability to move along and between these interconnected high conducting routes.

ZnO is considered a broad band gap semiconductor with an energy gap of 3.37 eV at ambient temperature. ZnO is widely used because of its electrical, optoelectronic, catalytic, and photochemical qualities. For a catalytic reaction process, ZnO nanostructures' strong catalytic activity and vast surface area are their greatest advantages

Table 5.1 Basic properties of CuO

Property	Value
Lattice parameters	a = 4.68 Å b = 3.42 Å c = 5.13 Å
Crystal structure	Monoclinic
Density	6.32 gm/cm^3
Melting point	1134^0C
Dielectric constant	18.1
Refractive index	1.4
Energy gap	1.2 eV, direct
Hole mobility	0.1–10 cm^2/Vs

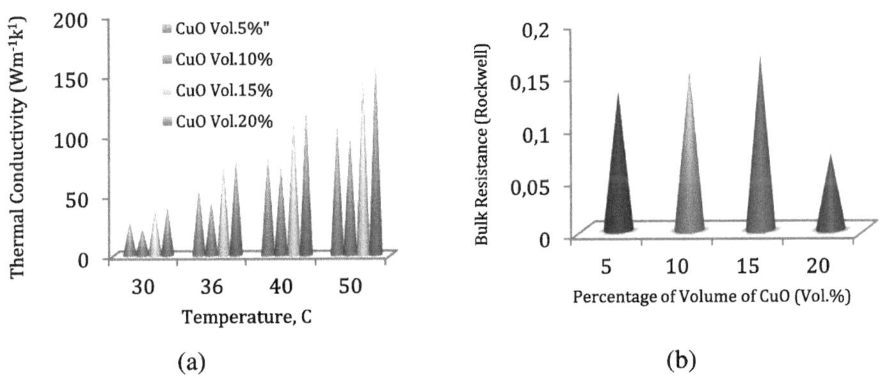

Fig. 5.4 Mechanical properties of CuO with ER Filler **a** thermal conductivity, **b** hardness [12]

[31–33]. ZnO's chemical and physical characteristics can change depending on the shape of its nanostructures. Because of its shape, the physical and chemical characteristics of generated zinc oxide must be examined in addition to different synthesis techniques [34].

Biocompatibility, high electron transfer rates, and improved analytical performance are some of ZnO's appealing qualities that make it appropriate for intracellular and extracellular sensing applications. Table 5.2 lists the basic physical characteristics of ZnO at room temperature. Metal oxides must be synthesized with carbon nanotubes, activated carbon, and conducting polymers in order to reach their full potential and be used as supercapacitor electrodes to improve supercapacitor performance. Therefore, this study proposes the synthesis of ZnO/CuO and filler conducting polymer.

Table 5.2 Basic properties of ZnO

Property	Value
Lattice parameters	a = b = 3.25 Å c = 5.21 Å
Stable crystal structure	wurtzite
Density	5.606 gm/cm^3
Melting point	1975 °C
Dielectric constant	8.66
Refractive index	2.008
Energy gap	3.4 eV, direct
Exciton binding energy	60 meV
Effective mass electron/Hole	$0.24\ m_o/0.59\ m_o$
Electron mobility	100–200 cm^2/Vs
Hole mobility	5–50 cm^2/Vs
Bulk Young modulus	111.2 ± 4.7 GPa

5.5 Synthesis of Electrodes Materials

Two or more electrically conducting plates that are separated by an insulator in a dielectric capacitor store electrostatic charge as electric charges. The capacitance of the dielectric capacitors depends on the location of the two conducting plates' overlap and separation. Two of the primary goals of a dielectric separator, as seen in Fig. 5.5, are to prevent any contact between two conducting plates and to obtain high capacitance from the dielectric polarization. Because the charging and discharging process solely includes electron transfer, dielectric capacitors have a faster recharge function and a longer cycle life than batteries [35].

Prior research [36–38] looked at two distinct dielectric capacitor designs. Glass fiber reinforced prepreg (GFRP) uses metallized polymer films (MPF) for conductive electrodes and GFRP as the dielectric material and supporting framework. In the instance of a polymer film dielectric capacitor, dielectric materials are created from various polymer thin films, and CFRP is used to provide conductive electrodes and structural support. Because of this type of lightweight design, the energy consumption is subsequently reduced during operations. Therefore, in sophisticated automotive structures, it could be utilized as a load-carry composite for energy storage applications.

However, structural dielectric capacitors have not been successful when using dielectric materials like epoxy and polyethylene terephthalate (PET) because of their poor dielectric permittivity, which leads to low energy density and multifunctional efficiency. Additionally, the interfacial connection between electrodes and dielectrics limits the mechanical performance of structural dielectric capacitors. To maximize the overall performance of structural dielectric capacitors, as seen in Fig. 5.6, the dielectric permittivity of the dielectric separator and the interfacial bonding between electrodes and dielectrics must be improved.

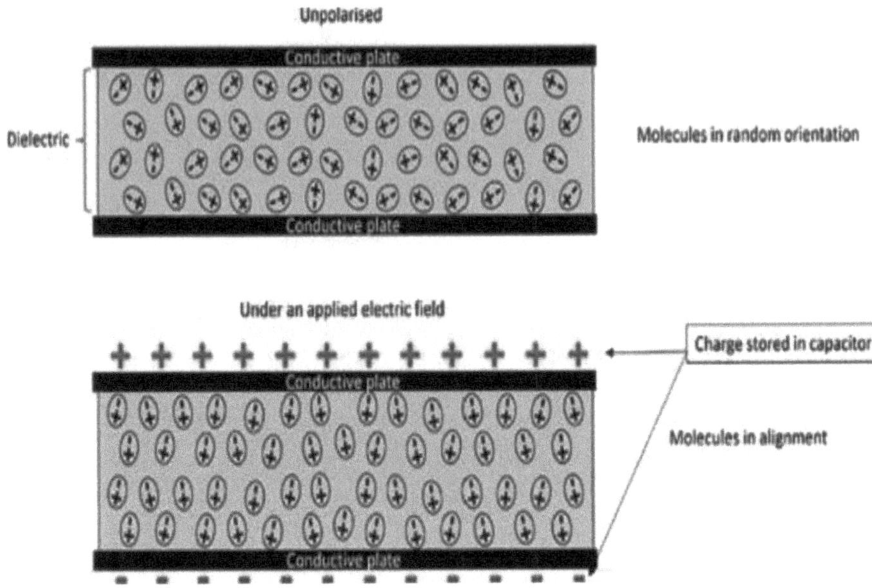

Fig. 5.5 Basic structure of a dielectric capacitor [28]

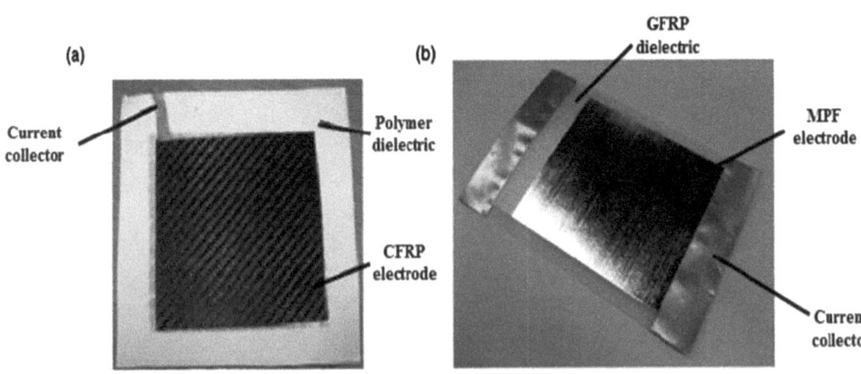

Fig. 5.6 Current collector configurations of dielectric capacitors **a** Polymer film dielectric capacitor **b** GFRP dielectric capacitor [28]

The materials used to construct the electrodes and dielectrics are among the aspects that determine the dielectric capacitors' mechanical and electrical characteristics. According to Carlson et al. [34], the parameters of dielectric separators, particularly electric breakdown strength (EBD), are additional significant factors influencing the mechanical and electrical characteristics of structural capacitors. When an electrical insulator experiences electrical breakdown (EBD), current flows through it and the voltage placed across it surpasses the breakdown voltage, making the insulator electrically conductive. Using CFRP for conductive electrodes and various materials for dielectric separators, the initial energy storage capacity experiments of structural dielectric capacitors were presented in the early phases of the research. The CFRP electrode layers were manufactured from 0.125 mm thick prepreg weaves as shown in Fig. 5.6

The average energy density of dielectrics made using impregnated sheets is 80 times lower than that of polymer film dielectric capacitors in terms of electrical properties. This is because EBD in polymer films is better than in impregnated sheets. The electrical characteristics of PET film dielectric capacitors in varying thicknesses were also examined in another study. As the thickness of the dielectric separators increases, the capacitances of the specimens are found to decrease (Table 5.3). In addition, Chen et al. [33] demonstrated that the EBD falls when epoxy-based dielectric thickness increases, as indicated in Table 5.4 245 g/m2 were the prepregs [42]. It is common practice to combine different grades of epoxy resin, utilize additives, plasticizers, or fillers to get the desired processing or final qualities, or to cut costs. Thus, numerous studies have been carried out to determine the impacts of epoxy resins packed with nanoparticles.

Nanoparticles with certain qualities can be created using a variety of techniques. A specific method of creating nanoparticles can be selected based on the intended conditions and use of the particles. Table 5.5 summarizes the benefits and drawbacks of different synthesis techniques used by different studies [41–43].

One way to increase capacitance is to maximize the surface area between the electrodes and the dielectric material. As seen in Fig. 5.7, the Sol–Gel technique yields

Table 5.3 Electrical properties of structural dielectric capacitors

Dielectric materials	Dielectric thickness (mm)	Capacitance (μF/cm^2)	EBD (kV/mm)	Energy (J/g)	Reference
Electrode material: CFRP prepreg					
Paper (40 g/m^2)	0.071	0.712	7.75	0.00013	Carlson et al. [32]
Paper (80 g/m^2)	0.089	2.466	8.76	0.00088	Carlson et al. [33]
Paper (150 g/m^2)	0.173	0.766	9.77	0.00120	Carlson et al. [33]
PA film	0.05	0.868	156	0.03400	Carlson et al. [32]
PET film	0.019	1.86	337	0.05200	Carlson et al. [32]
PC film	0.155	0.206	183	0.08900	Carlson et al. [32]
PET film	0.05	0.447	292	0.06	Chen et al. [28]
PET film	0.075	0.3	299	0.08	Chen et al. [28]
PET film	0.125	0.193	235	0.09	Chen et al. [28]

Table 5.4 Dielectric properties of epoxy-based nanoparticles

Nanoparticle	Concentration	Permittivity, ε	EBD (kV/mm)	Reference
Conductive nanoparticles				
Ag	0.03 wt%	4.6	36	Dorigato et al. [36]
Ag	23 vol%	300	1.25	Dorigato et al. [36]
Al	40 vol%	450	23	Dorigato et al. [36]
CNT	0.2 vol%	3.75	–	Ladani et al. [37]
CNT	0.5 wt%	1328	–	Ladani et al. [38]
CNT	1 wt%	5.5	–	Ladani et al. [37]
EG	2 wt%	12	–	Imran et al. [38]
Ferroelectric ceramic nanoparticles				
Al_2O_3	0.1 wt%	4.6	–	Tuncer et al. [39]
Al_2O_3	1 wt%	4.7	–	Tuncer et al. [39]
$xAl_2O_{3.y}$ SiO_2	40 wt%	6	–	Tuncer et al. [39]
$BaTiO_3$	5 phr	6	–	Hamdi et al. [40]
$BaTiO_3$	7 phr	7.7	–	Hamdi et al. [40]
$BaTiO_3$	30 vol%	10.9	200	Hamdi et al. [40]
SiO_2	0.1 wt%	4.5	–	Qu et al. [41]
TiO_2	1 wt%	4.8	–	Qu et al. [41]
TiO_2	5 wt%	5.3	30.15	Qu et al. [41]
TiO_2	5 wt%	6.5	–	Qu et al. [41]
ZnO	5 wt%	5	–	Rojas et al. [42]

Table 5.5 Comparison between different nanoparticle synthesis processes

Method	Advantages	Disadvantages
CVD	High deposition rate	High vapor pressure
PVD	No chemical substances	Low deposition rate
Mechanical grinding	Any metal	High impurities
Arc-discharge	High crystallinity	Broad distribution size

a pure and extremely fine powdered product at low temperatures (room temperature). Additionally, the procedure is quite easy to follow and straightforward. The synthesis of nanocrystalline ZnO powder using the Sol–Gel process was examined by the author Rani et al. [43]. Their findings revealed that the diameter of the crystallite size of the resulting ZnO particles ranged from 6 to 14 nm, making them the smallest among the various nanoparticle synthesis methods, including sonicating.

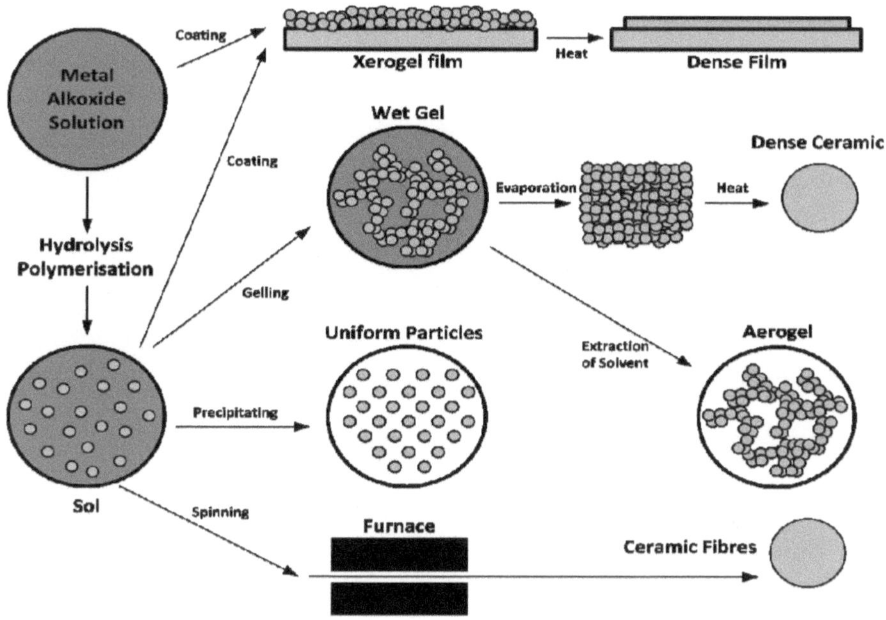

Fig. 5.7 Sol–gel process [43]

5.6 Body of EV as Structural Capacitor

The weight of batteries is one of the primary disadvantages of EVs. Replacing some of the steel body panels with carbon fiber composite materials that have battery-like power storage capabilities could be a game-changer for EV design. VOLVO'S project used innovative composite material to replace the bonnet, trunk lid, and door panels. The carbon fiber layers that make up the structural panels are separated from one another by fiber glass inserts. As structural supercapacitors, the cathode and anode are composed of layers of carbon fiber impregnated with polymers. Therefore, it can be charged like a battery of EV by renewable energy sources or external charging stations.

As seen in Fig. 5.8, the vehicle panel type nano-battery was created using high conductivity CuO filler epoxy resin encased in carbon fiber (CF) and lithium (Li) thin plates. Lithium thin plates serve as the cathode, CF as the anode, and material filler ER as solid electrolytes. In both charging and discharging modes, the body panel's permissible temperature has been regulated between 20 and 400C. 20A is the maximum charging current while 30A is the maximum discharge current. The electronic power management system regulates the SCE's rate of charging and discharging. The body panel is made of aluminum laminated polypropylene (ALP) material to stop electromagnetic interference (EMC) from entering the passenger cabin.

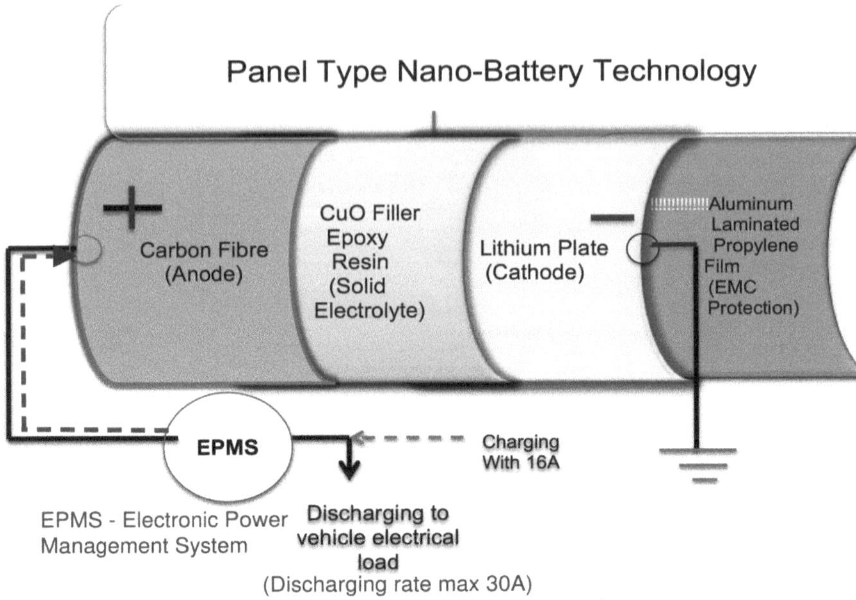

Fig. 5.8 Design of vehicle panel type battery, CuO with ER filler [1]

The filler material CuO powder was employed by the authors [1, 2] because of its greater capacity to promote electron transport along its surfaces. The epoxy resin composite for the electrolyte of the panel-type battery was constructed by the authors using nano-ZnO powder.

The James Dyson Foundation anticipated that the structural capacitor in the body panel will eventually take the role of the car's battery. The 30% reduction in steel body weight would result in a drop in the EV's weight. Made of carbon fiber, the structure can be employed as a structural element in the car body, such as the door panel or bumper. Furthermore, the powdered activated charcoal's high surface area and porosity result in a high power density and reduced internal resistance. A structure with high strength and good electrical density is produced by combining high electrical conductivity activated charcoal powder with high strength carbon fiber.

The anode and cathode components are formed of the same material, therefore experimental research is required to establish the James Dyson structure. It is possible to construct the composite body using carbon fibers, lithium foil, epoxy resin nanofiller, and extra ACP, an environmentally benign material that is based on the fundamental idea of a capacitor or battery. Utilizing carbon fiber as the EV's structural body would address the issue of battery packs and supercapacitors being too heavy.

Epoxy resin has not been used in any of the capacitor technologies described in the literature. Although there may be potential power losses, it is noted that the proposed technology [4, 8, 9], on the fabrication of structural capacitor in view of the increase in storage capacity of the renewable energy and discharge timing would prolong the discharge timing of the storage renewable energy. The timing of the structural body capacitor/battery's discharge further supports the findings [29, 43, 44]. According to the authors [29], as CuO nanoparticles are added to the polymer film, as seen in Fig. 5.9, the ionic conductivity of the polymer electrolyte drops.

The smart body panel is expecting to generate the renewable energy about 3.66 kW from solar power and vehicle deceleration from 125 to 25 km/h and also can save vehicle propelling power about 10% by cutting down the vehicle weight about 15%. Research done by Rahman et al. [3] highlighted that if the vehicle is equipped CuO filler ER solid electrolyte battery as the roof panel, the panel would able to save power due to the weight reduction 264W, 304W, 355W, and 289W for the filler concentration of 5%, 10%, 15%, and 20% respectively as shown in Table 5.6.

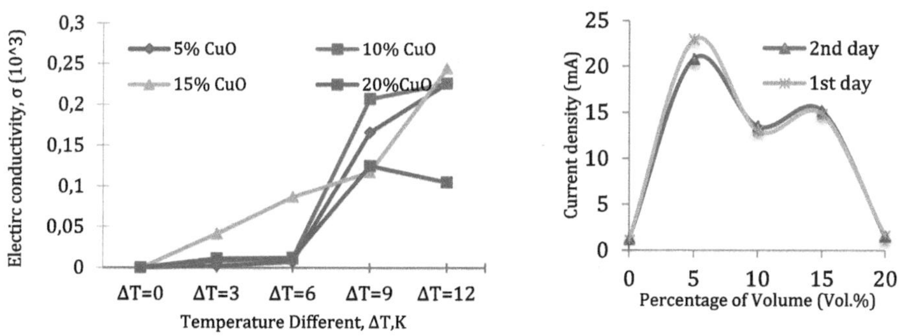

Fig. 5.9 Thermal conductivity of CuO with ER filler [1]

Table 5.6 Performance of the specimen as vehicle roof panel

Composition of CuO (wt%)	Vehicle weight saving (kg)	Fuel consumption (kg/s)	Power charge/ Discharge (W)		*Power saving (kW)
			A_p (m^2)	A_v(m^2)	
5	21	5.46×10^{-3}	105.6	264	2.384
10	20	5.58×10^{-3}	121.6	304	2.424
15	18	5.46×10^{-3}	142.1	355	2.475
20	17	5.47×10^{-3}	115.6	289	2.409

Notifications: A_p = Prototype size = $0.02m^2$, A_v = Vehicle roof size = 3 m^2*Power saving = Power from weight saving + 1.1 kW from heat insulation + 1.02 kW from electron mobility due to diffusion.

5.7 Conclusion

EV body panels may benefit from the use of structural organic solar supercapacitors (SOSCs). By retaining solar heat, it can increase EV energy generation by roughly 10% of storage battery energy, according to a study. It may also make the EV lighter and serves as insulation to prevent heat transfer to the passenger cabin.

Acknowledgements The authors express their gratitude to Eastern Michigan University for enabling this project to proceed and to the Research Management Centre (RMC), International Islamic University Malaysia, for providing funding. The Automotive Composite Laboratory Technician's assistance in creating the organic super-capacitor for the study's completion is also greatly appreciated by the authors.

References

1. Rahman, A., Mahbubur, R., Hawlader, M. N. A., & Mohiuddin, A. K. M. (2015). Nano-battery Technology for EV/HEV: A pioneering study. *IIUM Engineering Journal, 16*(2), 57–68.
2. Rahman, A., Mahbubur, R., & Mohiuddin A. K. M. (2015b). The prospects of panel style nano-battery technology for EV/HEV. In *2015 IEEE international conference on industrial engineering and engineering management, paper no. IEEM 2015-P-0509, December 6–9.* Singapore.
3. Ataur, R., Kyaw, M. A., & Shaheer, A. K. (2020). Organic photoelectric-supercapacitor power source of electric vehicle. ISSN: 0193–4120 Page No. 15175–15182.
4. Lee, M., Yang, R., Li, C., & Wang, L. (2010). *Journal of Physical Chemistry Letters, 1*(2010), 2929–2935.
5. Young-Min, K., Jang-Hee, L., Seok-Joon, K., & Daniel, F. (2012). Potential and evolution of compressed air energy storage: Energy and energy analyses. *Entropy, 2012*(14), 1501–1521.
6. Burke, A. & Zhao, H. (2015). Applications of supercapacitors in electric and hybrid vehicles. In *5th European symposium on supercapacitor and hybrid solutions (ESSCAP), Brasov, Romania, April 23–25, 2015.*
7. Vuorilehto, K., & Bautzen, M. N. (2014). Supercapacitors—basics and applications. *Skeleton-tech, Bautzen, 23*(1), 2014.
8. Conway, B. E. (1999). *Electrochemical supercapacitors: Scientific fundamentals and technological applications.* Kluwer-Plenum.
9. Burke, A. (2000). Ultracapacitors: Why, how, and where is the technology. *Journal of Power Sources, 91*(1), 37–50.
10. Kotz, R., & Carlen, M. (2000). Principles and applications of electrochemical capacitors. *Electrochimica Acta, 45*(15–16), 2483–2498.
11. Jurewicz, K., & Delpeux, S. (2001). Supercapacitors from nanotubes/polypyrrole composites. *Chemical Physics Letters, 347*(1–3), 36–40.
12. Rahman, M. A., & Golam, K. (2021). Al-ZnO/CuO_Cfrp organic structural auxiliary energy storage system for electric vehicle. *Journal of Solid State Technology, 28*(1 & 2), 103–116.
13. Chokri, M., Aymen, F., Lassaad, S. (2014). An overview of electric vehicle concept and power management strategies. In *Conference Paper 978–1–4799–7300–2/14IEEE.*
14. Hongjun, C., Fei, L., & Fujuan, G. (2012). Power management system design for small size solar-electric vehicle. In *2012 IEEE 7th international power electronics and motion control conference—ECCE Asia, 2012* (pp. 2658–2662).

15. Ganji, B. & Kouzani, A. Z. (2010). A study on look-ahead control and energy management strategies in hybrid electric vehicles. In *2010 8th IEEE International Conference on Control and Automation (ICCA)* (pp. 388–392).
16. Salmasi, F. R. (2007). Control strategies for hybrid electric vehicles: evolution, classification, comparison, and future trends. *IEEE Transactions on Vehicular Technology, 56*(2007), 2393–2404.
17. Xin, L., & Williamson, S. S. (2007). Assessment of efficiency improvement techniques for future power electronics intensive hybrid electric vehicle drive trains, in electrical power conference, 2007. EPC 2007. *IEEE Canada, 2007*, 268–273.
18. Frackowiak, E., & Beguin, F. (2001). Carbon materials for the electrochemical storage of energy in capacitors. *Carbon, 39*(6), 937–950.
19. Saliger, R., Fischer, U., Herta, C., & Fricke, J. (1998). High surface area carbon aerogels for supercapacitors. *Journal of Non-Crystalline Solids, 225*(1998), 81.
20. Choi, J. W., McDonough, J. R., Yang, Y., Mantia, F. L., Zhang, Y., & Cui, Y. (2009). Carbon nanofibre supercapacitors with large areal capacitances. *Applied PhysicsLetters, 95*(2009), Article 243109.
21. Cheah, L. W. (2010). *Cars on a diet: The material and energy impacts of passenger vehicle weight reduction in the U.S. PhD thesis, MIT, September 2010.*
22. Kim, I. H., & Kim, K. B. (2001). Ruthenium oxide thin film electrodes for supercapacitors. *Electrochemical and Solid State Letters, 4*(5), A62–A64.
23. Ryu, K. S., & Kim, K. M. (2002). Symmetric redox supercapacitor with conducting polyaniline electrodes. *Journal of Power Sources, 103*(2), 305–309.
24. Khomenko, V., Frackowiak, E., Jurewicz, K., Lota, K., & Béguin, F. (2006). Supercapacitors based on conducting polymers/nanotubes composites. *Journal of Power Source, 153*(2006), 413.
25. Jiang, J., Jiang, Q., cHEN, J., Zhou, X., Zhu, S., & Chen, T. (2021). Advanced power management and control for hybrid electric vehicles: A survey. *Wireless Communications and Mobile Computing, 2021*, Article ID 6652038, Hindawi.
26. Zheng, J. P., Cygan, P. J. & Jow, T. R. (1995). Hydrous ruthenium oxide as an electrodematerial for electrochemical capacitors. *Journal of Electrochemical Society, 142*(1995), 2699.
27. Chen, J. C., & Tang, C. T. (2007). Preparation and application of granular ZnO/Al2O3 catalyst for the removal of hazardous trichloroethylene. *Journal of Hazardous Materials, 142*, 88–96.
28. Li, J., Guo, Z., Chen, X., Liu, J.-H., & Huang, H.-J. (2011). ZnO/CuO hetero-hierarchical nanotrees array: Hydrothermal preparation and self-cleaning properties. *Langmuir, 27*(2011), 6193–6200.
29. Filipic, G., & Cvelbar, U. (2012). Copper oxide nanowires: a review of growth. *Nanotechnology, 23*(2012), 194001.
30. Tian, Y., Yan, J. W., Xue, R., & Yi, B. (2011). Capacitive properties of activated carbon in K4Fe (CN) 6. *Journal of Electrochemical Society, 158*(2011), A818.
31. Zhao, X. Y., Cao, J. P., Morishita, K., Ozaki, J. I., & Takarada, T. (2010). Electric doublelayer capacitors from activated carbon derived from black liquor. *Energy Fuels, 24*(2010), 1889.
32. Kumar, S. S., Venkateswarlu, P., Rao, V. R., & Rao, G. N. (2013). Synthesis, characterization and optical properties of zinc oxide nanoparticles. *International Nano Letters, 2013*(3), 30
33. Rahman, A., Nur, F., Hawlader, M., & Afroz, R. (2015). Fuzzy controlled evaporative battery thermal management system for EV/HEV. *International Journal of Electric and Hybrid Vehicles, 7*(1), 22–39.
34. Chen, X. (2017). Carbon-based supercapacitors for efficient energy storage. *National Science Review, 4*(3), 453–489. https://doi.org/10.1093/nsr/nwx009.
35. Carlson, T., Ordéus, D., Wysocki, M., & Asp, L. E. (2011). CFRP structural capacitor materials for automotive applications. *Plastics, Rubber and Composites, 40*(6–7), 311–316. https://doi.org/10.1179/174328911x12948334590286
36. Carlson, T., Ordéus, D., Wysocki, M., & Asp, L. E. (2010). Structural capacitor materials made from carbon fibre epoxy composites. *Composites Science and Technology, 70*(7), 1135–1140.

37. Dorigato, A., Giusti, G., Bondioli, F., & Pegoretti, A. (2013). Electrically conductive epoxy nanocomposites containing carbonaceous fillers and in-situ generated silver nanoparticles. *Express Polymer Letters, 7*(8), 673–682.
38. Ladani, R. B., Wu, S., Kinloch, A. J., Ghorbani, K., Zhang, J., Mouritz, A. P., & Wang, C. H. (2015). Improving the toughness and electrical conductivity of epoxy nanocomposites by using aligned carbon nanofibres. *Composites Science and Technology, 117*, 146–158. https://doi.org/10.1016/j.compscitech.2015.06.006
39. Imran, K. A., & Shivakumar, K. N. (2017). Enhancement of electrical conductivity of epoxy using graphene and determination of their thermo-mechanical properties. *Journal of Reinforced Plastics and Composites, 37*(2), 118–133. https://doi.org/10.1177/0731684417736143
40. Tuncer, E., Sauers, I., James, D. R., Ellis, A. R., Paranthaman, M. P., Aytuğ, T., & Goyal, A. (2006). Electrical properties of epoxy resin based nano composites. *Nanotechnology, 18*(2), Article 025703.
41. Hamdi, K., Aboura, Z., Harizi, W., & Khellil, K. (2017). Improvement of the electrical conductivity of carbon fiber reinforced polymer by incorporation of nanofillers and the resulting thermal and mechanical behavior. *Journal of Composite Materials, 52*(11), 1495–1503.
42. Qu, M., Nilsson, F., & Schubert, D. (2018). Effect of filler orientation on the electrical conductivity of carbon fiber/PMMA composites. *Fibers, 6*(1), 3.
43. Rojas, M., C. M., D., Flores, P., Peréz-Tijerina, E., & Meléndrez, M. F. (2016). Effect of ZnO nanoparticles obtained by arc discharge on thermo-mechanical properties of matrix thermoset nanocomposites. *Journal of Applied Polymer Science, 133*(30).
44. Rani, S., Suri, P., Shishodia, P., & Mehra, R. (2008). Synthesis of nanocrystalline ZnO powder via sol–gel route for dye-sensitized solar cells. *Solar Energy Materials and Solar Cells, 92*(12), 1639–1645.

Chapter 6
Development of AI Organic Ultra-Supercapacitor

Abstract Worldwide recently awares about the commercialization of EV to reduce the environmental threats about 90% which could be due to the internal combustion engine (ICE) engine vehicle greenhouse gas (GHG) emissions. Electricity is a potential clean energy substitute for ICE vehicles. This could be the biggest obstacle to charging electric vehicles in the future. Moreover, EVs are limited by range and speed. Most of the EVs have mileage range about 50–100 and low speed. EV's battery has longer charging time. While it only takes a couple of minutes to fill up the fuel tank of IC vehicle, EVs take about 4–6 h to get fully charged. Depending on the type and usage, batteries of all EVs must be replaced every 3–10 years. In addition, the main traction batteries of electric vehicles are overloaded by car accessories, radios, air conditioners and electrical systems. These applications drain battery faster and increase the frequency of battery charging, resulting in poor battery life. Artificial Intelligent (AI) control organic ultra-supercapacitor EV's body panel is one of the potential solution to generate energy from trapping renewable solar energy and it could be useful to meet the axially power demand of EV including dashboard.

Keywords Climate change · Artificial intelligent · GHG emission · EV mileage extended

6.1 Introduction

The innovation of electric vehicles, researchers have been investigating new technologies to overcome the limitations of electric such as mileage, battery life and maintenance cost. One of the promising technologies is the utilization of SOSC as body panel in EV. Conventional supercapacitor applications have already been utilized in EV as regenerative braking and power supply for start-stop system, etc. Limitations of conventional supercapacitors include expensive cost per Watt-hour and low specific energy (Wh/kg), which is only a fraction of regular battery. By utilizing SOSC as body panel of EV, energy density and quick discharging can be improved. Supercapacitor applications and solar applications have great potential

A. Rahman et al., *Autonomous Driveline System for Electric Vehicle*, Green Energy and Technology, https://doi.org/10.1007/978-981-96-8418-2_6

for advanced improvements. Combination of current activated carbon-based supercapacitors and lightweight solar panels would find its place in commercialization in the near future as structural body of EV. Furthermore, harvesting both electrical and solar energies from a single hybrid system is highly desirable and representing a new trend of all-in-one multiple energy harvesting technology [1, 2].

EV could potentially be improved by integrating organic panel type composite body with high conductive metal oxides (Graphene Oxide, Copper Oxide, Zinc Oxide, etc.). The Nano-micro filler epoxy resin could play a crucial role on the clean energy technology by reducing weight and functioning as a battery or capacitor which can store the renewable energy from various systems. The panel type battery has multifunctional characteristics, weight saving, energy storing/supplying on demand and protecting vehicle's body from external heat as an insulator. Based on the investigations, it can be summarized that downsizing of battery pack and capacitor replacement can be made by using light and efficient composite body panel [3, 4]. By utilizing solar supercapacitor as body panel of EV, energy density and quick discharging can be improved.

In recent years, many automobile manufacturers have invested in transforming the conventional vehicle into an EV that can provide renewable and green solution. EV demand is rising in terms of market share [5]. With advanced innovations and competitive price compared to commercial petrol vehicle, EV is a better choice not only for the users but also for the environment. Substantial research and innovation is required to enhance the range of EVs to overtake the vehicle market in the future [6]. The choice of the appropriate design parameters requires deep understanding of vehicle use purposes, study of driving cycles, vehicle size and weight, desired performance and type of application. Once the design parameters have been selected, the second step is the design of an energy management control strategy which is an essential key for an efficient electric vehicle [7].

Supercapacitors have several advantages over electrochemical batteries and fuel cells, including higher power density, shorter charging times, and longer cycle life and shelf life [48, 50]. The first group of materials, which is most frequently used in commercial supercapacitor electrodes, is carbon. Carbon has gained considerable interest due to its various properties including the existence of different allotropies, accessibility, low cost, high surface area, high conductivity, superior corrosion inhibition property, high temperature stability and environmental friendliness. So far the forms of carbon, which are well recognized in the use of the supercapacitor electrodes, are activated carbon, carbon aerogels, carbon nanotubes (CNTs), carbon fibers and grapheme [8]. Another attractive material for supercapacitor electrodes is carbon fiber. Carbon fiber has superior charge transportation property due to one dimensionality. It provides high adsorption capacities to ions due to the existence of large number of pores at the surface of the fiber and proves to be the best candidate for electric double-layer capacitor (EDLC) electrodes [9].

This paper presents modeling and developing of organic solar structural supercapacitor (OSSC) as EV body panel. The ultimate goal of this paper is to develop an OSSC system which will serve as a photovoltaic energy generator and store the renewable energy from the solar heat trapping and provide accurate power to EV's

Load (Accessories) and to charge EV's auxiliary battery. The SOSC model has been be developed by referencing experimental data from solar supercapacitor samples which are developed by sandwiching the CuO/ZnO and filler epoxy Resin (ER) with CF sheets. Simulation study is conducted to determine the performance of the SOSC and Power Management System (PMS) of SOSC using the adaptive neuro fuzzy intelligent system (ANFIS).

6.2 Development oF SOSC

The composite body panel will also act as a protection layer from external heat as insulator. CuO powder can facilitate higher electron transfer along their surfaces. ZnO nanostructures can have a great advantage to apply to a catalytic reaction process due to their large surface area and high catalytic activity. Therefore, synthesized CuO/ZnO material will result faster charging time. Presence of ER makes the nanostructures of CuO/ZnO compact, thus storing the charge longer without discharging too fast as a conventional capacitor. Photovoltaic thermal hybrid solar collectors, sometimes known as hybrid PV/T systems or PVT, are devices that can convert solar energy into thermal and electrical energy. A solar cell which converts sunlight into electricity is combined with a solar thermal collector which captures the remaining energy and removes waste heat from the PV module. Therefore, they are more energy efficient than standalone solar PV or solar thermal. Due to the presence of the conducting polymer, the panel can also harness solar energy and convert it to electricity.

The organic solar cell, electrical circuits, converter, controller, circuit breaker switch and batteries were all integrated for the modeling of OSSC. The OSSC was developed utilizing separator/Na_2SO_4 electrolyte, zinc oxide (ZnO), copper oxide (CuO), conductive polymer (ER) and carbon fiber (CF) as shown in Fig. 6.1. CuO doped polymer with CF reinforced is utilized to improve the concentration of electrons.

6.3 Development Process of OSSC

Carbon Fiber fabrics derived from PiCarbon (240 g/m^2) 0.25 mm thick is used for the development of cell. The CF acts as a rising electrode. Based on the compatibility with carbon fibers, epoxy resin and hardener was selected. The transparent epoxy resin and hardener came in a box with a recommended mixing ratio of 2:1 and a full 24 h curing period between 25 and 30 °C. The CF electrodes are reinforced with nano ZnO epoxy for the negative electrode in order to facilitate high electron mobility as n-type semiconductor with a bandage energy of 3.2–3.4 eV increases to 3.5 eV by adding 6% π-carbon [10]. The CF electrodes were reinforced with epoxy-filled nano-CuO as p-type semiconductor to facilitate bore/positive charging. They improve the

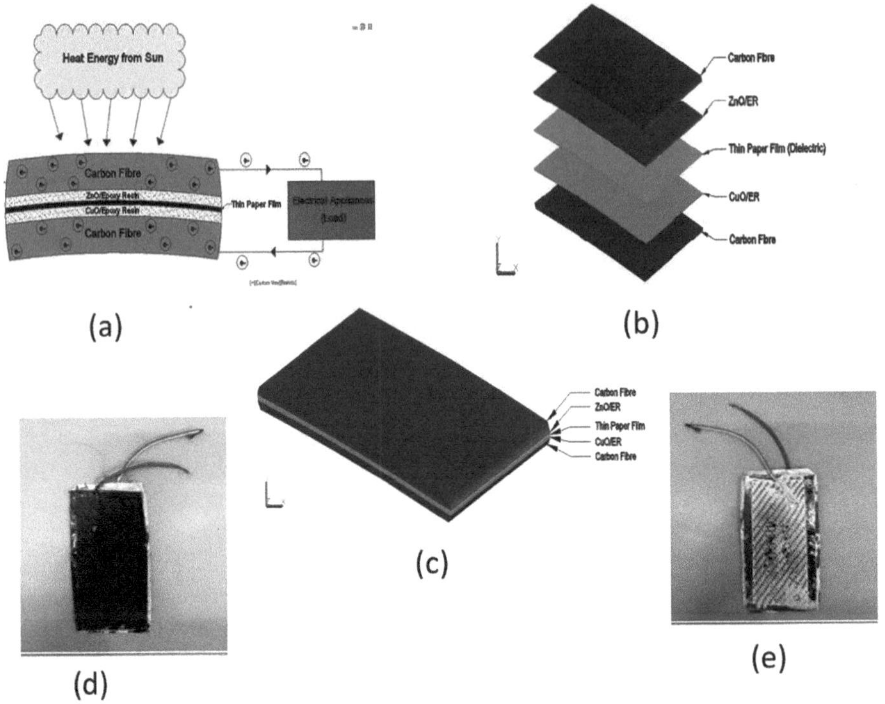

Fig. 6.1 Sample of OS-USC **a** Structure, **b** 3D-Auto CAD prototype, **c** Exploded View **d** Prototype with dielectric **e** Prototype with Na_2SO_4 aqueous dielectric [1]

conductivity of the OSSC. The dielectric separator film was placed between the ZnO doped ER and the CuO doped ER [11, 12]. The laboratory scale sample of solar supercapacitor is shown in Fig. 6.1.

6.4 Simulation Study

Based on experimental results of SSC samples, simulation study is conducted to investigate the full potential of SSC system integrated with EV battery. Full-sized SSC system will have 2.4 m^2 of roof area and be able to provide power to auxiliary loads of EV and charge EV main traction battery. Efficiency of the SSC from experiment will be converted for the simulations of full solar supercapacitor system with DC-DC converters and EV auxiliary load. Any physics-based system can be modeled and simulated by using software from MATLAB-Simulink. The simulation of IoT control systems and controllers to optimize the power management of structural solar capacitor is the main focus for this experiment. MATLAB-Simulink and ANFIS will be used to achieve the main objectives of this study.

6.5 Simulink Model of Structural SOSC

The system is designed as solar panel block, supercapacitor block, Li-ion battery block and voltage converters as shown in Fig. 6.2. EV accessories load will be supported by combination of Solar Supercapacitor (Structural Capacitor) and battery. Bi-directional converters are used to boost and buck the input voltage to required system voltage. Conventional capacitor is used as a DC-link between the supply side and load side.

Modeling of Battery

The Battery block in Simulink implements a generic dynamic model that represents most popular types of rechargeable batteries. Figure 6.3 shows the equivalent circuit of the battery model. The circuit parameters are modified to represent a specific battery type and its discharge characteristics. A typical discharge curve consists of three sections as shown in Fig. 6.4. The first section represents the exponential voltage drop when the battery is charged. Battery type determines the width of the voltage drop. The grey area represents the available charge that can be extracted from the battery until the voltage drops below the battery nominal voltage. Finally, the white area represents the total discharge from the battery when the voltage drops rapidly. When the battery current is negative, the battery recharges following a charge characteristic as shown in Fig. 6.5.

The lead–acid battery model is represented by the following equations:

- Discharge Model ($i^* > 0$)

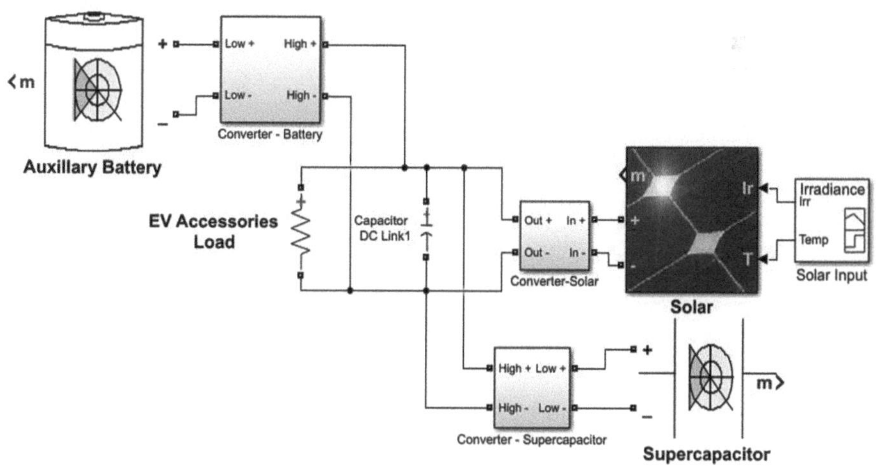

Fig. 6.2 Solar supercapacitor basic Simulink system [4]

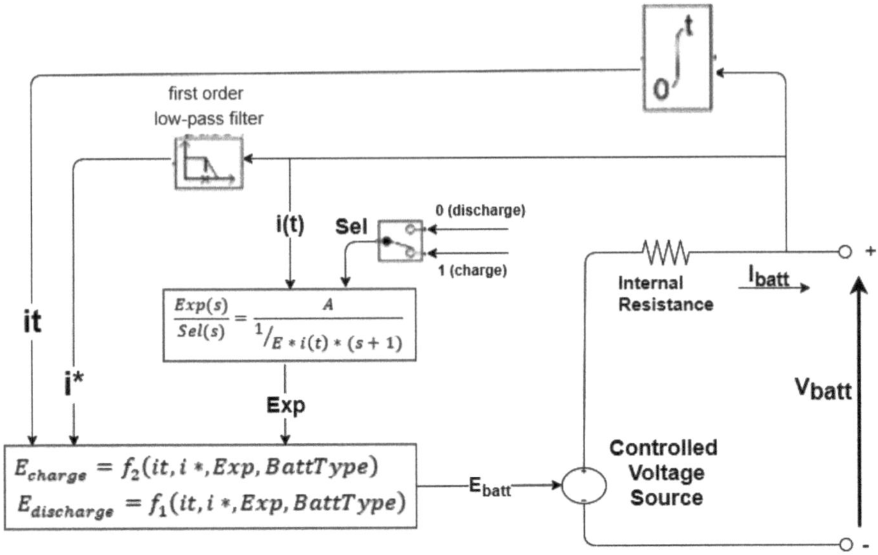

Fig. 6.3 Equivalent circuit of battery [1]

Fig. 6.4 Typical discharge characteristics of battery in Simulink [2]

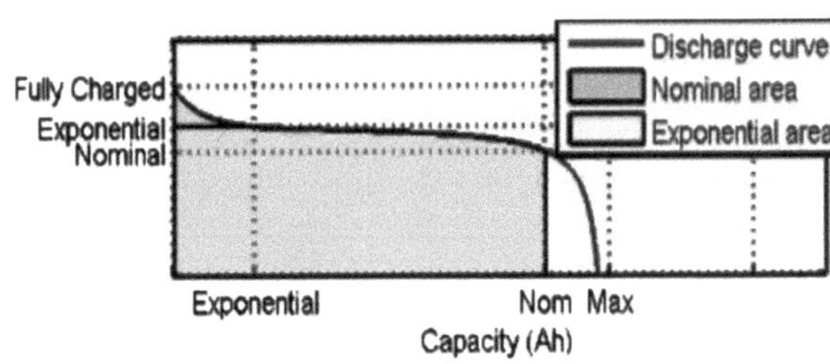

Fig. 6.5 Charge characteristic of lead–acid and Li-Ion battery in Simulink

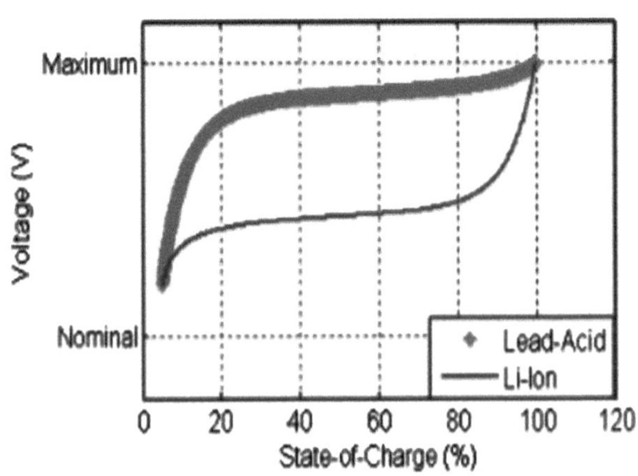

$$f_1(it, i*, i, Exp) = E_0 - K \cdot \frac{Q}{Q - it} \cdot i * - K \cdot \frac{Q}{Q - it} \cdot it + Laplace^{-1}\left(\frac{Exp(s)}{Sel(s)} \cdot 0\right) \tag{6.1}$$

- Charge Model (i* < 0)

$$f_2(it, i*, i, Exp) = E_0 - K \cdot \frac{Q}{it + 0.1.Q} \cdot i * - K \cdot \frac{Q}{Q - it} \cdot it + Laplace^{-1}\left(\frac{Exp(s)}{Sel(s)} \cdot \frac{1}{s}\right) \tag{6.2}$$

The lithium-ion battery model is represented by the following equations:

- Discharge Model (i* > 0)

$$f_1(it, i*, i, Exp) = E_0 - K.\frac{Q}{Q - it}.i * - K.\frac{Q}{Q - it}.it + A.exp(-B, it) \tag{6.3}$$

- Charge Model (i* < 0)

$$f_1(it.i * .i, Exp) = E_0 - K.\frac{Q}{it + 0.1Q}, i * - K.\frac{Q}{Q - it}, it + A.exp(-B, it) \tag{6.4}$$

where E_{Batt} is the nonlinear voltage (V), E_0 is the constant voltage (V), *Exp(s)* is the exponential zone dynamics (V), *K* is the polarization constant (V/Ah) or polarization resistance (Ohms), *i** is the low-frequency current dynamics (A), *i* is the battery current (A), *it* is the extracted capacity (Ah), *Q* is the maximum battery capacity (Ah), *A* is the exponential voltage (V), *B* is the exponential capacity (Ah^{-1}), and *Sel(s)* is the battery mode. *Sel(s)* is zero during battery discharge and is one during battery charging.

Limitation

Limitation of the Simulink Battery Model:

- The minimum no-load battery voltage is 0 V and the maximum battery voltage is equal to 2 × E0.
- The minimum capacity of the battery is 0 Ah and the maximum capacity is Qmax.

Assumption

Assumptions to be made to model the battery in Simulink:

- During the charge and discharge cycles, the internal resistance is assumed to be constant and does not vary with the amplitude of the current [13]
- The parameters of the model are derived from the discharge characteristics. The discharging and charging characteristics are assumed to be the same [14].

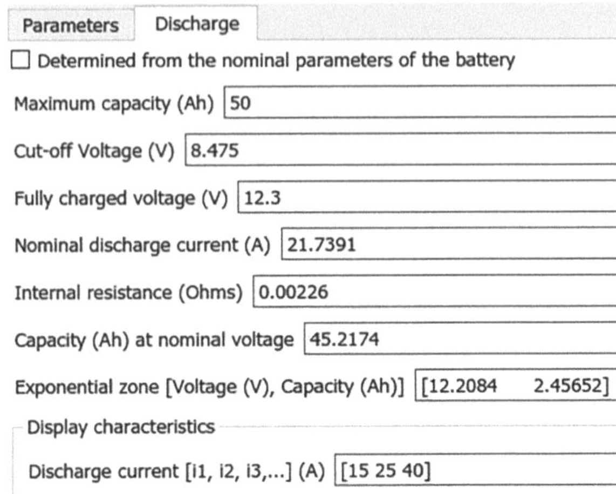

Fig. 6.6 Parameters and discharge properties of battery block in Simulink

- The amplitude of the current does not change with the capacity of the battery (No Peukert effect) [15]
- The self-discharge of the battery is not considered. It can be simulated by adding a large resistance in parallel with the battery terminals [16]
- There is no memory effect in the battery.

Battery is defined as 12 V and 50Ah as shown in Fig. 6.6 to represent the auxiliary battery of the EV to be charged by solar energy and to provide power to electronic accessories. The discharge profile of the designed battery is as shown in Fig. 6.7.

Modeling of Structural Organic Supercapacitor

The Supercapacitor block implements a generic model which can be parameterized to represent most popular types of supercapacitors. Figure 6.8 shows the equivalent circuit of the supercapacitor in Simulink.

The supercapacitor output voltage is expressed by Stern equation as

$$V_{SC} = \frac{N_s Q_T d}{N_P N_e \varepsilon \varepsilon_0 A_i} + \frac{2 N_e N_s RT}{F} sinh^{-1}\left(\frac{Q_T}{N_P N_e^2 A \sqrt{BRT \varepsilon \varepsilon_0 c}}\right) - R_{SC}.i_{SC} \tag{6.5}$$

with

$$Q_T = \int i_{SC} dt \tag{6.6}$$

The supercapacitor electric charge is modified as follows (when $i_{SC} = 0$) to represent the self-discharge phenomenon:

$$Q_T = \int i_{Self\_dis} dt \tag{6.7}$$

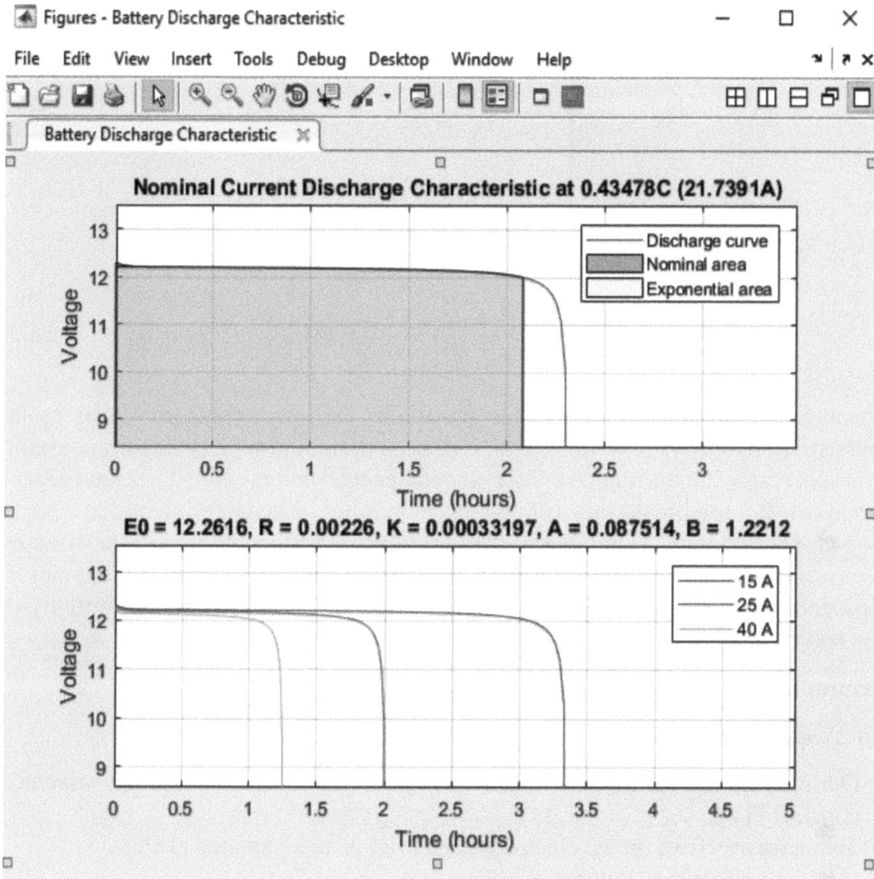

Fig. 6.7 Discharge profile of battery model in Simulink [2]

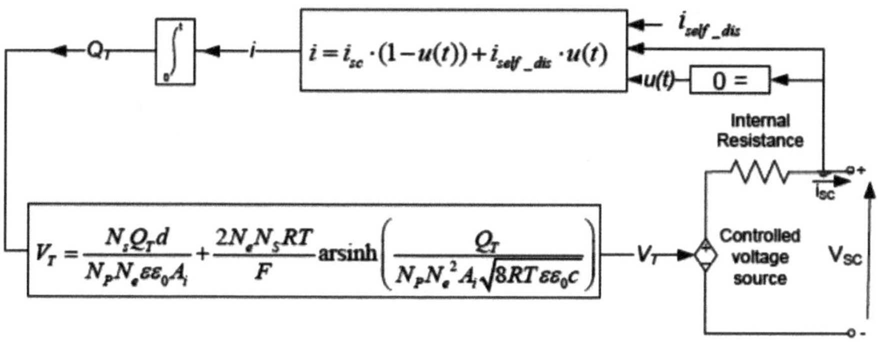

Fig. 6.8 Equivalent circuit of supercapacitor [2]

$$i_{self.dis} = \begin{cases} \frac{C_T \alpha_1}{1 + sR_{SC}C_T} if\ t - t_{oc} \le t_3 \\ \frac{C_T \alpha_1}{1 + sR_{SC}C_T} if\ t_3 < t - t_{oc} \le t_4 \\ \frac{C_T \alpha_1}{1 + sR_{SC}C_T} if\ t - t_{oc} > t_4 \end{cases} \quad (6.8)$$

The SOC is calculated as

$$SOC = \frac{Q_{in - \int_0^t i(\tau)d\tau}}{Q_T} x\ 100\% \quad (6.9)$$

where A_i is interfacial area between electrodes and electrolyte (m^2), c is molar concentration (mol/m^3, $c = 1/8N_A r^3$), r is molecular radius (m), F is Faraday constant, i_{sc} is supercapacitor current (A), V_{sc} is supercapacitor voltage (V), C_T is total capacitance (F), R_{sc} total resistance (ohms), N_e is number of layers of electrodes, N_A is avogadro constant, N_p is number of parallel supercapacitors, N_s is number of series supercapacitors, Q_T is electric charge (C), R ideal gas constant, d is molecular radius, T is operating temperature (K), ε is permittivity of material and ε_0 is Permittivity of free space.

Assumption

Supercapacitor is modeled in Simulink with following assumptions:

- During the charge and the discharge cycles, internal resistance is assumed constant [17].
- Temperature effect on the electrolyte material is no accounted [18].
- No aging effect is accounted.
- For all values of voltage, charge redistribution is the same [19].
- Cell balancing is not represented by the block.
- Current is continuous through the supercapacitor [20].

Parameters from the experiment (Table 6.1) is used as reference to construct the model of Solar Supercapacitor for optimal roof area of EV (2.4 m^2) in Simulink as shown in Fig. 6.9.

Table 6.1 Experiment parameters of SSC

SC dimension (cm^2)	45
Weight (gram)	9–12
Power generation (W/day)	2800
Power density (kW/kg)	33
Energy density (Wh/kg)	130
Capacitance ($\mu F/cm^2$)	20.15

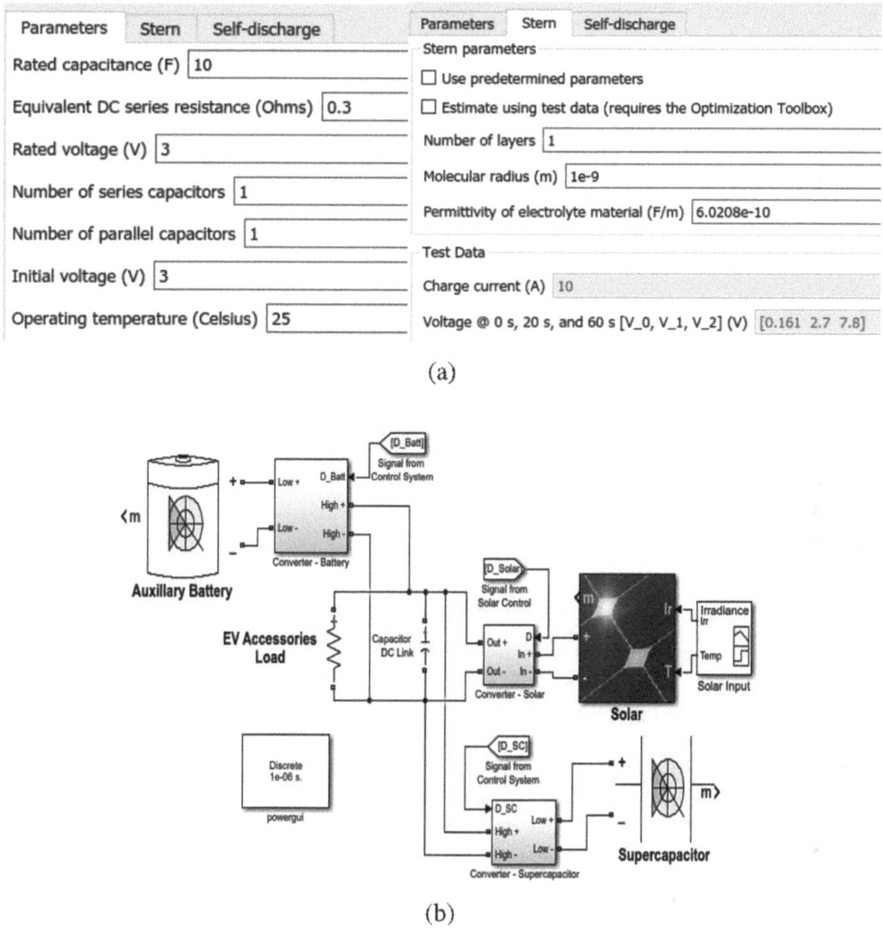

Fig. 6.9 **a** Parameters of supercapacitor in Simulink, **b** Simulink model of organic solar capacitor [4]

Modeling of Solar Energy Harvesting Using SOSC

The PV Array block implements an array of photovoltaic (PV) modules. The array is built of strings of modules connected in parallel, each string consisting of modules connected in series. Pre-set PV modules from the National Renewable Energy Laboratory (NREL) System Advisor Model [21] and also user-defined PV modules can be modelled [22]. The PV Array block is a five-parameter model using a light-generated current source (I_L), diode, series resistance (R_s), and shunt resistance (R_{sh}) to represent the irradiance and temperature dependent I-V characteristics of the modules [23–25], as shown in Fig. 6.10.

The diode I–V characteristics for a single solar module are defined by the equations:

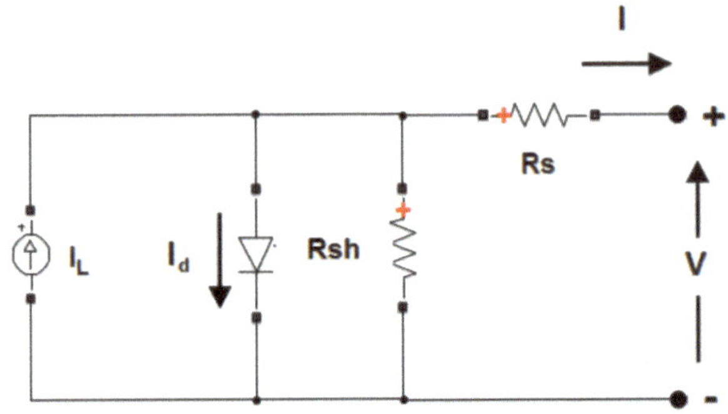

Fig. 6.10 Equivalent electrical circuit of solar module [2]

$$I_d = \left[exp\left(\frac{V_d}{V_T}\right) - 1\right] \quad (6.10)$$

$$V_T = \frac{k_T}{q}(nl \times N_{cell}) \quad (6.11)$$

where I_d is diode current (A), V_d is diode voltage (V), I_0 is diode saturation current (A), nI is diode ideality factor (close to 1.0), k is Boltzmann constant (1.3806e^{-23} J.K^{-1}), q is electron charge (1.6022e^{-19} C), T is cell temperature (K), and N_{cell} is number of cells connected in series in a module. The output current of PV module is given at equation as follows:

$$I = N_p.I_{ph} - N_p.I_s.\left[exp\frac{q(V + I_{RS})}{N_s.k.T.A} - 1\right] \quad (6.12)$$

Photocurrent of the PV module depends linearly with irradiance and temperature which is described at Eq. 6.2 as follows:

$$I_{ph} = [I_{SC} + K_i(T - T_r)].\lambda \quad (6.13)$$

The PV module current saturation varies with cell temperature and is defined by Eq. 3.14 as follows:

$$I_s = I_{rs}\left[\frac{T}{T_r}\right]^3 exp\left[q.Eg\frac{1/T_r - 1/T}{kA}\right] \quad (6.14)$$

The reverse current saturation of PV module in reference of temperature is given at Eq. 3.15 as follows:

$$I_{rs} = \frac{I_{SC}}{exp\left[\frac{q.V_{OC}}{N_s.k.AT}\right] - 1} \quad (6.15)$$

where I_{ph} is the photocurrent, I_s is saturation current of the PV module, q is electron charge ($1.6022e^{-19}$ C), k is Boltzman constant ($1.3806e^{-23}$ J.K^{-1}), T is temperature of the PV cell (°C), A is the ideal factor of PV module, R_{Sh} is the shunt resistance (Ohm), R_s is the series resistance, N_s is the number of cells connected in series, N_p is number of cells connected in parallel, I_{SC} is the cell short circuit current at 25 °C temperature and 1000 W/m2 irradiance, K; is the cell short circuit current temperature coefficient, T_r is the cell reference temperature, λ is the irradiance level of PV module in 1000 W/m2, I_{rs} is the reverse saturation current in reference of temperature and irradiance, E_g is band-gap energy of semiconductor which will be used by the PV module cell and V_{oc} is the PV module open cirüüf temperature [26, 27, 29].

Fill factor (FF) is another defining term in the overall behavior of a solar cell. The quality of a solar cell is measured by this factor. Fill factor is defined as

$$FF = \frac{p_m}{V_{OC}\, x\, I_{SC}} = \frac{\eta\, A_c\, G}{V_{OC}\, I_{SC}} \tag{6.16}$$

where FF is fill factor, P_m is maximum power (W), V_{OC} is open-circuit voltage, I_{SC} is short-circuit current (A), η is efficiency of solar panel, A_C is area of solar panel (m^2) and G is solar irradiance (W/m^2).

The fill factor is directly affected by the values of the cell's series, shunt resistances and diodes losses. Increasing the shunt resistance (R_{sh}) and decreasing the series resistance (R_s) lead to a higher fill factor, thus resulting in greater efficiency, and bringing the cell's output power closer to its theoretical maximum. Typical fill factors range from 50 to 82%. The fill factor for a normal silicon PV cell is 80%. Single-junction flat-plate terrestrial solar cells are fundamentally limited to about 30% solar-to-electricity conversion efficiency, but multiple junctions and concentrated light make much higher efficiencies practically achievable. Until now, four-junction III–V concentrator solar cells have demonstrated the highest solar conversion efficiencies. In latest technology in solar panel, 47.1% solar conversion efficiency can be achieved by using a monolithic, series-connected, six-junction inverted metamorphic structure operated under the direct spectrum at 143 Suns concentration [28–30]. Further reduction of the series resistance within this structure could realistically enable efficiencies over 50%. This efficiency and parameters from experiment results are used as reference to model solar panel with potential roof area 2.4 m^2 in Simulink Irradiance and temperature are set as constant 1000W/m^2 and 25 °C as shown in Fig. 6.11, to simulate the maximum performance of the system using the Simulink model as shown in Fig. 6.12.

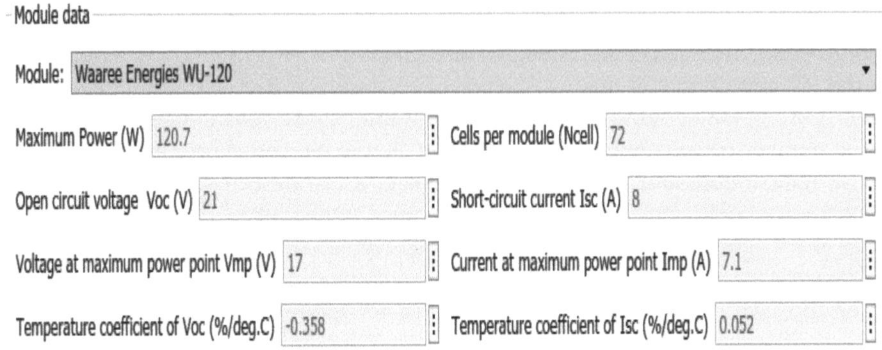

Fig. 6.11 Parameters of solar panel [2]

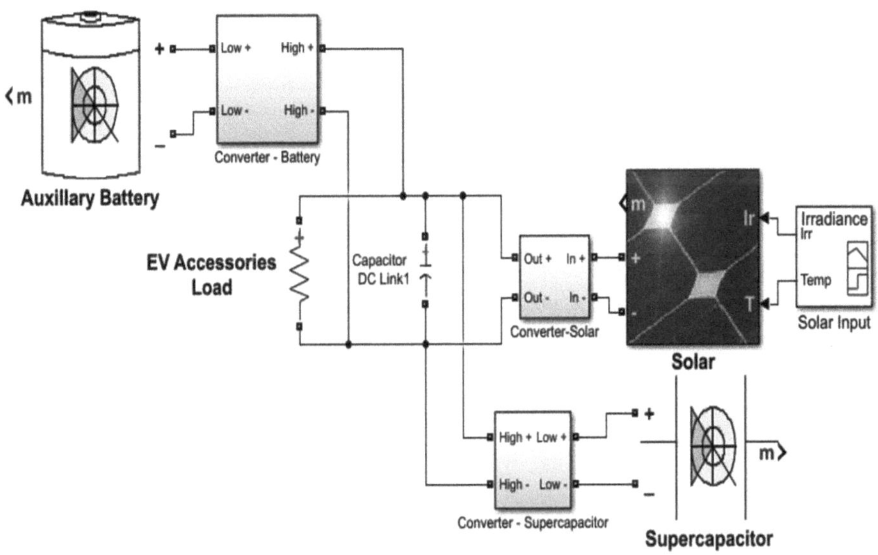

Fig. 6.12 Solar supercapacitor basic Simulink system [3]

6.6 Conclusion

Single-junction flat-plate terrestrial organic solar cells (OSC) are fundamentally limited to about 30% solar-to-electricity conversion efficiency, but multiple junctions and concentrated light make much higher efficiencies practically achievable. However, 47.1% solar conversion efficiency can be achieved by using a monolithic, series-connected, six-junction inverted metamorphic structure.

Innovations set to change the capacitor effectiveness include designing the electrodes with graphene to create lightweight OSSC with energy-storage capabilities between 150 and 550 F/g. There are also environmental and economic benefits of utilizing solar supercapacitor and EV such as reduction of GHG, reducing dependence on fossil fuel and creating economic development and jobs in new sector.

Acknowledgements The authors are grateful to the Research Management Centre (RMC), International Islamic University Malaysia for financing this study. Authors are also grateful to the Automotive Composite Laboratory Technician helps to develop the organic supercapacitor from this study.

References

1. Rahman, M. A., & Golam, K. (2021). Al-ZnO/CuO_Cfrp organic structural auxiliary energy storage system for electric vehicle. *Journal of Solid State Technology, 28*(1 & 2), 103–116.
2. Aung, K. M. (2022). *Development of solar structural supercapacitor system for EV*. PhD thesis. Library, International Islamic University Malaysia.
3. Ataur, R. & Kaw, M. A. (2021). Development of solar supercapacitor by utilizing organic polymer and metal oxides for subsystem of electric vehicle. *Journal of Material Research Express, 8*(2021), 125301.
4. Rahman, A. Kaw, M. A., Ihsan, S., Raja, A. S., Qubeissi, A. L. M., & Aljarrah, M. T. (2023). Solar energy dependent supercapacitor system with ANFIS controller for auxiliary load of electric vehicles. *Energies, 2023*(16), 2690.
5. Rahman, A., Mahbubur, R. & Mohiuddin, A. K. M. (2015b). The prospects of panel style nano-battery technology for EV/HEV. In *2015 IEEE international conference on industrial engineering and engineering management, paper no. IEEM 2015-P-0509, December 6–9*. Singapore.
6. Rahman, A., Mahbubur, R., Hawlader, M. N. A., & Mohiuddin, A. K. M. (2015). Nano-battery technology for EV/HEV. A Pioneering Study. *IIUM Engineering Journal, 16*(2), 57–68.
7. Ataur, R., Kyaw, M. & Shaheer, A.K (2020) Organic photoelectric-supercapacitor power source of electric vehicle. *ISSN: 0193–4120 Page No. 15175–15182*
8. Young-Min, K., Jang-Hee, L., Seok-Joon, K. & Daniel, F. (2012). Potential and evolution of compressed air energy storage: energy and energy analyses. *Entropy, 2012*(14), 1501–21.
9. Burke, A. & Zhao, H. (2015) Applications of supercapacitors in electric and hybrid vehicles. In *5th European symposium on supercapacitor and hybrid solutions (ESSCAP), Brasov, Romania, April 23–25, 2015*.
10. Tian, Y., Yan, J.W., Xue, R., & Yi, B. (2011). Capacitive properties of activated carbon in K4Fe (CN) 6. *Journal of Electrochemical Society, 158*(2011), A818.
11. Conway, B. E. (1999). *Electrochemical supercapacitors: Scientific fundamentals and technological applications*. Kluwer-Plenum.
12. Burke, A. (2000). Ultracapacitors: Why, how, and where is the technology. *Journal of Power Sources, 91*(1), 37–50.
13. Kotz, R., & Carlen, M. (2000). Principles and applications of electrochemical capacitors. *Electrochimica Acta, 45*(15–16), 2483–2498.
14. Jurewicz, K., & Delpeux, S. (2001). Supercapacitors from nanotubes/polypyrrole composites. *Chemical Physics Letters, 347*(1–3), 36–40.
15. Chokri, M., Aymen, F., & Lassaad, S. (2014). An overview of electric vehicle concept and power management strategies. *Conference Paper 978–1–4799–7300–2/14 IEEE*

16. Hongjun, C., Fei, L., & Fujuan, G. (2012). Power management system design for small size solar-electric vehicle. In *2012 IEEE 7th international power electronics and motion control conference—ECCE Asia* (pp. 2658–2662).
17. Ganji, B. & Kouzani, A.Z. (2010). A study on look-ahead control and energy management strategies in hybrid electric vehicles. In *2010 8th IEEE international conference on control and automation (ICCA)* (pp. 388–392).
18. Salmasi, F. R. (2007). Control strategies for hybrid electric vehicles: Evolution, classification, comparison, and future trends. *IEEE Transactions on Vehicular Technology, 56*(2007), 2393–2404.
19. Xin, L. & Williamson, S. S. (2007). Assessment of efficiency improvement techniques for future power electronics intensive hybrid electric vehicle drive trains. In *Electrical power conference, 2007. EPC 2007. IEEE Canada, 2007* (pp. 268–273).
20. Frackowiak, E., & Beguin, F. (2001). Carbon materials for the electrochemical storage of energy in capacitors. *Carbon, 39*(6), 937–950.
21. Saliger, R., Fischer, U., Herta, C., & Fricke, J. (1998) High surface area carbon aerogels for supercapacitors. *Journal of Non-Crystalline Solids, 225*(1998), 81.
22. Choi, J. W., McDonough, J. R., Yang, Y., Mantia, F. L., Zhang, Y., & Cui, Y. (2009). Carbon nanofibre supercapacitors with large areal capacitances. *Applied PhysicsLetters, 95*(2009), Article 243109.
23. Cheah, L.W. (2010). *Cars on a diet: The material and energy impacts of passenger vehicle weight reduction in the U.S.* PhD thesis, MIT, September 2010.
24. Kim, I. H., & Kim, K. B. (2001). Ruthenium oxide thin film electrodes for supercapacitors. *Electrochemical and Solid State Letters, 4*(5), A62–A64.
25. Ryu, K. S., & Kim, K. M. (2002). Symmetric redox supercapacitor with conducting polyaniline electrodes. *Journal of Power Sources, 103*(2), 305–309.
26. Zheng, J. P., Cygan, P. J., & Jow, T. R. (1995). Hydrous ruthenium oxide as an electrodematerial for electrochemical capacitors. *Journal of Electrochemical Society, 142*(1995), 2699.
27. Chen, J. C., & Tang, C. T. (2007). Preparation and application of granular ZnO/Al2O3 catalyst for the removal of hazardous trichloroethylene. *Journal of Hazardous Materials, 142*, 88–96.
28. Filipic, G., & Cvelbar, U. (2012). Copper oxide nanowires: a review of growth. *Nanotechnology, 23*(2012), 194001.
29. Tian, Y., Yan, J.W., Xue, R. & Yi, B. (2011). Capacitive properties of activated carbon in K4Fe (CN) 6. *Journal of Electrochemical Society, 158*(2011), A818.
30. Zhao, X. Y., Cao, J. P., Morishita, K., Ozaki, J. I., & Takarada, T. (2010). Electric doublelayer capacitors from activated carbon derived from black liquor. *Energy & Fuels, 24*(2010), 1889.

Chapter 7
AI Power Management System Organic Ultra-Supercapacitor

Abstract The EV auxillary load's power and energy needs are satisfied by the organic solar supercapacitor (OSSC). For the simulation using Simulink, the solar irradiance profile was adjusted to 1000 W/m^2 for 30% from 10:00 am to 3:00 pm, 600 W/m^2 for 30% from 3:00 pm to 5:00 pm, 200 W/m^2 for 5:00 pm to 7:00 pm, and 0 W/m^2 (no sunlight). The EV load receives 500 W from the supercapacitor at 1000 W/m^2 of solar radiation. The OSSC's output might cut the weight of an EV battery by approximately 7.5% and its size by 10%. Furthermore, taking into account the greenhouse gas emissions of 2.31 kgCO$_2$/liter of gasoline, it would be able to lower them by 25%. When comparing the system response under tuned PID and FLC, the FLC performs better than the PID controller in terms of reference point tracking. While the tuned PID provides a respectable response with overshoot, the fuzzy controller retains great tracking of the reference condition.

Keywords AI power management system · Solar capacitor · EV auxially power source · Solar power tracking

7.1 Introduction

The effectiveness of a solar supercapacitor power management system using PI and fuzzy logic controllers is examined under various load scenarios and irradiance levels. Solar irradiance profile settings include 1000 W/m^2 at 30% for 10 a.m. to 3 p.m., 600 W/m^2 at 30% for 3 p.m. to 5 p.m., 200 W/m^2 at 5.00 p.m. to 7 p.m., and 0 W/m^2 (no sunlight). Simulink signal statistics indicate that the average solar power during a given time period is 495.4 W. The solar supercapacitor system provides 500 W to the EV load at 1000 W/m^2. At 500 W/m^2, the EV load cannot be supplied with 500 W of solar electricity alone.

As a result, the extra battery provides the EV load with 4 A. The extra battery must provide greater power to the EV load at 28 A at 200 W/m^2. In order to give the converter system precise signals, the power management system with FLC is working more effectively. When comparing the system response under tuned PID and FLC, the

A. Rahman et al., *Autonomous Driveline System for Electric Vehicle*, Green Energy and Technology, https://doi.org/10.1007/978-981-96-8418-2_7

FLC performs better than the PID controller in terms of good reference point tracking, which in turn affects system energy consumption and comfort standards. While the tuned PID provides a respectable response with overshoot, the fuzzy controller retains great tracking of the reference condition. The PID controller must therefore be adjusted because some of its resilience is lost. For HESS applications, the FLC has a better robustness characteristic than the adjusted PID.

To regulate the voltage and current of each device, realistic models of solar supercapacitor systems and power management systems are necessary. Buck–boost converters, both one-directional and bi-directional, are made to control the system's power flow. A DC/DC power converter that steps up or steps down voltage from its input [1–4] to its output (load) is called a buck–boost converter.

7.2 Mathematical Modeling

The following equations establish the theoretical transfer function of the buck–boost converter and component selection in continuous conduction mode, when current flowing through the inductor never drops to zero:

$$D = 1 - \frac{V_{n.\min}\eta}{V_{\text{out}}} \tag{7.1}$$

where, D is Duty Cycle, $V_{in.min}$ is minimum input voltage (V), η is efficiency of converter (%) and V_{out} is output voltage (V).

$$di = I_{\text{ripple}} + I_{\text{out}}\frac{V_{\text{out}}}{V_{\text{in}}} \tag{7.2}$$

where, di is input current and I_{ripple} is inductor ripple current (20–40% of output current).

$$f_s = di\frac{L(v_{\text{out}} - v_{\text{in}})}{V_{\text{out}}V_{\text{in}}} \tag{7.3}$$

$$L = \left[v_{\text{in}}\frac{v_{\text{out}} - v_{\text{in}}}{di f_s v_{\text{out}}}\right] \tag{7.4}$$

where, L is inductance (H) and f_s is switching frequency.

$$C = \frac{Q}{f_s dv} \tag{7.5}$$

where, C is the capacitance, Q is the charge store in the panel (organic photovoltaic cell), dv is the output voltage of the ripple in volt.

Table 7.1 Difference between organic and inorganic solar cells

	Organic PV	In-organic PV
Lifetime	10 000 h	10 years
Efficiency	~10%	~25%
Cost	Low	High
Transparency	Transparent	Opaque
Integration	Easy	Difficult
Flexibility	Flexible	Solid
Weight	Very light	Heavy

The switching frequency determines the size of the inductor. A larger switching frequency will result in a smaller inductor but will increase the switching losses in the circuit. A typical switching frequency for low input voltage applications is 15 kHz which is a good compromise [4, 5].

A PI Controller is a feedback control loop that determines an error signal by calculating the difference between a system's output, in this case the power management between a load, battery, and solar supercapacitor. It is possible to manually or automatically adjust the PI controller coefficients. There are two ways to manually adjust the controllers. Integral gain (Ki) and differential gain (Kd) are initially set to zero in the Ziegler–Nichols method. After that, the proportional gain (Kp) is raised until the oscillations in the system become steady and reliable. The oscillation period is known as "Tu," and this Kp value is known as the ultimate gain, or "Ku." Finally, for each controller type, Kp, Ki, and Kd are configured as indicated in Table 7.1.

7.3 Theoritical Performance of OSSC

A physical and chemical process known as the photovoltaic effect transforms solar cells' energy from sunlight. A physical and chemical phenomena known as the photovoltaic effect directly transforms solar energy into electrical power. The photovoltaic effect is how a solar cell or photovoltaic cell produces electricity. When exposed to light, the PV cell's electrical properties—such as current, voltage, and resistance—change. Blocks of solar cells make up PV modules, sometimes referred to as solar panels. At the moment, one of the greatest technologies for capturing solar energy is solar panel.

The different uses of solar cells are determined by their cost per watt, practicality, and solar energy conversion efficiency (85). A clear picture of the applicability of the two types of photovoltaics or solar cells available today for various application scenarios can be obtained by comparing them. Table 7.1 provides an overview of the distinctions between inorganic and organic solar cells [6–8].

7.4 OSSC with Anfis Controller

A subfield of computer science called artificial intelligence (AI) is capable of analyzing vast amounts of data. Artificial Intelligence approaches include ANNs, ANFIS, and FLC [9]. There are two approaches to ANN training: unsupervised training and supervised training. While unsupervised training simply needs the network's input and an artificial neural network (ANN) is meant to correctly classify the data, supervised training requires a set of inputs as well as the network's intended output. The Takagi–Sugeno fuzzy inference system serves as the foundation for the ANFIS type of ANN. It has the ability to combine the advantages of neural networks and fuzzy logic concepts into a single framework.

In order to estimate nonlinear functions, its inference system translates into a collection of fuzzy IF–THEN rules with learning capabilities. ANFIS is therefore regarded as a universal estimator [10]. ANFIS control is a hybrid approach that uses the gradient method to determine the parameters of the input membership function and the least squares method to determine the parameters of the output function. Figure 7.1 depicts the ANFIS structure.

ANFIS is a straightforward machine learning method that uses fuzzy logic to convert random inputs into a predetermined output through the processing of information and connections between highly interconnected neural networks. The numerical inputs are then mapped into an output by weighting these. ANFIS uses neural network learning techniques to control a fuzzy inference system's (FIS) parameters. A first-order Sugeno model with two fuzzy 'If–Then' rules can express the ANFIS configuration with following rules and equations:

$$\begin{aligned} &\text{Rule (1)}: \ \text{IF}\, x \,\text{is}\, a_1 \ \text{AND}\, y \,\text{is}\, b_1, \ \text{THEN} \\ &f_1 = p_1 x + q_1 y + r_1 \end{aligned} \tag{7.6}$$

$$\begin{aligned} &\text{Rule (2)}: \ \text{IF}\, x \,\text{is}\, A_1 \ \text{AND}\, y \,\text{is}\, B_1, \ \text{THEN} \\ &f_2 = p_2 x + q_2 y + r_2 \end{aligned} \tag{7.7}$$

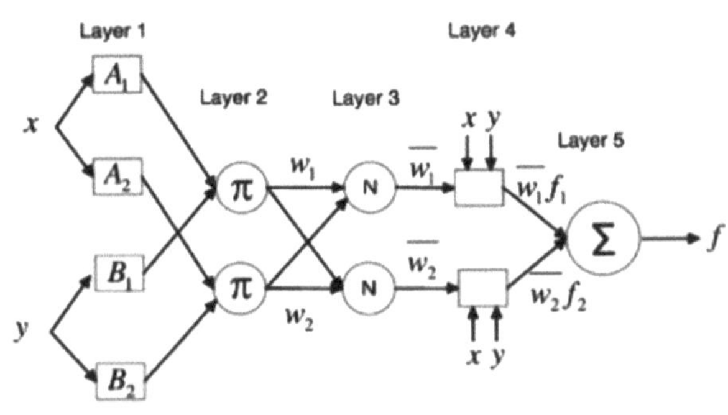

Fig. 7.1 Architecture of ANFIS system [1]

where x_i and y_i are the inputs, a_i and b_i are the fuzzy sets, f_i are the outputs specified by the fuzzy rule and p_i, q_i, and r_i are the design parameters that are determined during the training process.

ANFIS's fixed configuration has a lot of parameters, which makes it prone to overfit the training data, particularly when there are a lot of training epochs. If overfitting takes place, the trained FIS might not adjust to other independent data sets well. In the best-case scenario, the converters' efficiency is 0.9.

For solar power, it is represented as equation.

$$\text{Ideal}\, P_{\text{solar}} = [V_{\max} * N_{\text{series}}]\left[I_{\max} * N_{\text{parallel}}\right] \tag{7.8}$$

where P_{solar} is power generated by solar panel, $V_{\max}$ is maximum voltage, N_{series} is number of solar modules in series, $I_{\max}$ is maximum current, and N_{parallel} is number of solar modules in parallel.

Ideal power of the battery is determined by following equation and ideal power of supercapacitor is determined by equation.

$$\text{Ideal}\, P_{\text{battery}} = V_{\text{battery}} * I_{\text{battery}} \tag{7.9}$$

$$\text{deal}\, P_{sc} = V_{sc} * I_{sc} \tag{7.10}$$

As seen in Fig. 7.2, power simulation has been carried out using the Simulink model, and Fig. 7.3 displays the findings. The product of load voltage and load current from load sensors simulates the power consumption of EV loads, such as starters, lighting, instrumentation, dashboard, wiper motors, power windows, air conditioning power, and so on. Supercapacitor voltage and current from sensors are multiplied to imitate the supercapacitor's power supply and charging power. The product of solar panel voltage and solar panel current simulates solar power. To test the efficiency of the solar supercapacitor at real life cycle, input irradiance is set as variable irradiance in the range of 200–2000 W/m^2.

The Simulink model based on the organic structural solar supercapacitor panel and a constant temperature of 25 °C, with solar input irradiance in the range of 200–1000 W/m^2, produced the simulated data shown in Table 7.1. Table 7.2 was created using the ANFIS structure as the output parameter, as seen in Fig. 7.3.

ANFIS training data is loaded with battery power (P_{battery}) simulation results from Table 7.3, as seen in Fig. 7.4, and ANFIS solar power simulation results are displayed in Fig. 7.5. Irradiance and load are the two inputs, while P_{battery} is the only output. There are five membership functions for the irradiance input (constant 1000 W/m^2) and five membership functions for the load input (depending on the electrical system load of the EV, 500–100 W).

Figure 7.6 displays the ANFIS structure for Pbattery. Figure 7.7 displays the solar power results at various irradiance settings trained by ANFIS. With different input values, ANFIS is operating more effectively to deliver correct results. Through the

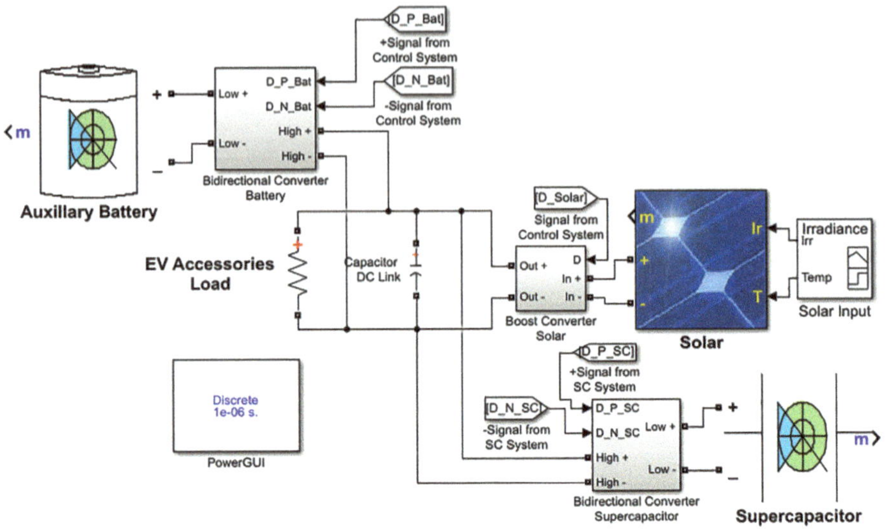

Fig. 7.2 Simulation model of ANFIS [2]

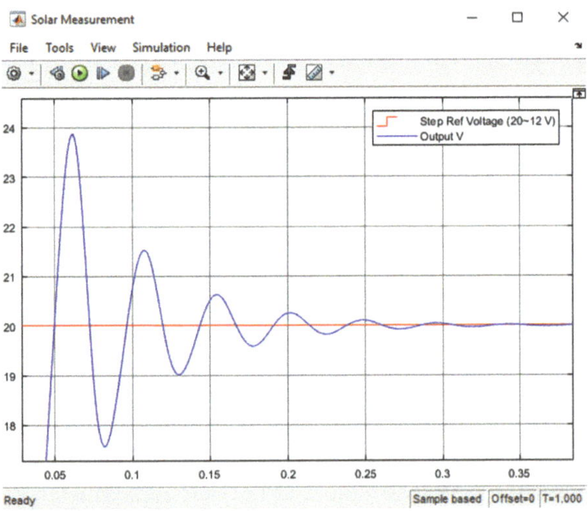

Fig. 7.3 Output voltage profile of solar panel without controller at 1000 W/m^2

Table 7.2 Solar power simulation results from simulink (OSSC size of 9.75 m^2)

Input parameter		Output parameter
Temperature(°C)	Irradiance (W/m^2)	P_{solar} (W)
25	1000	957.1
25	800	729.1
25	600	584.1
25	400	388.4
25	200	180.8

Table 7.3 Battery power simulation results from simulink

Input parameters		Output
Irradiance (W/m^{2})	Load (W)	$P_{battery}$ discharge (W)
1000	500	−375.6
1000	400	−453.4
1000	300	−525.4
1000	200	−610.8
1000	100	−705.3

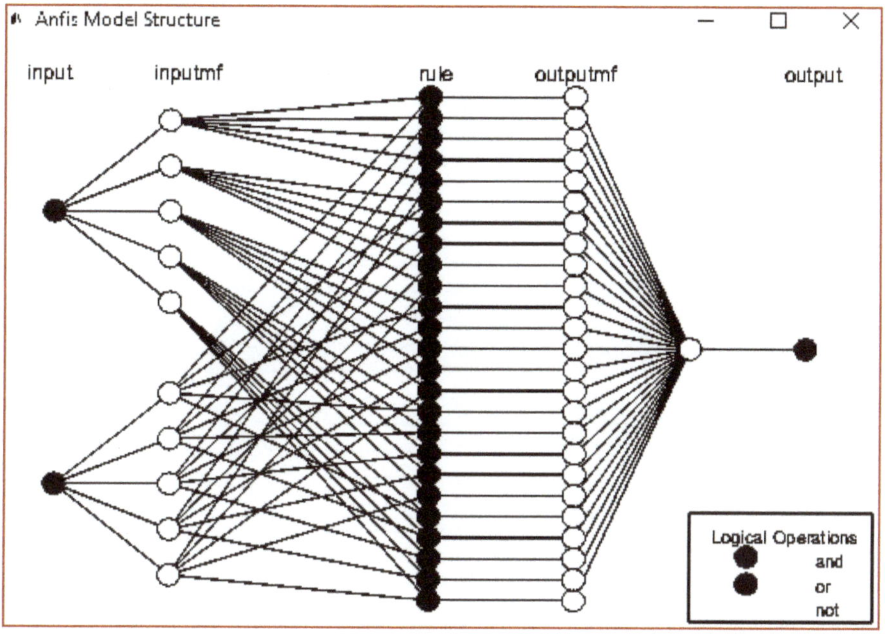

Fig. 7.4 ANFIS network structure for P_{solar} [2]

Fig. 7.5 Simulation of P_{solar} ANFIS system

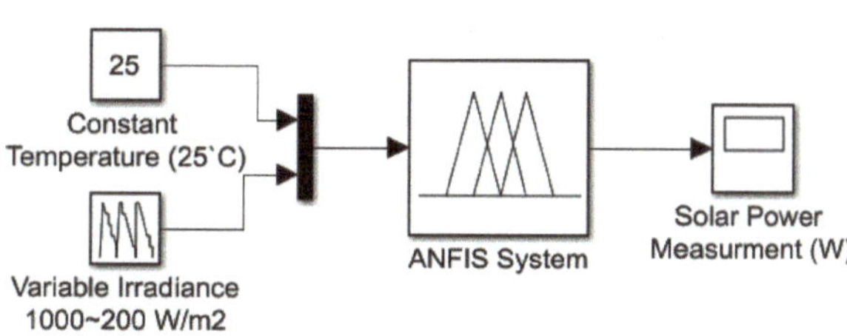

training of the fuzzy interference system's input data, ANFIS has produced effective solar output values.

In real-world solar supercapacitor applications, input irradiance will differ depending on the application's location. For 30% of the day, the solar irradiance profile is set at 1000 W/m^2, for 30% of the day, 600 W/m^2, for 30% of the day, 200 W/m^2, and for 10% of the day, 0 W/m^2 (no sunlight). As seen in Fig. 7.8, Simulink signal statistics indicate that the average solar power during a certain time period is 495.4 W.

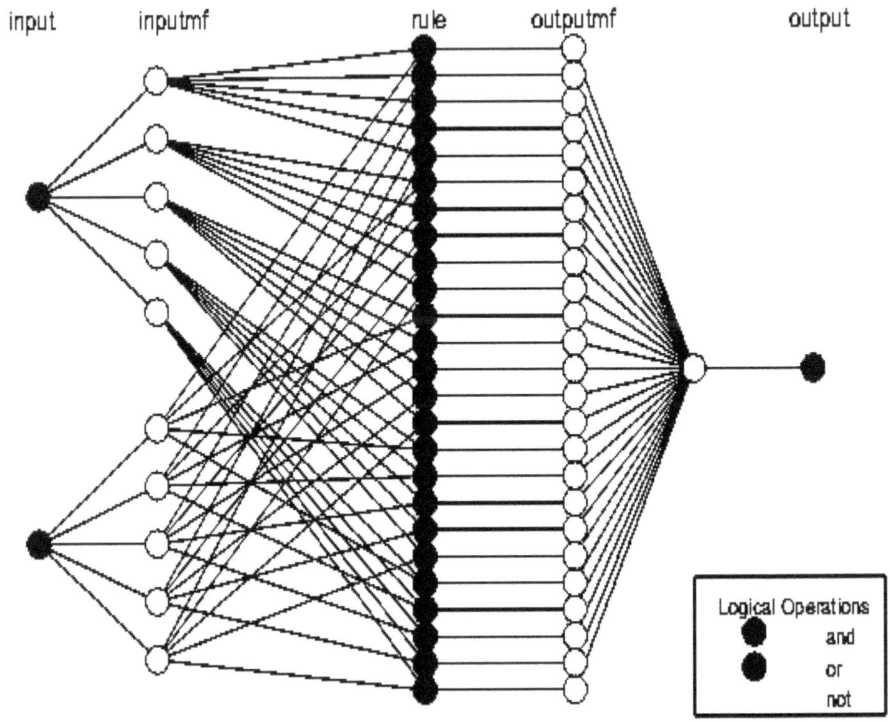

Fig. 7.6 ANFIS network structure for $P_{battery}$ [2]

Fig. 7.7 Simulation of $P_{battery}$ ANFIS system

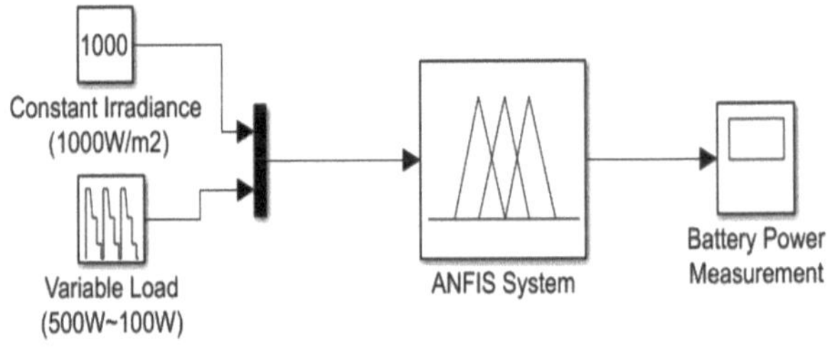

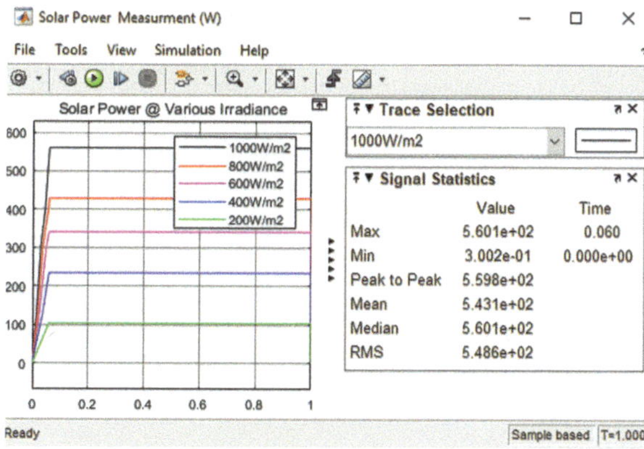

Fig. 7.8 Solar power at different irradiance setting by ANFIS

The potential performance of solar supercapacitor system is estimated by the following equation and calculation:

$$P_{\text{system}} = P_{\text{solar}} + P_{sc} + P_{\text{battery}} \tag{7.11}$$

The Tesla Model 3 has a 50 kWh battery with a 400 km range. With 1000 W/m^2 of solar irradiance, a fully charged solar supercapacitor system can generate 2.3 kWh of electricity (18.24 km of additional range per hour). As a result, the system can supply up to 4.56% of the power and range of a traditional EV each hour.

Investigations have been conducted into the system's overall efficiency, fuzzy controllers, and solar supercapacitor performance. The suggested OSSC provides precise power distribution among several storage systems while satisfying the energy and power needs of the EV accessory load. The OSSC's output might cut the weight of an EV battery by approximately 7.5% and its size by 10%. Additionally, by saving $1,000 by reducing the battery size, EVs with OSSC might help reduce manufacturing costs by 10%. According to data on greenhouse gas emissions, it would be possible to cut emissions by 25% (2.31 kgCO_2/liter of gasoline).

Variable Load and Irradiance

As seen in Fig. 7.9, the effectiveness of the organic solar supercapacitor power management system is examined under various irradiation conditions and EV electrical load demands. The 2 m^2 solar supercapacitor system is delivering 1122 W to the EV's batteries and charging the battery at 88 A at sun temperature of 25 °C. When the solar temperature is between 25 °C and 500 W/m^2, the OSSC can produce 885 W of power and charge batteries with a current of up to 70 A, which charges batteries at a very slow rate. One could draw the conclusion that solar energy is insufficient for battery charging.

The OSSC receives 5 A from an extra battery to increase the current needed to charge the EV's batteries. 200 W/m^2 can only supply 35 A at solar temperatures of 25 °C, which is significantly less than what is needed to meet the load requirement

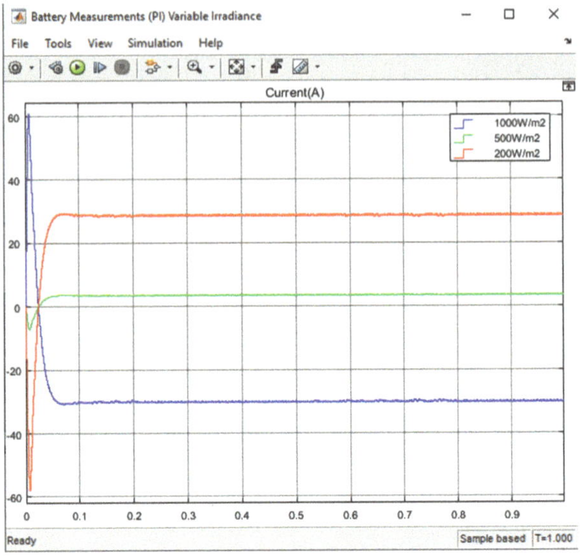

Fig. 7.9 Current profile of battery at different irradiance settings

of roughly 65 A. In order to meet the EV's load demand of 65 A, the auxiliary battery must provide OSSC with an additional 30 A of power. The contribution of the auxiliary battery to the total solar output power required to meet the electrical load demand of the EV is depicted in Fig. 7.4.

If the sun temperature is 32 °C or higher, it can be said that the organic solar structure capacitor (OSSC) is highly effective. The data comparison between the simulation and experimental outcomes is displayed in Table 7.4. There is a 0.271% error between experiment data and Simulink and a 1.296% error with ANFIS. Table 7.4 displays a comparison between ideal data, Simulink data, and ANFIS data. In a perfect world, the converters' efficiency would be 93%. The real converter component settings are used to simulate the results in Simulink and ANFIS.

The error percentage of the solar supercapacitor is less than 7%, as shown in Table 7.5. Because there are more converters for the current to flow through, the error percentage of power delivered to the battery is larger. Higher current in real-world applications will result in a drop in the converter's efficiency. As a result, designing the system with a high operating voltage and low operational current is efficient.

Table 7.4 Comparison of experiment data and simulation (solar power)

Irradiance (W/m$^{2)}$	Estimation from experiment (W)	Simulink data (W)	ANFIS data (W)	Variation of experimental result over	
				Simulink	ANFIS
1000	959.976	957.1	972.3	0.271%	1.296%

Table 7.5 Comparison of simulation data

Irradiance (W/m^2)	Ideal power solar (W)	Simulink data (W)	ANFIS data (W)	Error % (compared with ideal)	Error % (compared with ANFIS)
1000	976	957.1	972.3	5.83%	1.72%
600	584	584.5	594.1		
200	192	180.8	182.6		
Load (W)	Ideal power battery (W)	Simulink (W)	ANFIS data (W)	Error % (compared with ideal)	Error % (compared with ANFIS)
500	−397.17	−375.6	−377.1	6.85%	1.35%
300	−564.03	−525.4	−532.6		
100	−730.85	−705.3	−703.4		

7.5 OSSC with PI Controller

Both charging and discharging must be possible using the solar supercapacitor technology. Thus, at continuous 1000 W/m^2 irradiance (ideal condition), the performance of the bi-directional converter with PI controllers is examined. Figure 7.10 illustrates the voltage and power delivered to the load. The load is receiving 12.4 V of voltage, and 524 W of electricity is being used. If the OSSC size is taken to be 2 m^2, the result is greater than the specified values of 65 A to fulfill the electrical load demand of the EV, which is 69 A. Table 7.6 displays the bi-directional PID controller coefficient parameters.

Figure 7.10 depicts the model of SC's converter control system. The reference voltage range is adjusted to 16–8 V, and the PI controller receives an error signal from the difference between the reference and battery voltages (Fig. 7.11a). As seen in Fig. 7.11b, the PI controller's output is converted to a pulse width modulation

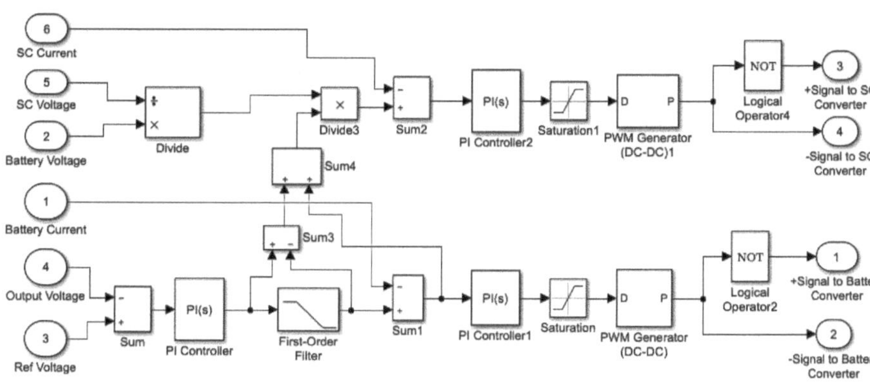

Fig. 7.10 Control system with PI controllers for battery and supercapacitor

Table 7.6 Parameters of Bi-directional controllers

PI controller (Solar ref)	$K_p = 1.45$, $K_i = 3099$
PI controller (Battery signal)	$K_p = 4.5$, $K_i = 60$
PI controller (SC signal)	$K_p = 0.5$, $K_i = 15{,}000$
Switching frequency (PWM)	15 kHz

(PWM) signal and transmitted to control switches. Table 7.5 displays the parameters of the components of the converter circuit. The OSSC panel, which is 1 m^2 in size and has an irradiance of 1000 W/m^2, provides the voltage and current. As seen in Fig. 7.12, the solar panel's voltage and current reach their maximum power points of 23 V and 38 A at 0.03 s. The Maximum Power Point tracking system's effective operation is the cause of this. Nevertheless, at 0.06 s, the OSSC's voltage and current stabilize at 13 V and 7 A. However, as Fig. 7.12 illustrates, if there is solar irradiation, the auxiliary battery drain will decrease dramatically in 0.05 s.

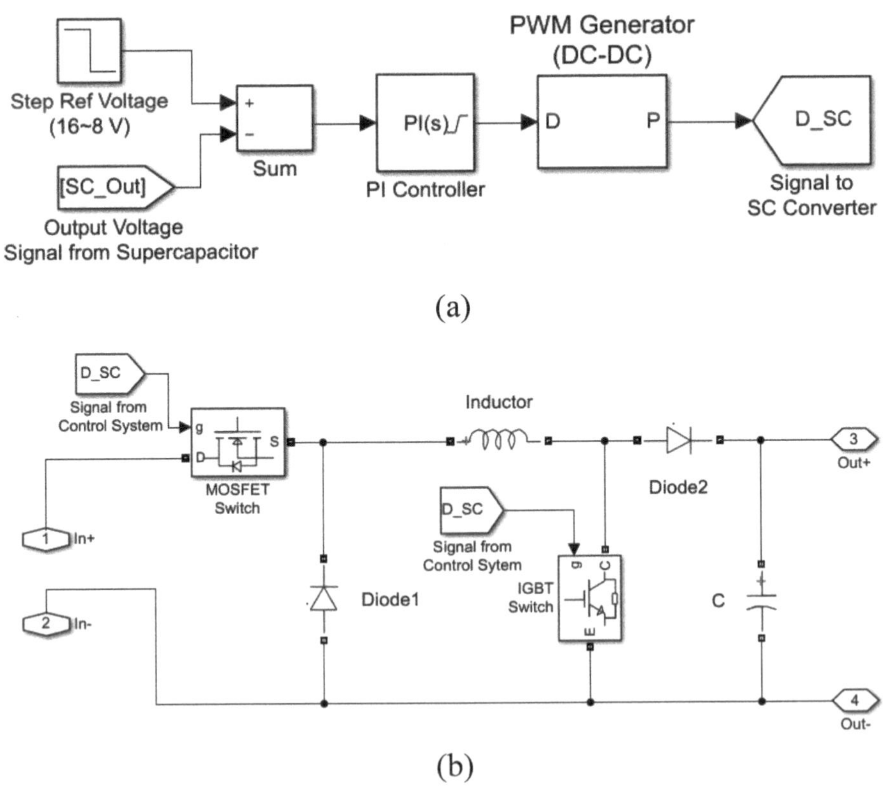

Fig. 7.11 One-directional PI PMS for supercapacitor **a** control system **b** converter system [1]

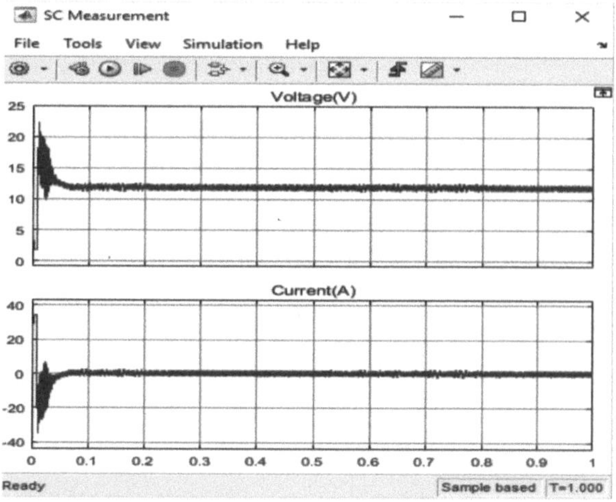

Fig. 7.12 Voltage and current profile of supercapacitor

7.6 SOSC with Fuzzy Logic Controller

Lotfi Zadeh created fuzzy logic (FL) at the University of California, Berkeley in 1965. Instead of using binary numbers, FL can use natural values to carry out computer operations. FL depends on the input's state levels, and the output is reliant on the input's state and rate of change. The basic idea behind a FL system is to assign a specific output according to the likelihood of the input's condition. FL also operates on the tenet that knowledge-based data determines the outcome. A multi-input system's variables are distinct inputs, and the output is the possible outcome of the variables' "AND" operation. The Simulink model for the OSSC with the FL controller as shown in Fig. 7.13 has been developed.

The OSSC's power management system is controlled using a fuzzy controlling system. An error signal is sent when the output voltage and reference voltage disagree. The battery's FLC is configured to accept the error signal and battery current as input signals. FLC generates "Duty Cycle" signals, which are then transformed into pulse width modulation (PWM) signals and transmitted as opposing signals to converter systems. The mode of the storage devices and the quantity of current flowing to the load are determined by fuzzy control signals. The SC control system receives an error signal based on the difference between the battery's error signal and battery current. The FLC of the supercapacitor is configured to receive the error signal and the supercapacitor current as input signals (Fig. 7.14).

Figures 7.15 and 7.16 illustrates the architecture of the solar panel's maximum power point tracking (MPPT) control system and converter system for both PI and fuzzy controllers. The P&O technique, which varies the duty cycle until the maximum voltage and current are reached, is used to produce MPPT for solar panels. Table 7.7 displays the PMS parameters.

As seen in Fig. 7.17, the FLC's inputs are "Ref Current" (error signal) and "Battery Current," and its output is "Duty Cycle." The Battery FLC's input membership

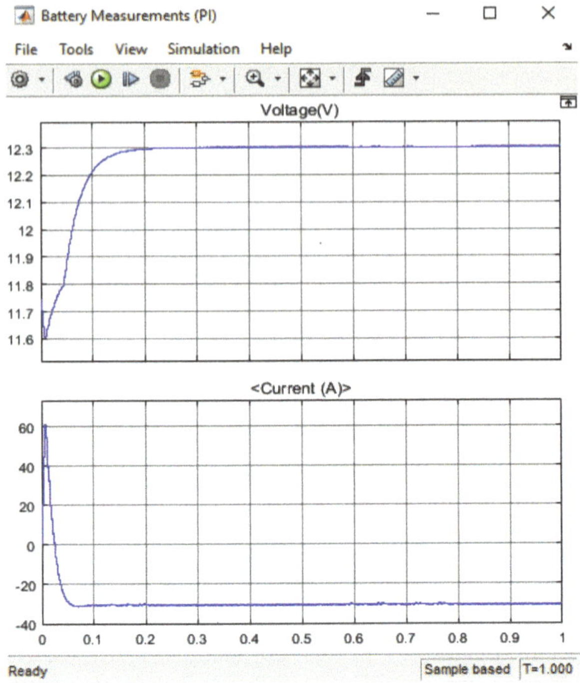

Fig. 7.13 Auxillary battery profile

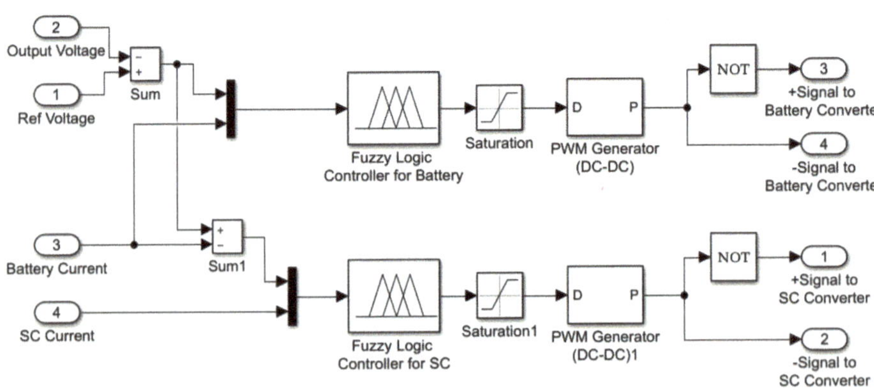

Fig. 7.14 Control system with FLC for battery and supercapacitor in simulink [1]

functions are configured as depicted in the figure. According to Fig. 7.17 a, the membership functions of “Ref Current” are set as follows: “Very Low (VL)” (−15 to −7.5), “Low (L)” (−15 to 0), “Zero (Z)” (−7.5 to 7.5), “High (H)” (0 to 15), and “Very High (VH)” (7.5 to 0). “High Charging (HC)” (−30 to −15), “Low Charging (LC)” (−30 to 0), “Zero (Z)” (−15 to 15), “Low Discharging (LD)” (0 to 30), and

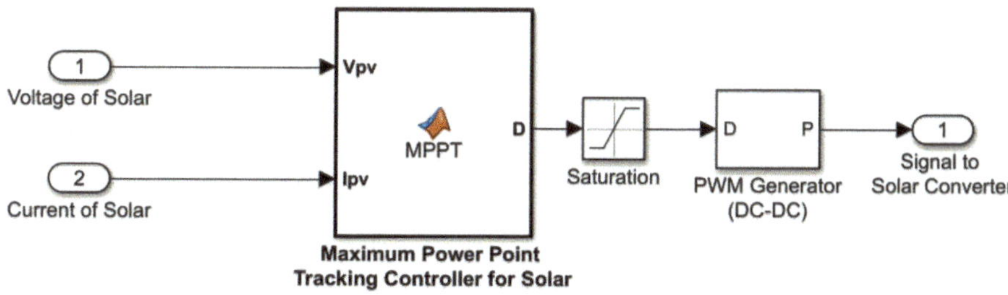

Fig. 7.15 MPPT control system for solar

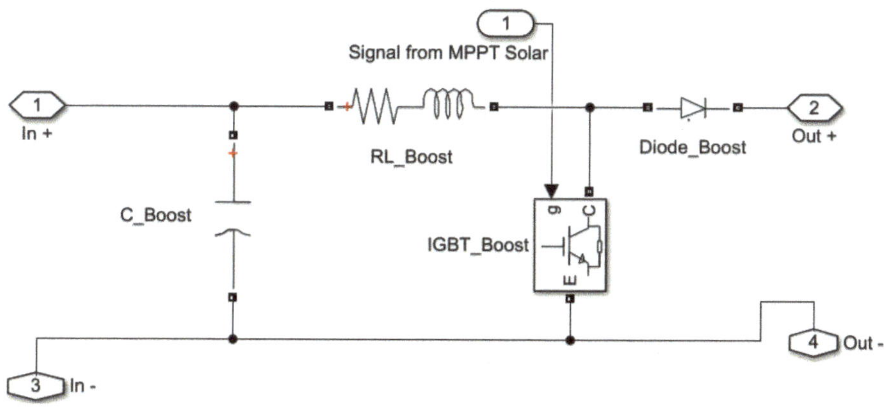

Fig. 7.16 DC-DC converter system of solar MPPT (P&O) [1]

Table 7.7 Parameters of solar PMS

Switching frequency	Values (for 5 kHz)
Converter inductance	0.352 mH
Converter capacitance	2200 μF
Converter resistance	50 mOhm
Diode resistance	0.1 mOhm
IGBT resistance	1 mOhm

"High Discharging (HD)" are the membership functions of "Battery Current" (15 to 30), as Fig. 7.17b illustrates. Membership functions of "Duty Cycle" are set as "Negative Big (NB)" (−1 to −0.5), "Negative Small (NS)" (−1 to 0), "Zero" (−0.5 to 0.5), "Positive Small (PS)" (0 to 1), "Positive Big (PB)" (0.5 to 1) as shown in Fig. 7.17c.

As seen in Fig. 7.18, the FLC's inputs are "Error Ref" (error signal) and "Supercapacitor Current," and its output is "Duty Cycle." Fig. 7.18a illustrates how the SC FLC inputs' membership functions are configured. As seen in Fig. 7.18b, the membership functions of "Error Ref" are set as "Very Low" (−15 to −7.5), "Low"

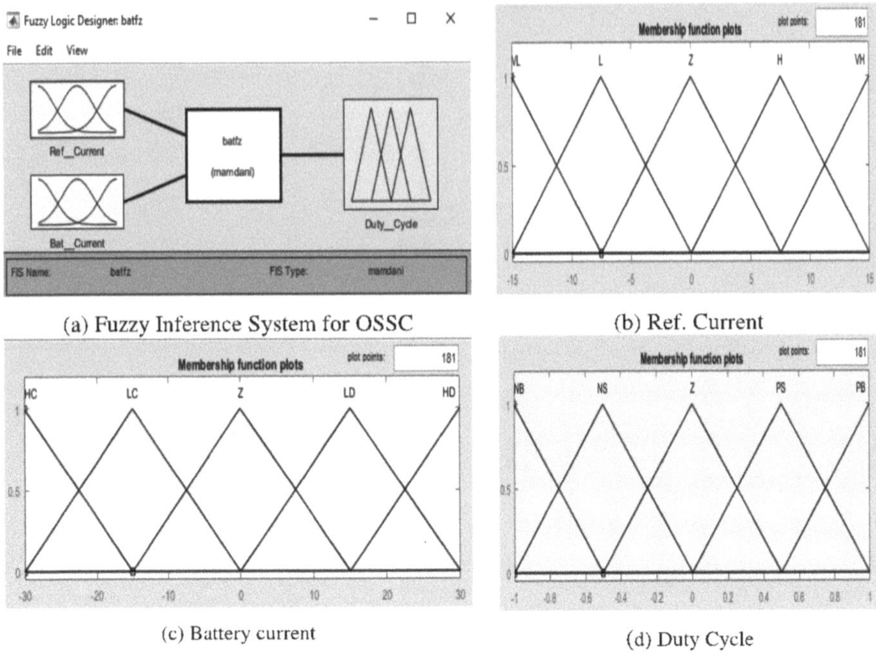

(a) Fuzzy Inference System for OSSC (b) Ref. Current

(c) Battery current (d) Duty Cycle

Fig. 7.17 Fuzzy intelligent system for auxiliary battery

(−15 to 0), "Zero" (−7.5 to 7.5), "High" (0 to 15), and "Very High" (7.5 to 0). As illustrated in Fig. 7.18c, the membership functions of "Supercapacitor Current" are "High Charging" (−30 to −15), "Low Charging" (−30 to 0), "Zero" (−15 to 15), "Low Discharging" (0 to 30), and "High Discharging" (15 to 30). Figure 7.18d illustrates the membership functions of "Duty Cycle": "Negative Big" (−1 to −0.5), "Negative Small" (−1 to 0), "Zero" (−0.5 to 0.5), "Positive Small (PS)" (0 to 1), and "Positive Big (PB)" (0.5 to 1). Tables 7.8 and 7.9 display fuzzy rules for the battery and supercapacitor, respectively.

At a constant 1000 W/m^2 (ideal condition), the performance of the bi-directional converter with fuzzy logic controllers is examined. Figures 7.19 and 7.20 display the OSSC's voltage and current curve. The effectiveness of a solar supercapacitor power management system using fuzzy logic controllers is examined under various load scenarios and irradiance levels. The solar supercapacitor system is delivering 500 W to the EV load and charging the battery at −30 A at 1000 W/m^2. At 500 W/m^2, the EV load cannot be supplied with 500 W of solar electricity alone. As a result, the extra battery provides the EV load with 4 A. The extra battery must provide greater power to the EV load at 28 A at 200 W/m^2.

The solar supercapacitor system can only charge the battery at 8 A when the load is 800 W. The solar supercapacitor system is charging the battery at 48 A when the load is 300 W. The solar supercapacitor system is charging the battery at 70 A when

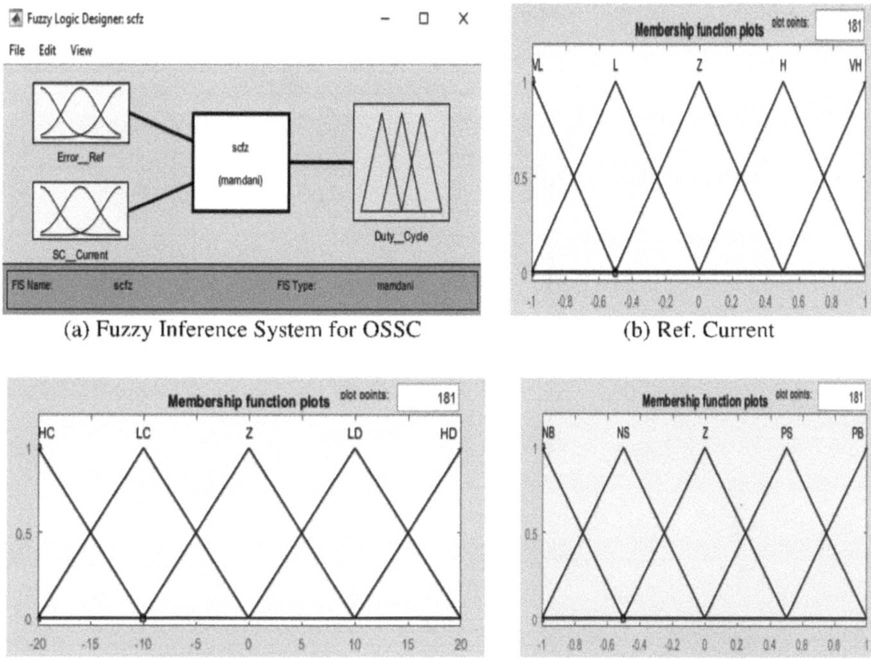

(a) Fuzzy Inference System for OSSC (b) Ref. Current

(c) OSSC current (d) Duty Cycle

Fig. 7.18 Fuzzy intelligent system for supercapacitor (OSSC) [2]

Table 7.8 Fuzzy rules for battery

		Battery_Current									
		HC		LC		Z		LD		HD	
Ref_current	VH	Z	1	PS	2	PS	3	PB	4	PB	5
	H	NS	6	Z	7	PS	8	PS	9	PB	10
	Z	NS	11	NS	12	Z	13	PS	14	PS	15
	L	NB	16	NS	17	NS	18	Z	19	PS	20
	VL	NB	21	NB	22	NS	23	NS	24	Z	25

Table 7.9 Fuzzy Rules for Supercapacitor

		Rate of error (Δe)									
		HC		LC		Z		LD		HD	
Error (e)	VH	Z	1	PS	2	PS	3	PB	4	PB	5
	H	NS	6	Z	7	PS	8	PS	9	PB	10
	Z	NS	11	NS	12	Z	13	PS	14	PS	15
	L	NB	16	NS	17	NS	18	Z	19	PS	20
	VL	NB	21	NB	22	NS	23	NS	24	Z	25

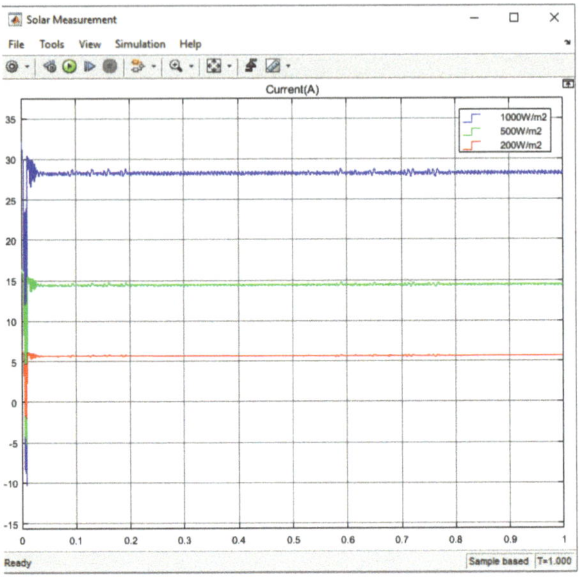

Fig. 7.19 OSSC performance using fuzzy [1]

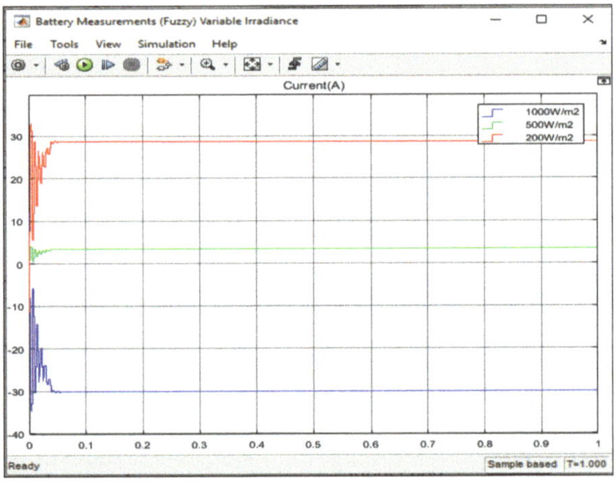

Fig. 7.20 Current profile of battery at different load setting (fuzzy) [1]

the load is 100 W. In order to give the converter system precise signals, the power management system with FLC is working more effectively.

Based on simulation results, Table 7.10 compares ANFIS's performance to those of FL and PID. By only monitoring the battery terminal voltage and preventing excessive voltage from damaging the battery, the OSSC 12.5 V output voltage control was used for comparison. When generating current from the OSSC at an irradiation of 1000 W/m^2, the ANFIS controller performs 16% better than the PID controller and 29% better than the fuzzy logic controller. The present stabilizing performance

Table 7.10 ANFIS best for OSSC power management system at irradiance 1000 W/m^2

OSSC		ANFIS	PID controller	FL controller	Advantage of ANFIS over (%)	
					PID	FLC
Current	Maximum	45	38	32	16	29
	Stabilized	45	10	27	77	40
	Overshooting ms	0.03	0.07	0.03	133	0

Table 7.11 Solar power output by different systems at 12.5 V of EV's electrical system

Temperature (°C)	Irradiance (W/m^2)	Power (W)		Current (A)	
		Fuzzy	ANFIS	Fuzzy	ANFIS
25	1000	450.5	560.1	36	45
25	800	424.94	430.2	33.9	36.81
25	600	342.07	345.49	27.4	27.63
25	400	230.48	234.69	18.43	18.72
25	200	105.71	107.93	8.46	8.63

for charging the auxiliary battery during overshooting time is 0% better than FL and 133% better than PID.

As seen in Fig. 7.20, the outcome shows that ANFIS can deliver improved performance to satisfy vehicle load requirement and regulate battery charging performance with 0% mistake. Table 7.11 allows for a similar conclusion to be drawn between the ANFIS and FLC. In terms of output power generation to satisfy the vehicle load requirement for the irradiance in the 200–1000 W/m^2 range, the ANFIS performs better than the FL. This rationale places a strong emphasis on using the ANFIS controller in conjunction with the OSSC to regulate power flow in order to satisfy load demands and charge auxiliary batteries.

7.7 Experimental Study of OSSC with ANFIS

Fluorescent lights set at 32 °C were used to test each sample's energy storage capabilities. To be certain that there is nothing left in the sample, the terminals were first shorted out. The OSSC was then tested at 320 C in the sun on December 8, 2019, Malaysian time, at 13:20. The ANFIS controller's output port had a millimeter connected to it. For five minutes, readings of voltage and current were taken at one-second intervals. The OSSC samples are displayed in Fig. 7.21.

At the time, the OSC's performance was investigated for the highest values of SoC(t), W/kg, and energy density W/Ah while both charging and discharging. The OSSC's SoC(t) and SoD(t) for varying percentages of activated carbon (AC) are

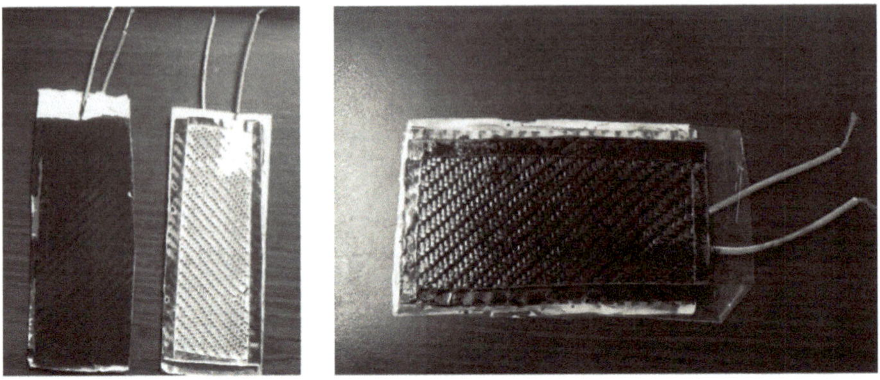

Fig. 7.21 Organic solar cell with the composition of ZnO/CuO: AC [2]

displayed in Figs. 7.22 and 7.23. The weight percentage of ZnO for the P-type has been shown to be constant, while the weight percentage of AC varies. The outcome demonstrates the variation in charging and discharging for the AC 6–20%. A higher proportion of AC particles results in a larger surface area, more pores, and the use of PVDF polymer chains to encase the CFs.

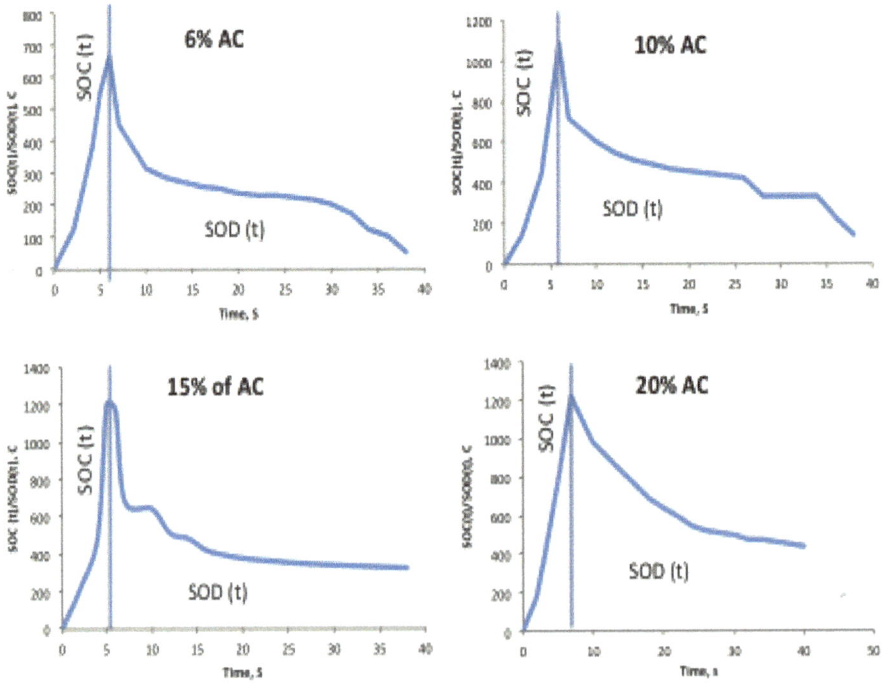

Fig. 7.22 State of Charging (SoC(t)] and State of discharging (SoD(t)) [3]

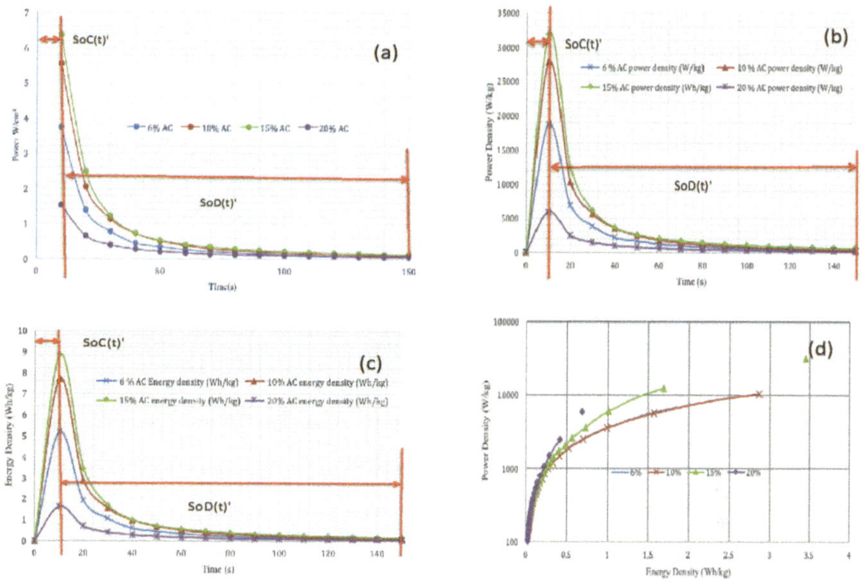

Fig. 7.23 Performance of OSC for the different % of AC [3]

However, an overabundance of AC may cause the OSSC to form a thin coating and obstruct electron transport. As a result, the OSSC has a longer charge-holding capacity than the lower AC %.

Figure 7.22a displays the power stored at 6.7 W/kg for 15% AC_OSSC at 5 s of charging and 1.8 W/kg for 20% AC. The power density of the OSC, as determined by the digital multimeter, is displayed in Fig. 7.22b for both charging and discharging. The findings indicate that a 15-weight percent AC sample may yield the highest power density of 33,000 W/kg, while a 20 weight percent AC sample can yield the lowest power density of 5700 W/kg. This is because the AC is tightly mixed, which makes the micropore area form a film and seals the interstitial gap between the PVDF and the ZnO filler. See Fig. 7.23c for OSSC's energy density. According to the results, the 15% AC sample had the maximum energy density (8.9 Wh/kg), while the 20% AC sample had the lowest energy density (1.7 Wh/kg).

The overall performance of AC-ZnO/CuO conductive polymer carbon fiber reinforced OSC is displayed in Table 7.12. The 15% AC sample yields the best results, with 10% coming in close. The 20% AC sample's performance was subpar due to the ZnO's light film formation, which may have an impact on overall performance because of the lowest interstitial space. For the EDLC supercapacitor, 10–15 weight percent AC is a very suitable choice.

The optimal sample's solar power is 133.33 W/m^2. The ideal supercapacitor has a 13.33% solar energy conversion efficiency. Table 7.13 summarizes the promise of this when compared to other metal oxide solar panels. Solar supercapacitors with ideal experimental conditions and improved nanoparticle synthesis can generate 2800 W

Table 7.12 Overall performance of the AC-ZnO/CuO-ER_CF EDLC

AC (%)	Voltage (V_{oc})	Capacitance $C(\mu F/cm^2)$	Current density $J_{sc}(A/cm^2)$	Power density $P_{ms}(KWkg^{-1})$	Energy Density $E_{ms}(W.hkg^{-1})$
6	2.414	13.356	2.234	18.691	5.198
10	2.649	16.356	3.013	27.747	7.707
15	2.438	20.156	3.584	31.859	8.85
20	2.154	11.27	1.312	5.863	1.623

Table 7.13 Comparison of performance for different types of solar cells

PEC	V_{oc} (mV)	J (A/m^2)	Efficiency (η) (%)	Reference
CuO/ZnO	599	222.22	13.33	This research
CuO/ZnO	630	180	6.12	[67]
ZnO/SnS	120	0.4	0.003	[67]
CdS/SnS	260	96	1.3	[67]

of power per day, have a solar energy conversion efficiency of 19–22%, a power density of 33 kW/kg, an energy density of 13 kWh/kg, and a capacitance of 11.17 $\mu F/cm^2$ at 25–30 °C and 1000 W/m^2 of irradiance.

7.8 Conclusion

The advancement of supervisory control logic and converter power efficiency have a significant impact on the performance of solar supercapacitors. The performance of this technology can be more accurately estimated under real-world driving situations and testing conditions. Because the solar supercapacitor device uses renewable energy and supplements electricity, it can extend the battery life of electric vehicles.

Using materials that are efficient and lightweight for the solar supercapacitor can lower the system's cost. Reducing the size of EV batteries has the potential to result in lower costs and lighter vehicles by improving the hybrid energy storage system's efficiency and dependability. In order to lower the system's weight and cost, the solar supercapacitor's characteristics can also be adjusted to the consumer's needs.

Acknowledgements The Research Management Centre (RMC) at International Islamic University Malaysia provided funding for this study, and Eastern Michigan University let the authors to proceed with their project of turning the car body into a portable energy source. The automotive composite laboratory technician who assisted in the development of the organic supercapacitor from this study is also appreciated by the authors.

References

1. Aung, K. M. (2022) Development of solar structural supercapacitor system for EV. PhD thesis, Library, International Islamic University Malaysia.
2. Rahman, A. Kaw, M. A., Ihsan, S., Raja, A. S., Qubeissi, A. L. M., & Aljarrah, M. T. (2023) Solar energy dependent supercapacitor system with ANFIS controller for auxiliary load of electric vehicles. *Energies, 16*, 2690 (2023).
3. Rahman, M. A., & Golam, K. (2021). Al-ZnO/CuO_Cfrp organic structural auxiliary energy storage system for electric vehicle. *Journal of Solid State Technology, 28*(1 & 2), 103–116.
4. Salman, S., Xin, A. I, & Zhouyang, W. U. (2018) Design of a P-&-O algorithm based MPPT charge controller for a stand-alone 200 W PV system, Vol. 3, p. 25.
5. Mustafa, E., Sahin, H., & Okumus, I (2011) Fuzzy logic controlled buck-boost DC-DC converter for solar energy-battery system. Conference Paper, June 2011.
6. Tong, Y., Wu, G., & Jin, X. (2013). Topological study of bidirectional dc/dc Converters. *Proceeding of the CSEE, 27*(13), 81–86.
7. Lia, B. Y., Xua, C., Lib, C., & Guan, Z. (2017). Working principle analysis and control algorithm for bidirectional DC/DC converter. *Journal of Power Technologies, 97*(4), 327–335.
8. Muhammad, A. K., Kamran, Z., Sathishkumar, P., Muhammad, U., Ali Waqar, A. U., Hussain, S., Ishfaq, M., Imran, K., Hwan-Gue, C., & Hee-Je, K. (2018). A novel supercapacitor/lithium-ion hybrid energy system with a fuzzy logic-controlled fast charging and intelligent energy management system. *Electronics, 2018*(7), 63.
9. Steve, W. (2017) Power supplies for LED driving (2nd edn., pp. 43–71).
10. Patel, D., Patel, S, & Shah, M. (2022) Solar radiation and solar energy estimation using ANN and Fuzzy logic concept: A comparative and systematic study. *Environmental Science and Pollution Research, 29*(22), 32428–32442.

Part III
Intelligent Battery Thermal Management System for EV

Chapter 8
Revolutionary Battery Cooling System

Abstract Keeping LIBs safe is the largest challenge facing EV development. The LIB may sustain damage or perhaps catch fire as a result of thermal output at −20 °C and 60 °C. It is advised to keep the temperature range between 20 °C and 40 °C to preserve peak performance. Independent evaluations were conducted on the performance of LIB's thermal management system (ThMS) utilizing air-cooled (AC), liquid-cooled (LC), and evaporative-cooled (EC) systems. The outcome demonstrates that BThMS with EC performs better at temperature control than AC and LC. Furthermore, the microchannel heatsink turns into a useful cooling technique for lowering battery temperature. The LIB's heat pipe cooling is constrained by weight, gravity, and passive control.

Keywords Battery thermal management system · Battery life safety and life span · EV acceptances

8.1 Introduction

Vehicle electrification requires an optimized component sizing such as battery size and propulsion electric motor to make it more sustainable and cost-effective, particularly for small and medium segment vehicles. For traction, these vehicle segments usually need a compact 25 kWh lithium-ion battery (LIB) pack, where the DC link voltage must vary dramatically depending on the load. In order to meet the vehicle's power demands during brief peak acceleration, LIB must produce a high current output in a short amount of time. Internal resistance causes considerable LIB heating during these brief peak load periods, which need thermal regulation. For example, LIBs may not be able to supply enough power or energy for the vehicle below −20 °C, and their operation may result in internal structural damage. At 60 °C, LIB performance may be subpar, or internal module damage may occur, which could cause thermal runaway [1–3].

Because of its high energy density, lack of memory effect, and extended cycle life, LIB is frequently used in electric and hybrid vehicles. However, because of

A. Rahman et al., *Autonomous Driveline System for Electric Vehicle*, Green Energy and Technology, https://doi.org/10.1007/978-981-96-8418-2_8

its increased high power consumption, driving range, and fast charging, it may pose safety risks [4–6]. Additionally, to avoid accelerated cell deterioration, the temperature gradient inside LIB cells must be maintained between 2 and 5 °C [7–9].

Battery thermal management systems (BTMS) can be used to regulate the thermal behavior of LIBs through the use of various cooling medium strategies, including liquid- or air-cooled systems. Air-cooled BTMS is a lightweight, inexpensive, and less complicated technology [10, 11]. However, because of the air's low heat transfer coefficient, the air-cooled system is impractical under certain working situations, such as high discharge/charge rates and high ambient temperatures [12–14]. Consequently, in order to improve the cooling effectiveness of an air-cooled system's design, a high heat transfer device must be used. For instance, Bergveld [15] have designed air-cooled BTMS using a double silica cooling plate with copper mesh. This design is generally used in many air-cooled applications to increase heat transfer but has a drawback in terms of additional space and weight of the system.

The most efficient and economical cooling method for LIB in EV applications is thought to be liquid-cooled BTMS. Because liquid-cooled BTMS releases the heat created from the chemical reaction of LIB during current discharge/charge, it has been demonstrated to improve LIB deterioration and produce high efficiency [15]. The high cost of production, space utility rate, structural complexity, and potential for leakage are some of the drawbacks of liquid-cooled BTMS. A microchannel liquid-cooled system has been suggested for BTMS in order to address these problems [16]. Increasing the rate of heat removal and lowering the LIB temperature can be accomplished with the use of a microchannel heat sink.

For heat dissipation rate, the impacts of coolant inlet velocity, discharge rate, and contact area between LIB and the microchannel liquid-cooled tube were investigated [17–19]. By increasing coolant flow or plate wall thickness at the expense of pumping power, a number of stacks integrated into liquid-cooled BTMS have preserved the temperature gradient and optimal operating temperature (OOT).

Utilizing a heat pipe (HP), which has a very efficient heat transfer capability, is an additional option [20–22]. HP-cooled technology has been widely used in LIB in recent years. According to the authors [23, 24], the HP plate could regulate the LIB ($LiFePO4$) temperature below 50 °C while generating less than 30 W of heat. Additionally, it can keep LIB's temperature differential within 5 °C. A cooling system and sophisticated control methods can work together to maximize BTMS's goals. One possible technique in BTMS to regulate the thermal behavior of LIB in EV applications is the artificial intelligence (AI) controlled area network (CAN) bus.

Higher performance and longer battery life are provided by the battery's intended operating temperature range, which also presents no safety risks. A module that experiences excessive temperature changes during charging and discharging may use the active material unevenly, which could result in decreased performance and a shorter lifespan. In order to maximize cycle life and charging and discharging capacity, the battery must be maintained within a specific operating temperature range. Therefore, a battery pack's thermal management system starts to be designed and developed in order to guarantee the proper operating temperature throughout both charging and discharging modes and to extend the battery's lifespan. It could be mentioned that developed prototype is not only enhanced the battery life span but it also contributes on the emission control of the country and improve the air quality.

8.2 Battery Power Pack for EV

A variety of anode and cathode material combinations are used in the lithium-ion battery family of battery chemistries. Regarding performance, cost, safety, and other factors, each combination has unique benefits and drawbacks. Lithium-nickel–cobalt-aluminum (NCA), lithium-nickel-manganese-cobalt (NMC), lithium-manganese-oxide (LMO), lithium titanate (LTO), lithium ferro-phosphate (LFP), and lithium iron phosphate (LiFePO4) are the most widely used technologies for automotive applications, according to Fig. 8.1's storage capacity. Regarding this investigation, the Proton Saga EV is powered by LiFePO4 Li-ion batteries.

Comparing LiFePO4 battery cells to other lithium-ion batteries reveals a number of benefits. In comparison to the nickel-based batteries that have been in use for a long time, LiFePO4's energy density, quicker charge/discharge periods, and high current or peak power rating are its main advantages, according to authors [25]. This conclusion is supported by writers [26–28], who emphasize that LiFePO4 batteries have enhanced abuse tolerance, high power density, and good cycle life.

8.2.1 *Battery Pack Packaging*

Figures 8.2 and 8.3 demonstrate the superior performance of the LiFePO4 battery, which makes it suitable for use in modern automobiles, especially electric vehicles. Table 8.1 displays the specifications of a typical lithium battery cell. A automobile with 21 kg of these cells can travel up to 160 km on a single charge because of their energy density of 0.38 MJ/kg, which is equivalent to 47 MJ/kg of gasoline. According

Fig. 8.1 Maximum stored energy per unit of battery mass [1]

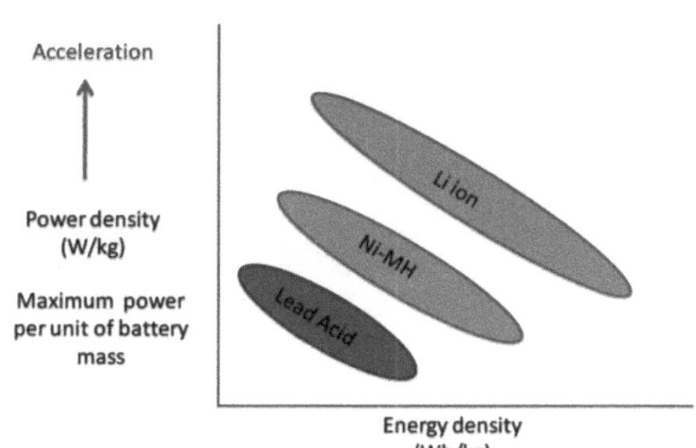

to the scientists [29, 30], these cells could travel over 320,000 km without needing to be recharged because they maintained 70% of their capacity even after 2000 cycles. However, the primary obstacle to the integration of current EV technologies is the relatively short lifespan and high cost of the current LiFePO4 battery per unit of energy storage. Therefore, a lot of research currently being done on increasing its performance.

Fig. 8.2 Lithium-ion battery cell, module, and pack [8]

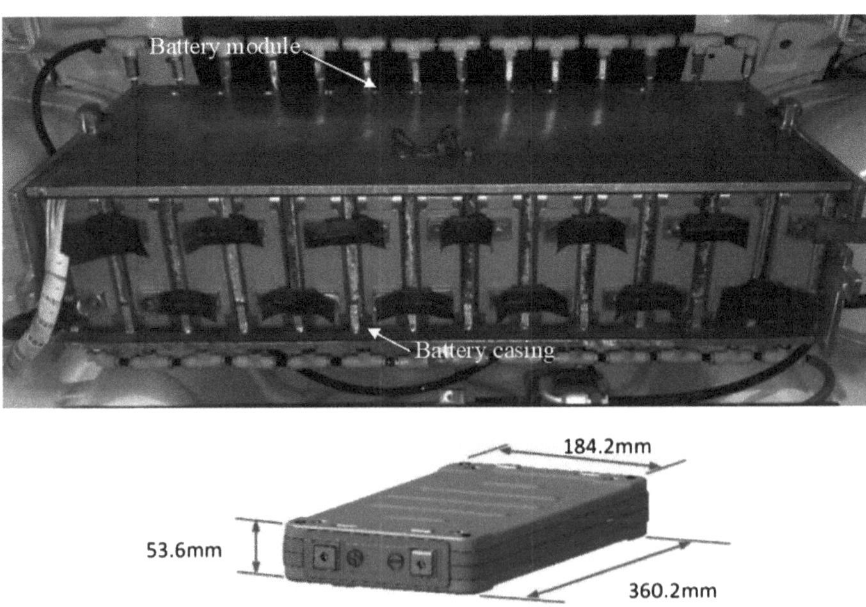

Fig. 8.3 Battery casing with 13 units of modules [31]

Table 8.1 Battery module specification

Item	Specification
Weight	≤4.8 kg
Normal capacity	86 Ah @ 0.5 C
Capacity range	SOC10 ~ SOC95 (85% Usable)
Normal voltage	(1 C) 7.5 V
Charging cut-off voltage	8.30 ± 0.05 V
Discharging cut-off voltage	6.00 ± 0.05 V
Standard charging current	1 C
Standard discharging current	1 C
Max continuous charging current	172 A
Max continuous discharging current	460 A @ SOC50, 30 s, 25 °C
Max plus discharging current	600 A
Heat capacity	925 J/kg·K
Energy density	160 Wh/kg
Power density (SOC50, 10 s, 25 °C)	1.813 W/kg
Internal resistance	1.64 mΩ (SOC50, 30 s discharge)
Charge working temperature	−20 ~ 50 °C
Discharge working temperature	−20 ~ 50 °C
Storage temperature	−40 ~ 60 °C

8.3 Battery Thermal Management System

The performance of EVs and HEVs is influenced by LIBs, whose discharge/charge behavior can be greatly controlled by the operating temperature and temperature gradient. Concerns over LIB's thermal behavior have been raised by the aggressive torque demand trends in EVs and HEVs [32]. The BTMS's main goals are to keep LIB functioning within the ideal temperature range of 30–40 °C, avoid going over high temperature limits that could harm or shorten LIB life [33], and keep LIB temperature variations at low levels to avoid high imbalanced voltages [34, 35].

Heat will be produced by LIB's discharge/charge processes. Particularly at high discharge rates, a considerable amount of heat is produced. The LIB temperature will rise dramatically in a short amount of time if this heat cannot be released right away [36, 37]. According to the authors [35, 36], the majority of the heat was produced during quick cycles of charging and discharging, which caused the LIB temperature to rise sharply above the permitted level. Heat control and management are necessary to enhance LIB degradation, lower power losses, and avoid thermal runaway in order to address the thermal problem related to the quick charge and discharge of LIB [37, 38].

Additional research has also been done to examine LIB's thermal behavior. For example, Soldo et al. [39] created a lumped thermal model based on experimental

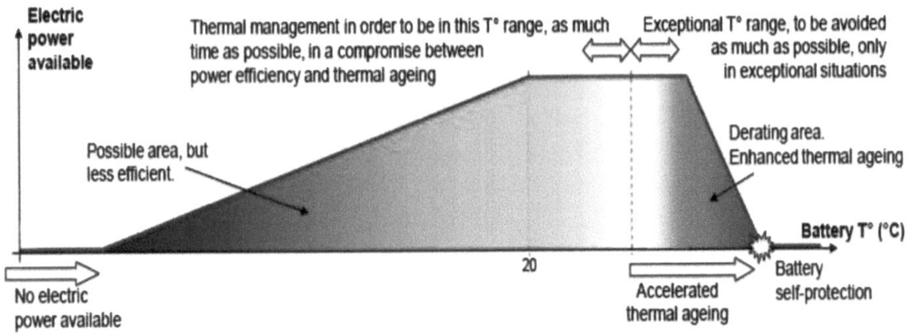

Fig. 8.4 Temperature range, regarding available performances, efficiency, and thermal aging [40]

data after studying the thermal properties of a cylindrical LiFePO4/graphite LIB. The electrochemistry and structures of LiFePO4 material have been investigated by Dominko et al. [20] and Duan and Naterer [21]. According to their findings, the LIB capacity decreases significantly at low working temperatures—by 17–22% at −20 ºC. When the temperature rises above 25 °C, the internal impedance is nearly constant; nevertheless, when the temperature falls from 25 °C to −20 °C, it rises by 25%.

The solid electrolyte interface membrane decomposes as a result of heat dissipating rapidly between 40 and 72 °C and producing five times as much when heat reaches 80 to 119 °C. At a temperature of 172 °C, the anode material begins to break down and react with the electrolyte. Operating beyond 40 °C will cause LIB to deteriorate somewhat, and operating above 60 °C will cause LIB to deteriorate considerably [40]. The ideal operating temperature range for LIB is between 20 and 40 °C, which offers the highest power capability and tolerable rates of thermal aging [41]. Figure 8.4 illustrates the performance, efficiency, and temperature-dependent thermal aging characteristics of LIB as determined by author Lefebvre [40].

Operating at a temperature lower than 15 °C can significantly reduce LIB's capacity because it increases the electrolyte's density and disrupts the ion transition between the positive and negative electrodes. However, LIB can have an adverse effect on performance [41–43] and increase the danger of causing thermal runaway [39] if it works at a temperature higher than 40 °C. From these data, it can be inferred that a small EV model's ideal LiFePO4 LIB operating temperature falls between 20 and 40ºC [44].

8.4 Sustainabile ThMS of LIB

Thermal management of LIB is usually achieved by combining two methods, according to Zolot et al. [45]: (i) a cooling system is made to absorb heat from LIB, and (ii) the LIB controller modifies the vehicle's load demand in response to

LIB circumstances. Air, liquid (coolant and water), refrigerant, phase change material (PCM), or any combination of these could be used as the heat transfer and dissipation medium for BTMS.

8.4.1 Air-Cooled

In EV applications, air-cooled BTMS is a popular LIB cooling technique. In an air-cooled system, heat transmission and dissipation are accomplished by guiding the air across LIB, as seen in Figs. 8.5 and 8.6. According to Smith and Wang [46], in order to meet cabin thermal comfort and noise consideration, the maximum flow rate for cold air extracted from the passenger compartment is normally restricted between 100 and 250 m^3/h (depending on the kind of vehicle).

Cooling efficiency can be increased by utilizing a fan or blower to force air-cooled convection. Heckenberger [34] asserts that the primary benefits of air-cooled BTMS are simpler, as EV and HEV manufacturers typically build the system by extracting cold air from the passenger compartment, using a second blower, and sending the cool air to LIB. This approach minimizes the cost of the system by requiring few changes or extra parts.

This method's primary flaw is its reliance on the temperature of the car's cabin. Additionally, because of the low heat transfer coefficient of air, it is challenging to provide LIB (OOT and temperature gradient) with a sufficient amount of cooling when the ambient temperature is high, like in high-climatic regions.

As seen in Fig. 8.7, this approach has been used by a number of EV manufacturers. The Plug-in Hybrid (PHEV) Prius model's 5.2 kW·h nickel-metal hydride (NiMH)

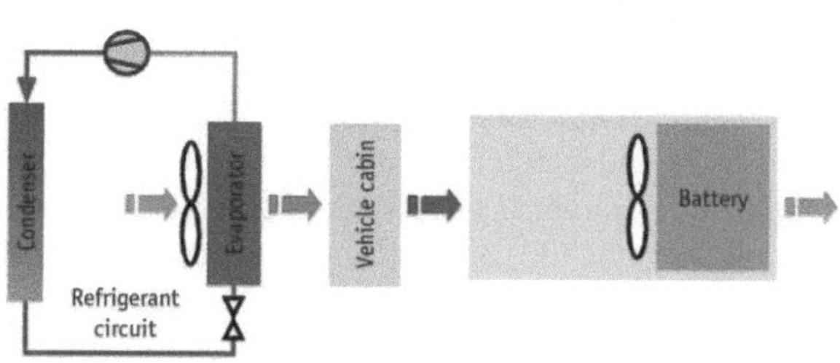

Fig. 8.5 Air cooling system for LIB [34]

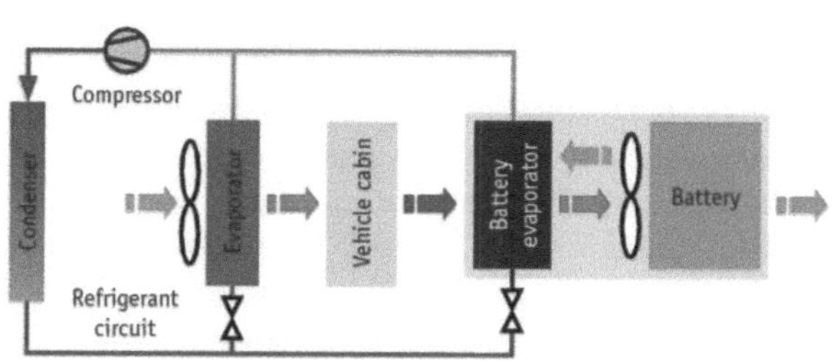

Fig. 8.6 Independent evaporator to blow cool air to the LIB [34]

Fig. 8.7 Force air-cooled convection BTMS of Toyota Prius [45]

battery pack has been cooled by Toyota using cold air extracted from the passenger cabin. The Mitsubishi i-MiEV and Nissan Leaf are two more EV models that have adopted the same BTMS approach.

8.4.2 Liquid-Cooled

Three techniques are available for liquid-cooled BTMS: (i) immersing modules in a dielectric fluid for direct contact; (ii) encircling each module with discrete tubing, passing it through a liquid-filled pipe with high thermal conductivity; and (iii) enclosing LIB in a jacket or setting LIB on a liquid-cooled cold plate, as illustrated in Fig. 8.8. The inlet temperature and coolant flow rate determine the LIB's average coolant temperature. The average LIB temperature will drop because to the increased coolant flow rate [33, 38]. LIB must be in close contact with the cold plate to create efficient cooling.

Larger, more sophisticated LIBs respond better to temperature regulation using liquid-cooled BTMS. The uniform temperature distribution within LIB, which results from the strong cooling capability of liquid, is another benefit of liquid-cooled BTMS [46]. The primary issue with liquid-cooled BTMS, however, is the potential for liquid leakage, which could result in an electrical short circuit. In extreme situations, such as car crashes, it may result in fire. Several EV manufacturers have implemented liquid-cooled BTMS such as Ford Focus, Chevrolet Volt, Tesla Roadster, BYD e6, Smart EV, and Volvo C30 [MIT Technology Review 2012].

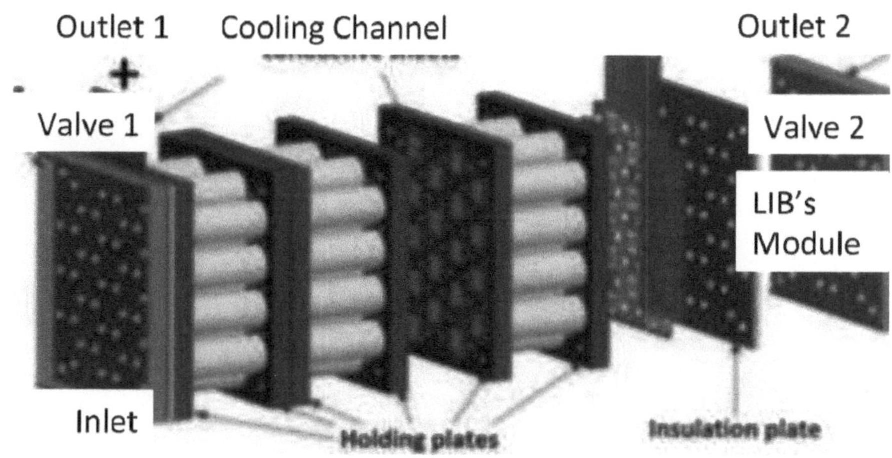

Fig. 8.8 Liquid-cooled BTMS for Porsche Panamera S E-hybrid's battery pack [37]

8.4.3 Evaporative-Cooled

Additionally, BTMS can regulate the temperature of LIB by using refrigerant, usually R134a. In a closed-loop circulation managed by a pump or compressor, the refrigerant travels via tubing or a hose to a heat exchanger (evaporator/condenser). Refrigerant will absorb and carry away the heat generated by LIB before dissipating it to a condenser that is cooled by either air or water. This technique is often referred to as evaporative cooling or two-phase cooling. As seen in Fig. 8.9, Heckenberger [34] created an evaporator plate positioned between LIB to create an evaporative-cooled BTMS.

Parker Hannifin [6] asserts that evaporative-cooled BTMSs that employ a refrigerant as a cooling medium are more dependable, lighter, and safer than traditional cooling systems that use liquid or air because of their higher heat transfer coefficient. The intricacy of the system, requiring the integration of LIB with active components like the pump/compressor, condenser, and evaporator, is the drawback of evaporative-cooled BTMS. As a result, the car will be more expensive, heavier, and use more energy. Evaporative-cooled BTMS is not a desirable technology because of these problems, and only a small number of manufacturers have included it in their EV vehicles, including the Mercedes S400 blue hybrid.

8.4.4 Phase Change Material Cooling

The integrated-cooled phase change material (PCM) shown in Fig. 8.10 illustrates the BTMS cooling technique, which involves placing LIB in direct contact with PCM. A composite known as PCM stores and releases thermal energy as it melts

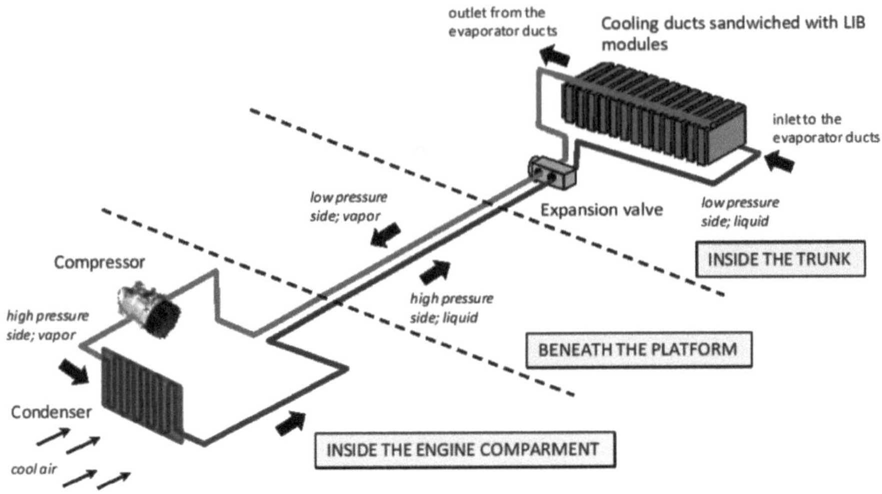

Fig. 8.9 Schematic diagram of evaporative-cooled BTMS [31]

and freezes. The way that PCM stores and releases latent heat has been covered by author Nellis [47]. According to them, the graphite serves as a thermal conductor and disperses the heat throughout the LIB as it heats up, preventing hot spots and guaranteeing temperature homogeneity across the LIB. All extra thermal energy is absorbed by the melting process rather than being used to raise the temperature once it reaches the PCM's melting point.

In this sense, the PCM's melting point establishes a maximum temperature that cannot be raised until the PCM has completely melted. Even at very high temperatures, the composite material itself stays solid and nothing seeps out, even while the PCM is melting.

Kim et al. [42] described that LIB utilizing PCM can control temperature excursions and maintain temperature uniformity with or without reduced use of active cooling components such as blowers, or pump/compressor in air, liquid and evaporative-cooled BTMSs. These attributes will allow for a compact, lightweight and energy-efficient system design.

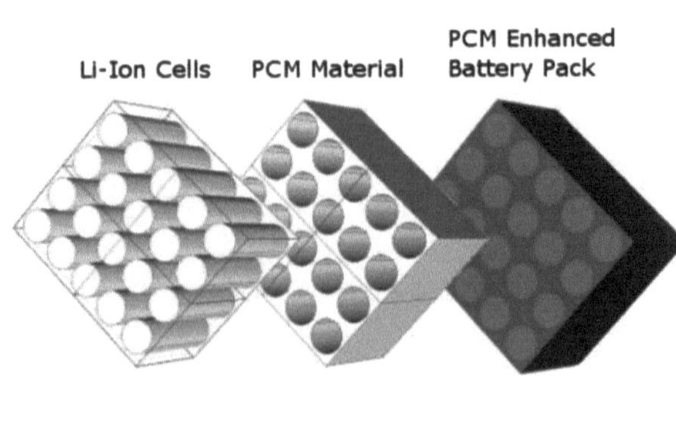

Fig. 8.10 PCM directly integrated into LIB [42]

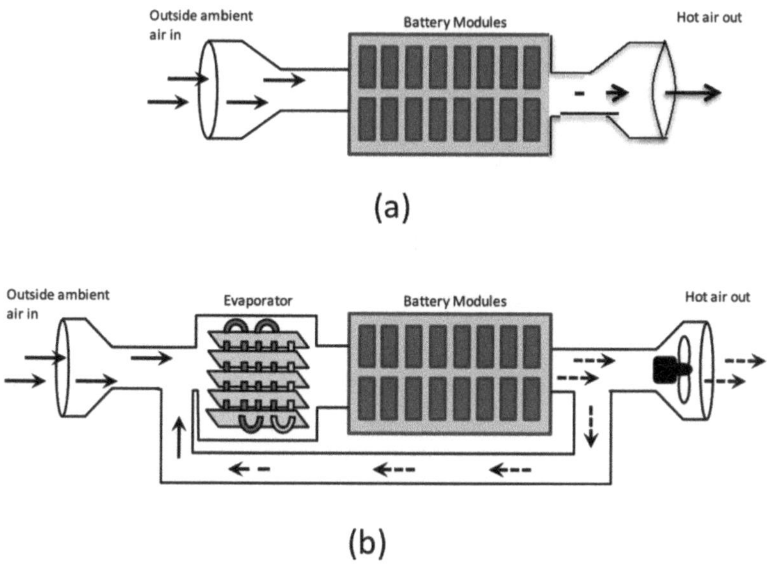

Fig. 8.11 Cooling system **a** passive cooling **b** active cooling [49]

8.4.5 *Passive and Active-Cooled*

Figure 8.11a illustrates how the free convection air-cooled passive-cooled BTMS regulates the LIB temperature. The heat from LIB is absorbed by either ambient or air-conditioned air. Only when the ambient air temperature is between 10 and 35 °C can BTMS function properly for the passive system [48]. Conversely, active-cooled BTMS generates a cooling cycle through LIB and dissipates heat from the cooling medium using active-controlled components.

The evaporative cooling system used in electric car battery cooling systems, as seen in Fig. 8.12, can save a substantial amount of battery energy because of its passive manner of operation. In inactive mode, it can save roughly (700–200)*15/60*2 = 250 Wh of energy for 15 min. Reducing the frequency of battery charging would help extend battery life, improve vehicle mileage, and save battery energy.

8.4.6 *Cooling System Effectiveness*

The cooling medium's density, velocity, viscosity, and thermal conductivity all affect how quickly heat moves between the LIB's walls. The most practicable direct-contact cooling medium, such oil, has a substantially faster heat transfer rate than air for the same flow rate due to its thinner boundary layer and higher fluid thermal conductivity. However, a lower flow rate is typically employed due to oil's higher viscosity and

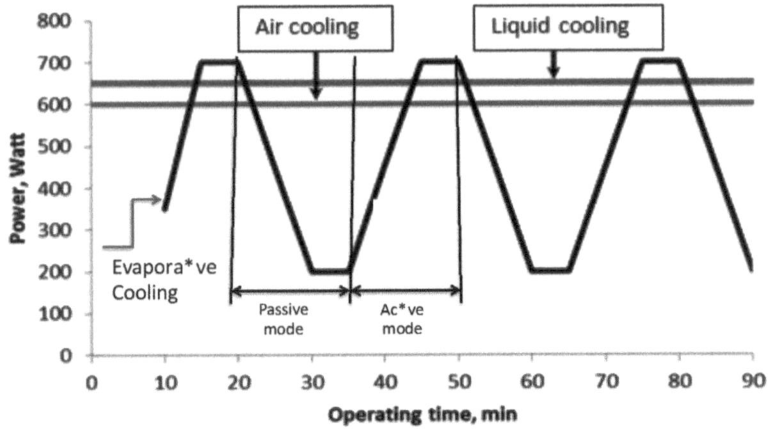

Fig. 8.12 Energy saving by a passive mode battery cooling system [31]

correspondingly higher pumping power, resulting in an oil heat transfer coefficient that is only 1.5 to 3 times higher than air.

Ethylene glycol water solutions and other indirect-contact heat transfer cooling media often have a higher heat transfer coefficient than most oils because of their higher thermal conductivity and lower viscosity. However, the efficiency of indirect contact cooling media is reduced since the heat must be transferred through the fins or the walls of the cooling jacket/container. According to research by Sabbah et al. [50], air-cooled BTMS is unable to maintain the LIB temperature within the ideal range without using a lot of fan power. According to Alaoui [7], BTMS may need to be more complex for EV applications, where a liquid-cooled system may need to be included. HEVs, which typically have a lower LIB capacity, can be fitted with air-cooled BTMS.

Kizile al. [38] employed HP and a metallic aluminum fin to reduce the temperature increase that occurred during LIB discharge. HP is a particularly helpful heat pump for transferring heat from LIB to the surrounding environment since it can operate independently and transmit heat at a high rate over long distances without the need for external pumping power. The temperature distribution when discharging was examined using a test sample, a 12 Ah cylindrical LIB. The findings demonstrated that the aluminum fins' increased surface area for heat transmission, which helped remove the generated heat, resulted in a lower temperature profile under forced convection.

Additionally, compared to the outcome prior to inserting HP, the temperature dropped by 7 °C at the end of the discharging procedure after HP was inserted into aluminum fins. According to their findings, HP and metallic aluminum fins work very well together to improve heat dissipation from various battery types. The testing findings were based on employing a single LIB alone, and there was no more research done for big capacity LIB, therefore the application of this approach in large-scale LIB is unclear.

A Peltier-effects thermoelectric heat pump module (dissimilar semiconductor material, P-type and N-type) for BTMS was studied by Alaoui and Salameh [7] in terms of airflow. Their findings indicated that the LIB temperature increased linearly for the first six minutes before beginning to saturate as a result of losses to the surrounding air temperature. The temperature within LIB reached a maximum of 29 °C, while the thermal unit reached 37 °C. The outcome indicates that this approach is applicable to a small-scale LIB. Since there are no active components in the system, it can provide great reliability and minimal maintenance.

A study by Mousavi et al. [51] examined the best air-cooled BTMS architecture for EVs. To let air to flow through LIB, 150 tiny air passageways were created. The purpose of LIB was to be placed in front of a car's dash panel. The number of heat transfer units (NTU) was determined; if NTU is 5, heat transmission has reached its limit. The outcome, which had an NTU value of 4.85 (out of 5), indicated that the tube's maximum diameter could not exceed 50 mm. They have, however, assumed three things about the parameters: i) LIB's maximum surface temperature of 90 °C remains constant; ii) no heat transfer process takes place from the surface of LIB; and (iii) LIB surface convective heat transfer is negligible. These assumptions are likely to affect the results of their study. The method of placing LIB in front of the vehicle is not a preferable and suitable method to be implemented in EV. Kawa [52] stated that comparatively, liquid-cooled is more thermally efficient and has better heat transfer qualities than air-cooled, but it is more expensive and more difficult to integrate into LIB.

To absorb the heat produced by LIB, Yeow et al. [53] employed a PCM integrated between LIB that served as a heat sink. The PCM's high latent heat will keep the LIB temperature from rising significantly when it beyond the PCM's melting point. According to their findings, the temperature of LIB without PCM rose quickly from 30 to 90 °C at a discharge current of 10 amps during the 15-min testing period, whereas the temperature of LIB with PCM climbed rather slowly from 50 to 60 °C. According to their findings, PCM can be used in BTMSEV, where a number of advantages, including its efficiency in absorbing heat and the compactness of LIB, which eliminates the need for active components like a manifold or pump, add to the method's advantages. However, as the target temperature range during operation is between 30 and 40 °C, this method is not appropriate for our investigation.

Furthermore, Kim et al. [54] used PCM and graphite matrix modules attached surrounding of the LIB, together with airflow for the cooling. The graphite matrix impregnated with wax absorbs thermal energy through both sensible heat and latent heat. The results showed that PCM effectively limits the peak temperature rise, especially in high ambient temperature conditions. It can be concluded that combining some degree of forced air-cooled with the PCM would help accelerate the heat removal process from LIB. An analysis of continuous cycling conditions suggests that the potential role of PCM in BTMS is not to replace the active cooling system, but rather to allow the active cooling system to be downsized or used less frequently.

Optimization of the loop heat pipe (LHP) consists of an evaporator, compensation chamber, vapor/liquid line, and condenser as main components to cool LIB onboard a military aircraft [55]. They carried out a heat transfer analysis of a thin liquid film model. The result of their work showed that the design optimization of LHP for the fixed and varied heat load cases can sustain LIB temperature near 10 °C.

8.5 Conclusion

A thorough examination of the thermal management system from the standpoint of battery stability reveals that the evaporative battery thermal management system (eBThMS) is better suited for EV battery cooling since it dissipates heat more rapidly than air and liquid cooling. It is unsafe to operate BThMS without BMS technology, nevertheless. Lithium-ion battery (LIB) cooling is the technique of extending battery life by regulating the battery's temperature via an AI thermal management system (AI_ThMS).

Acknowledgements The authors are grateful to the PROTON, Automotive Manufacturer Malaysia to finance this project.

References

1. Liu, H., Wei, Z., He, W., Zhao, J. (2017). Thermal issues about Li-ion batteries and recent progress in battery thermal management systems: A review. *Energy Conversion and Management, 150*, 304−330.
2. Ye, Y., Saw, L. H., Shi, Y., Tay, A. A. O. (2015). Numerical analyses on optimizing a heat pipe thermal management system for lithium-ion batteries during fast charging. *Applied Thermal Engineering, 86*, 281−291.
3. Huo, Y. Rao, Z. Liu, X. Zhao, J. (2015) Investigation of power battery thermal management by using mini-channel cold plate. *Energy Conversion and Management, 89*, 387−395.
4. Jarrett, A., Kim, I. Y. (2014) Influence of operating conditions on the optimum design of electric vehicle battery cooling plates. *Journal of Power Sources, 245*, 644−655.
5. Rahman, A., Helmi, A., & Hawlader, M. N. A. (2017). Two-phase evaporative cooling battery thermal management technology for electric vehicle. *International Journal of Automotive Technology, 18*(5), 875–882.
6. Rahman, A., Farhana, S., & Hawlader, M. N. A. (2015). Two phase fuzzy controlled evaporative thermal management system for EVs/HEVs. *International Journal of Automotive Technology, 28*(4), 297–304.
7. Alaoui, C., & Salameh, Z. M. (2003). Modeling and simulation of a thermal management system for electric vehicles (pp. 887–890). IEEE.
8. Albright, G. (2012). Making lithium-ion safe through thermal management. AllCell Technologies, Chicago, IL, (pp. 6(1)–6(10)). Retrieved November 26, 2013. http://www.battcon.com/PapersFinal2012/GregAlbright-MakingLithium-ionsafethroughthermalmanagement.pdf.
9. Allcelltech (2013). Thermal management with phase change material. Retrieved November 30, 2013. http://www.allcelltech.com/technology/thermal-management-with-phase-change-material.

10. Al-Hallaj, S., Kizilel, R., Lateef, A., Sabbah, R., Farid, M., & Selman, J. R. (2005). Passive thermal management using phase change material (PCM) for EV and HEV Li-ion Batteries. In *Vehicle Power and Propulsion, 2005 IEEE Conference*
11. Anderson, D. L. (2009) An evaluation of current and future costs for lithium-ion batteries for use in electrified vehicle powertrains. Master Dissertation, Nicholas School of the Environment of Duke University (p. 53).
12. Benger, R., Wenzl, H., Beck, H., Jiang, M., & Ohms, D. (2009). Electrochemical and thermal modeling of lithium-ion cells for use in HEV or EV application. *The World Electric Vehicle Journal, 3*(3), 1–10.
13. Bergveld, H. J. (2001). Battery management systems design by modelling. Ph.D. dissertation, Universiteit Twente (p. 61).
14. Boyd, S. J. (2006). Hybrid electric vehicle control strategy based on power loss calculations. Master dissertation, Virginia Polytechnic Institute and State University.
15. Campbell, J. B., Tolbert, L. M., Member, S., Ayers, C. W., Ozpineci, B., & Lowe, K. T. (2007). Two-phase cooling method using R134a refrigerant to cool power electronic devices. *Industry Applications, IEEE Transactions, 43*(3), 648–656.
16. Cengel, Y. A. (2007). *Heat and mass transfer; a practical approach* (3rd ed.). McGraw-Hill.
17. Chen, Y., & Evans, J. W. (1994). Thermal analysis of lithium polymer electrolyte batteries by a two dimensional model—thermal behaviour and design optimization. *Electrochimica Acta, 39*(4), 517–526.
18. Dangelmayr, G., & Kirby, M. (2005). Mathematical modeling. In *Mathematical modeling; A comprehensive introduction* (pp. 1–12). Upper Saddle, New Jersey: Prentice Hall.
19. Dinger, A., Martin, R., Mosquet, X., Rabi, M., Rizoulis, D., & Sticher, G. (2009). Batteries for electric cars; Challenges, opportunities, and the outlook to 2020. The Boston Consulting Group. Retrieved January 20, 2013. http://www.bcg.com/documents/file36615.pdf.
20. Dominko, R., Bele, M., Gaberscek, M., Remskar, M., Hanzel, D., Pejovnik, S., & Jamnik, J. (2005). Impact of the carbon coating thickness on the electrochemical performance of LiFePO4/C composites. *Journal of The Electrochemical Society, 152*(3), A607–A610.
21. Duan, X., & Naterer, G. F. (2010). Heat transfer in phase change materials for thermal management of electric vehicle battery modules. *International Journal of Heat and Mass Transfer, 53*(23–24), 5176–5182.
22. Dwiggins, B. H. (2002). *Automotive air conditioning* (8th ed.). New York: Delmar.
23. Erjavec, J., & Arias, J. (2007). *Hybrid, electric & fuel-cell vehicles*. Clifton Park, NY: Thomas Delmar.
24. Freitas, G. C. S., Peixoto, F. C., & Vianna, A. S. (2008). Simulation of a thermal battery using Phoenics®. *Journal of Power Sources, 179*(1), 424–429.
25. Kim, J., & Lee, H. (2007). Lithium ion battery. U.S. Patent, 0 154 795, A1, July 5, 2007.
26. Young, K., Wang, C., Wang, L. Y., & Strunz, K. (2013). Electric Vehicle Battery Technologies. In R. Garcia-Valle & J. A. Peças Lopes (eds.), Electric Vehicle Integration into Modern Power Networks (pp. 15–56). New York, NY: Springer New York. https://doi.org/10.1007/978-1-4614-0134-6
27. Swierczynski, M., Teodorescu, R., & Rodriguez, P. (2011). Lifetime investigation of a lithium iron phosphate (LFP) battery system connected to a wind turbine for forecast improvement and output power gradient reduction. In *Proceedings of the 15th Battcon Stationary Battery Conference and Trade Show 2011* (pp. 1–8).
28. Ghosh, D., Maguire, P. D., & Zhu, D. X. (2009). Design and CFD simulation of a battery module for a hybrid electric vehicle. SAE Technical Paper 2009-01-1386. https://doi.org/10.4271/2009-01-1386.
29. Gu, W. B., & Wang, C. Y. (2000). Thermal-electrochemical modeling of battery systems. *Journal of The Electrochemical Society, 147*(8), 2910–2922.
30. Hadhazy, A. (2009). A better battery? *The Lithium Ion Cell Gets Supercharged*. Scientific American. Retrieved September 25, 2012. http://www.scientificamerican.com/article.cfm?id=better-battery-lithium-ion-cell-gets-supercharged.

31. Helmi, A. (2014). Development of evaporative cooling battery thermal management system for electric vehicles. M.Sc Thesis dissertation, International Islamic University Malaysia.
32. Headway (2007). Technical specifications for Li-ion battery. Ecocarforum.com. Retrieved September 25, 2012. http://www.scientificamerican.com/article.cfm?id=better-battery-lithium-ion-cell-gets-supercharged.
33. Haines, R. W., & Wilson, C. L. (2003). *HVAC systems design handbook* (4th ed.). McGraw-Hill.
34. Heckenberger, T. (2009). Li-ion battery cooling: more than just another cooling task. *Technical Press Day, 2009*, 4–9.
35. Hussain, I. (2011). *Electric and hybrid vehicles: Design fundamentals* (2nd ed.). CRC Press.
36. Jurgen, R. K. (Ed.). (2002). *Electric and hybrid-electric vehicles*. Society of Automotive Engineers.
37. Karimi, G., & Dehghan, A. R. (2012). Thermal management analysis of a lithium-ion battery pack using flow network approach. *International Journal of Mechanical Engineering and Mechatronics, 1*(1).
38. Kizilel, R., Sabbah, R., Selman, J. R., & Al-Hallaj, S. (2009). An alternative cooling system to enhance the safety of Li-ion battery packs. *Journal of Power Sources, 194*(2), 1105–1112.
39. Soldo, V., Grozdek, M., & Curko, T. (2006). Two-phase heat transfer during refrigerant evaporation in a solar collector tube. In *61th ATI National Congress-International Season "Solar Heating and Cooling"* (pp. 2–7).
40. Lefebvre, L. (2011). HV battery thermal management for PHEV & EV. In *International Scientific Conference on HEV and EV (RHEVE 2011)*, Rueil Malmasion.
41. Khankari, K. K., Patankar, S. V., & Morey, R. V. (1995). A mathematical model for natural convection moisture migration in stored grain. *American Sosiety of Agricultural and Biological Engineers, 33*(6), 1777–1787.
42. Kim, G., Gonder, J., Lustbader, J., & Pesaran, A. A. (2008). Thermal management of batteries in advanced vehicles using phase-change materials. *The World Electric Vehicle Journal, 2*(2), 46–59.
43. Kim, G., & Pesaran, A. (2007). Battery thermal management design modeling. *The World Electric Vehicle Association Journal, 1*, 126–133.
44. Krüger, I., Sievers, M., & Schmitz, G. (2009). Thermal modeling of automotive lithium ion cells using the finite elements method in Modelica. In *Proceedings 7th Modelica Conference* (pp. 20–22). COMO, Italy
45. Zolot, M., Pesaran, A. A., & Mihalic, M. (2002). Thermal evaluation of Toyota Prius battery pack. SAE Technical Paper 2002-01-1962, 2002. 10.4271.
46. Smith, K., & Wang, C.-Y. (2006). Power and thermal characterization of a lithium-ion battery pack for hybrid-electric vehicles. *Journal of Power Sources, 160*(1), 662–673.
47. Mi, C., Li, B., Buck, D., & Ota, N. (2007). Advanced electro-thermal modeling of lithium-ion battery system for hybrid electric vehicle applications. In *Vehicle Power and Propulsion Conference* (pp. 107–111). IEEE.
48. Smyshlyaev, A., Krstic, M., Chaturvedi, N., Ahmed, J., & Kojic, A. (2011). PDE model for thermal dynamics of a large Li-ion battery pack. In *2011 American Control Conference* (pp. 959–964).
49. Pesaran, A. A., & Keyser, M. (2001) Thermal characteristics of selected EV and HEV batteries. In *National Renewable Energy Laboratory, 16th Annual Battery Conference: Advances and Applications*. Long Beach, California.
50. Sabbah, R., Kizilel, R., Selman, J. R., & Al-Hallaj, S. (2008). Active (air-cooled) vs. Passive (phase change material) thermal management of high power lithium-ion packs: Limitation of temperature rise and uniformity of temperature distribution. *Journal of Power Sources, 182*(2), pp. 630−638.
51. Onda, K., Ohshima, T., Nakayama, M., Fukuda, K., & Araki, T. (2006). Thermal behavior of small lithium-ion battery during rapid charge and discharge cycles. *Journal of Power Sources, 158*(1), 535–542.
52. Park, S. (2010). A comprehensive thermal management system model for hybrid electric vehicles. Ph.D. Dissertation, University of Michigan, 2011, 78.

53. Park, Y., Jun, S., Kim, S., & Lee, D. (2010). Design optimization of a loop heat pipe to cool a lithium ion battery onboard a military aircraft. *Journal of Mechanical Science and Technology, 24*(2), 609–618.
54. Ramotar, L., (2010). Second law analysis of a liquid cooled battery thermal management system for hybrid and electric vehicles (p. 73). Master dissertation, University of Ontario Institute of Technology.
55. Shannon, R. E. (1975). *Systems simulation—The art and science.* Prentice-Hall.

Chapter 9
Battery Thermal Management System for Electric Vehicle

Abstract The efficiency of the battery thermal management system (ThMS) can be used to assess the performance and safety of electric vehicles. Heat is produced by high battery discharge/charge current. Controlling battery heat quickly and effectively is crucial to preventing battery deterioration or even fire. Different BThMS were experimentally investigated in both cruising and vehicle acceleration modes. By keeping an eye on the battery's temperature between 30 and 40 °C in both the state of charge (SoC) and the state of discharge (SoD), smart evaporative BThMS is advised for maximizing battery performance. To regulate battery temperature and extend battery life, an algorithm was created based on the LIB temperature range.

Keywords Battery heat gain · Evaporative battery thermal management system · Battery maintenance

9.1 Introduction

Strong electrical currents are produced during the brief peak load times, and because of internal resistance, the Li-ion cells warm up significantly. The ideal operating temperature range for lithium-ion batteries is between 25 and 35° C. This could result in balanced active material use, enhance performance, and extend battery life. A battery generates heat throughout its charging and draining processes. Particularly during the high-rate discharges, a tremendous amount of heat is produced. In the event that this heat cannot be released right away, the battery temperature quickly increases [1]. Research has been done on the heat behavior of Li-ion batteries during cycles of rapid charging and discharging [2, 3]. According to their findings, quick cycles of charge and discharge result in significant heat buildup, which makes the battery temperature rise significantly.

The modeling and simulation for the BCThMS for the 100 V, 86 Ah, 8.6 kWh battery of the Proton Saga EV are presented in this chapter. The maximum current that an AC induction motor powered by a Li-ion battery pack may consume from the battery is 270 A for traction and 120 A for continuous mode. Because of internal

A. Rahman et al., *Autonomous Driveline System for Electric Vehicle*, Green Energy and Technology, https://doi.org/10.1007/978-981-96-8418-2_9

resistance and ohomic heat, these brief peak load times produce strong electrical currents that significantly warm the Li-ion cells. The cooling load and refrigerant mass flow rate have been taken into consideration in order to create a prototype model of EC-BThMS.

9.2 Mathematical Modeling

The mathematical model has been developed for the simulation of heat generation and temperature distribution within the battery pack due to the fast charging and high rate of discharging as shown in Fig. 9.1, using the equations of [1, 2, 4].

9.2.1 LIB Heat Generation

As previously mentioned, heat is produced in a battery cell during operation (the discharging or charging process) when an electric current *I* is removed from or introduced into the cell because:

(i) Entropy changes from electrochemical reaction
(ii) Joule's effect (Ohmic heating)
(iii) Polarization resistance.

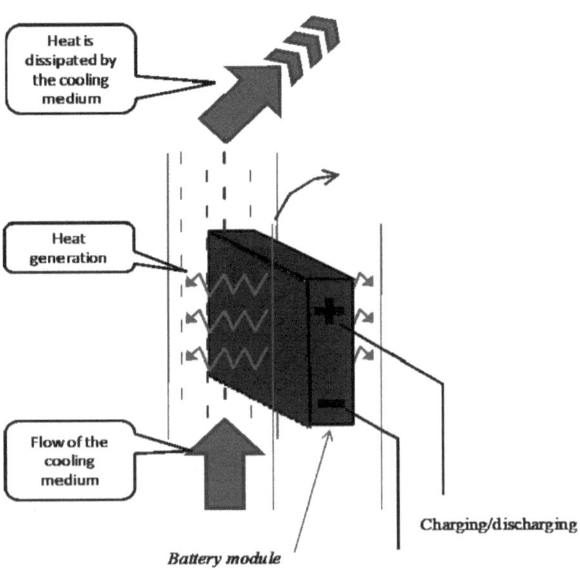

Fig. 9.1 Heat generation and dissipation [1]

Total heat rate Q_{gen} generated inside a battery cell due to electrochemical reactions Q_{ele}, polarization resistance Q_{polar}, and Joule's effects Q_{joule} can be expressed as follows [1].

$$Q_{gen} = Q_{elec} + Q_{polar} + Q_{joule} \tag{9.1}$$

According to [4, 5], each of the factors of heat generation are described

$$Q_{elec} = |I| \cdot \sum_i \frac{T_{batt(initial)} \Delta S_i}{n_i F} \tag{9.2}$$

$$Q_{polar} = \sum_i I \eta_i \tag{9.3}$$

$$Q_{joule} = I^2 R \tag{9.4}$$

Therefore,

$$Q_{gen} = |I| \cdot \sum_i \frac{-T_{batt} \Delta S_i}{n_i F} + \sum_i I \eta_i + I^2 R \tag{9.5}$$

where I is the current flow, F is the Faraday constant (9.64122×10^4 C/mol), $T_{batt(initial)}$ is the battery initial temperature, ΔS_i donates the entropy change associated with the electrochemical reaction i, n_i is the number of electrons exchanged in electrochemical reaction, η_i is the total over-potential due to kinetic aspects and mass transport associated with electrochemical reaction i and R is the total ohmic series electrical resistance inside the battery [4, 5].

9.2.2 Heat Dissipation

The battery cell generates heat during operation, which causes heat to be transferred to the cell's surface and either removed or dissipated into the surrounding environment. Through the process of convective heat transfer, heat is transferred and dissipated from the cell to the surrounding environment. Convection heat transfer happens from the battery modules' surface to the coolant as a result of fluid (refrigerant) motion within the modules' boundary layer. The following Eq. (9.5) can be used to explain the heat removal:

$$Q_{conv} = h_m A_s \left(T_{batt(initial)} - T_\infty \right) \tag{9.6}$$

where h_m is termed the convection heat transfer coefficient of the cooling medium, A_s is the area of the modules' surface, and T_∞ is the temperature of the cooling medium.

The total heat generation into the battery can be estimated as

$$Q_{\Delta \text{energy}} = Q_{\text{in}} + Q_{\text{gen}} - Q_{\text{out}} \tag{9.7}$$

where $Q_{\Delta \text{energy}}$ is the rate of energy increment stored within the battery cell, Q_{in} is the rate of energy transfer into the cell, Q_{gen} is the rate of energy generation, and Q_{out} is the rate of energy transfer out of the cell. It is assumed that there is no heat transfer energy transfer from cell to cell, $Q_{\text{in}} = 0$, therefore the overall energy balance on a battery module can be expressed as

$$Q_{\Delta \text{energy}} = Q_{\text{gen}} - Q_{\text{out}} \tag{9.8}$$

Considering Q_{out} equals to Q_{conv}, Eq. 9.8 can be expressed using the equation [2] as

$$Q_{\Delta \text{energy}} = \left(|I| \cdot \sum_i \frac{T_{\text{batt(initial)}} \Delta S_i}{n_i F} + \sum_i I\eta_i + I^2 R \right) - h_m A_s \left(T_{\text{batt(initial)}} - T_\infty \right) \tag{9.9}$$

The change in energy storage due to temperature change over time t when the vehicle is operated, or current is drawn from the battery can be expressed using the equation [4].

$$Q_{\Delta \text{energy}} = Q_{\text{accu}} = \frac{dU_t}{dt} = \frac{d}{dt}\left(\rho V c_p T \right) \tag{9.10}$$

The term Q_{accu} is called heat accumulation associated with the rate of change in the internal thermal energy of the battery module, whether the energy is stored or released over time. ρ and c_p are the mass density and the specific heat capacity, respectively of the battery module material and V is the volume of the module.

The above overall energy equation used to calculate the average module temperature T_{batt} over time has been expressed the equation [2, 4, 5].

$$\begin{aligned} \frac{d}{dt}\left(\rho V c_p T \right) = & \left(|I| \cdot \sum_i \frac{T_{\text{batt(initial)}} \Delta S_i}{n_i F} + \sum_i I\eta_i + I^2 R \right) \\ & - h_m A_s \left(T_{\text{batt(initial)}} - T_\infty \right) \end{aligned} \tag{9.11}$$

Therefore, by replacing the battery module mass m into the above equation, the overall energy equation used to calculate the average module temperature T_{batt} can be expressed as follows [2, 4]:

$$\begin{aligned} \frac{dT_{\text{batt}}}{dt} = \frac{1}{mc_p} & \left[\left(|I| \cdot \sum_i \frac{T_{\text{batt(initial)}} \Delta S_i}{n_i F} + \sum_i I\eta_i + I^2 R \right) \right. \\ & \left. - h_m A_s \left(T_{\text{batt(initial)}} - T_\infty \right) \right] \end{aligned} \tag{9.12}$$

Equation 9.12 can be written by the simplified equation [5] as

$$T_{\text{batt(final)}} = T_{\text{batt(initial)}} + \frac{\Delta t}{mc_p}\left[\left(|I| \cdot \sum_i \frac{-T_{\text{batt(initial)}}\Delta S_i}{n_i F} + \sum_i I\eta_i + I^2 R\right)\right.$$
$$\left. -h_m A_s\left(T_{\text{batt(initial)}} - T_\infty\right)\right] \tag{9.13}$$

Equation 9.13 can be defined as

$$T_{\text{batt(final)}} = T_{\text{batt(initial)}} + \frac{\Delta t}{mc_p}\left(Q_{\text{gen}} - Q_{\text{out}}\right) \tag{9.14}$$

where ΔT, Q_{gen}, and Q_{out} are the variable parameters and $T_{\text{batt(initial)}}$ and mc_p are the constant parameters, are considered to simulate the battery heat gain both in charging and discharging.

9.2.3 Refrigerant Mass Flow Rate

The refrigeration cycle is presented by the line 1-2-3-4 in Fig. 9.2. In the cycle 1-2 is gas compressed which equals the energy input into the gas by the compressor, the line 2-3 represents the vapor gas is condensed to a liquid, 3-4 represents the liquid is passed through an expansion device, where the liquid changes the phase 20% vapor and 80% liquid. However, the proportion of gas and liquid depends on the temperature of the evaporator (battery pack with hollow ducts), and 4-1 represents the liquid is evaporated completely to a gas by the compressor and enthalpy is extracted from the battery's generated heat (evaporating process at evaporator region).

Based on the diagram, cooling capacity of the evaporator Q_{evap} can be calculated by Eq. (9.15):

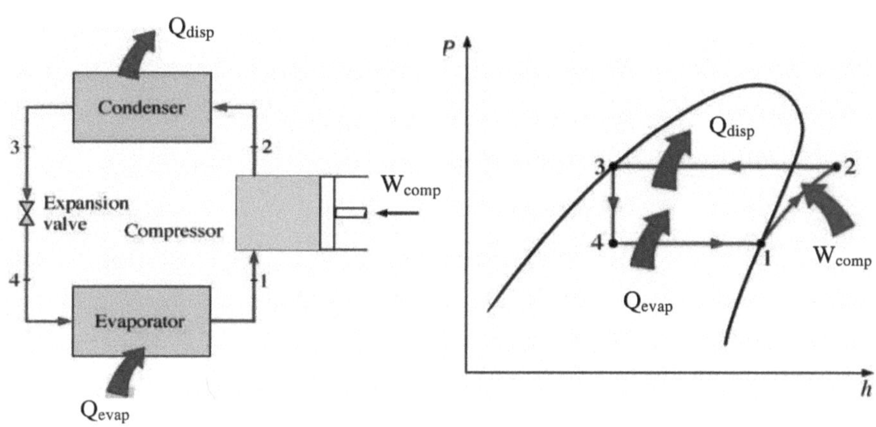

Fig. 9.2 p-H diagram of the ideal refrigeration cycle [1]

$$Q_{evap} = \dot{m}_{ref}(h_1 - h_4) \tag{9.15}$$

where $\dot{m}_{ref}$ is the refrigerant mass flow rate, h_1 is the enthalpy at point 1, and h_4 is the enthalpy at point 4. $(h_1 - h_4)$ is called change of enthalpy during evaporation process. It can be calculated using the equation [2, 4]:

$$\Delta h = (h_1 - h_4) = c_{p_{ref}}(T_{in} - T_{out}) \tag{9.16}$$

where $c_{p_{ref}}$ is the heat capacity of refrigerant, T_{in} is the inlet refrigerant temperature, and T_{out} is the outlet refrigerant temperature.

Work input from the compressor W_{comp} given to the refrigerant during compression process was obtained by Eq. (9.15).

$$W_{comp} = \dot{m}_{ref}(h_2 - h_1) \tag{9.17}$$

where h_2 is the enthalpy at point 2.

Assuming that the heat received by the refrigerant is equal to the heat generation from the battery, Eq. (9.5) can now be rewritten using the equation [2]:

$$Q_{evap} = Q_{gen} \tag{9.18}$$

$$|I| \cdot \sum_i \frac{-T_{batt}\Delta S_i}{n_i F} + \sum_i I\eta_i + I^2 R = \dot{m}_{ref}(h_1 - h_4) \tag{9.19}$$

Hence, from Eq. 9.19, the refrigerant mass flow rate can be expressed as follows [4, 5]:

$$\dot{m}_{ref} = \frac{\sum_i \frac{-T_{batt}\Delta S_i}{n_i F} + \sum_i I\eta_i + I^2 R}{(h_1 - h_4)} \tag{9.20}$$

9.3 Simulation and Considerations

Battery temperature has been simulated using mathematical Eq. 9.13 by varying the refrigerant's mass flow rate and discharge current. Optimizing the cooling system for the EV battery is the goal of this simulation. In the simulation, three different kinds of cooling media have been considered:

(i) Force air cooling,
(ii) Liquid cooling,
(iii) Refrigerant cooling.

The battery pack has a capacity of 8.6 kWh, and its starting temperature is set at 30 °C. According to Soldo et al. [6], Rahman et al. [7] and Bergveld [4], the cooling mediums' respective temperatures and heat transfer coefficients are 30 °C and 40 W/m^2·K for air, 28 °C and 200 W/m^2·K for water, and 15 °C and 500 W/m^2·K for refrigerant [4]. Li-ion batteries have an entropy change/module of 5–12 J/mol·K and a discharge current range of 50–350A.

9.4 Performance of EC-BTHMS

The purpose of this simulation is to examine the performance of developed smart BThMS theoretically for the Proton Saga EV:

(i) Capacity of the battery pack is 8.6 kWh.
(ii) Initial temperature of the battery is 30 °C, which is similar to the surrounding temperature.
(iii) Input parameters for each running mode during PGMC 2012

Figure 9.3 illustrates how the battery pack's internal heat generation rises as speed and discharge current increase. Furthermore, when the temperature gradient grows, so does the discharge current. For instance, if the car is traveling at 120 km/h on a level road with no gradient, 208 A of discharge current is needed, producing 924 W of heat. Similarly, if the car is driven at the same speed on a slope with a maximum gradient of 3.67%, the necessary discharge current is roughly 335 A, which will generate roughly 2500 W of heat [2, 5].

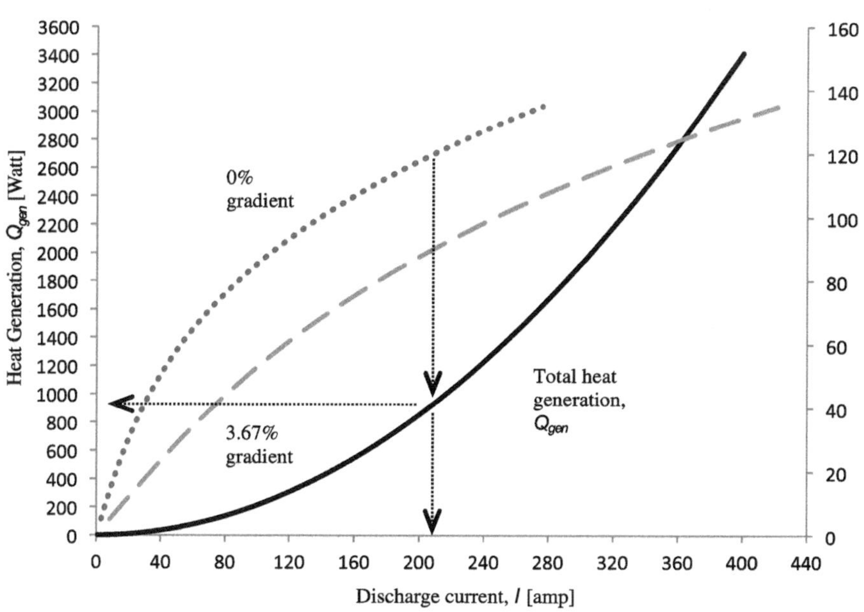

Fig. 9.3 Heat gain during discharging [4, 5]

9.5 Conclusions

The battery's temperature gradient is increased by high charging and discharge currents. The cooling media's respective temperatures and heat transfer coefficients are 30 °C and 40 $W/m^2 \cdot K$ for air, 28 °C and 200 $W/m^2 \cdot K$ for water [4], and 15 °C and 500 $W/m^2 \cdot K$ for refrigerant. According to a simulation study, a discharge current of roughly 335 A generates roughly 2500 W of heat. It takes a very effective cooling system to keep the battery temperature between 30 and 35 °C.

Acknowledgements The experiment was financed by PROTON, an automotive manufacturer in Malaysia, for which the authors are thankful. The authors express their gratitude to the International Islamic University Malaysia's Research Management System (RMC) for providing money for the creation of this project's intelligent system and Eastern Michigan University, Ypsilanti, MI, USA allowing to continue with the project work.

References

1. Haines, R. W., & Wilson, C. L. (2003). *HVAC systems design handbook* (4th ed.). McGraw-Hill.
2. Helmi, A. (2014). Development of evaporative cooling battery thermal management system for electric vehicles. MSc Thesis, International Islamic University Malaysia.
3. Li, M., & Wang, F. (2010). Thermal performance analysis of the lithium-ion batteries. In *The 11th International Conference on Parallel and Distributed Computing, Application and Technologies* (pp. 483–486).
4. Bergveld, H. J. (2001). Battery management systems design by modelling (p. 61). Ph.D. dissertation, Universiteit Twente.
5. Rahman, A., Farhan, S., Ahmed, H., & Hawlader, M. N. A. (2014). Fuzzy controlled evaporative battery cooling system for EVs/HEVs. *International Journal of Electric and Hybrid Vehicle System, Inderscience, 8*(2), 233–240.
6. Soldo, V., Grozdek, M., & Curko, T. (2006). Two-phase heat transfer during refrigerant evaporation in a solar collector tube. In *61th ATI National Congress-International Season "Solar Heating and Cooling"* (pp. 2–7).
7. Rahman, A., Helmi, A., & Hawlader, M. N. A. (2017). Two-phase evaporative cooling battery thermal management technology for electric vehicle. *International Journal of Automotive Technology, 18*(5), 875–882.

Chapter 10
Theoretical Modelling of Battery Thermal Management System

Abstract The battery's temperature rises when the traction and cruise modes use too much current from the battery, which might lead to cell damage or fire. Providing dependable battery protection against short circuits and quick heat dissipation, the BThMS shields cells from damage, overheating, and even fire. The BThMS regulates the mass flow of refrigerant depending on the heat of the battery modules, which are surrounded by hollow heat-conducting tubes. The temperature of the battery was regulated between 25 and 35 °C. By doing this, the modules' state of health (SoH) is maintained and they are able to equally distribute the load in order to satisfy the vehicle's load demand.

Keywords Evaporative battery thermal management system · Battery state-of-health · Safety · Maintenance

10.1 Introduction

The assessment of vehicle mobility power demands is presented in this chapter. The Proton Saga EV's design parameters, including load (W), wheel base (L), CGx position (lfr), CG height (h), front projection area (Af), drag coefficient (CD), and traveling speed (V), as well as road conditions, including air density (ρair), road slop (θg), and road frictional coefficient (μ), have been used to estimate it. The battery module and motor rating voltage of 100 V has been used in the construction of battery packs. The vehicle's traction power and cruising speed need an excessive current drain from the battery, which raises the battery's temperature and may harm or even kill the cells. As a result, the cooling system is necessary to keep the battery's temperature between 20 and 40 °C in order to avoid the battery from 20–40 °C for preventing the battery from damage or fire and increasing battery life span [1–4].

A. Rahman et al., *Autonomous Driveline System for Electric Vehicle*, Green Energy and Technology, https://doi.org/10.1007/978-981-96-8418-2_10

10.2 Mathematical Model

Vehicle weight, wheelbase, wheel radius, CG position, rolling motion resistance, aerodynamic resistance, slope angle, air density, drag coefficient, and traveling speed were all taken into consideration when estimating the vehicle's traction torque and propulsion power using the vehicle dynamics equation.

10.2.1 Traction Torque

The traction torque required by a vehicle to propel forward at a desired velocity can be described as follows:

$$T_t = F_t \times r_{\text{wheel}} \tag{10.1}$$

where T is the traction torque in Nm, F_t is the traction force in N and r_{wheel} is the radius of the drive wheel in m. While traction force F_t can be calculated as follows:

i. For initial condition (during stand still to start moving) [5]

$$F_{t\text{initital}} = m_a g \left(\mu_R \frac{\left(l_{fr} + f_r h\right)/L_{wb}}{1 + \mu_R h/L_{wb}} + \sin\theta_g \right) \tag{10.2}$$

where m_a is the mass of the vehicle in kg, g is the gravitational acceleration constant equal to 9.81 m/s^2, μ_r the adhesion coefficient of the road, l_{fr} distance from the front wheel to CG in m, f_r the coefficient of rolling motion resistance, h is the height of CG in m, L_{wb} the wheel base in m, θ_g the slope angle with respect to the horizon in deg.

ii. For moving condition (dynamic component) [5]:

$$F_{t\text{dynamic}} = f_r m_a g + m_a g \sin\theta_g + \frac{1}{2}\rho_{\text{air}} c_D A_f v^2 \tag{10.3}$$

where ρ_{air} is the air density in kg/m^3, C_D is the coefficient of aerodynamic resistance, A_f is the frontal area of the vehicle in m^2 and v is the travelling speed of the vehicle relative to the wind in m/s.

10.2.2 Propulsion Power

The traction power required to propel the vehicle at maximum speed, $P_{p(\max)}$ in Watt or W can be calculated from Eq. 4.3, and based on maximum travelling speed ($v_{\max}$):

Table 10.1 Vehicle traction corresponding heat generation [6–8]

Vehicle operation mode	Traction (Nm)	Power required (kW)	Discharge current (A)	Heat generation (Watt)
For speed: 60 km/h				
0% grade	85	5	49	100
3.67% grade	190	12	110	210
For Speed: 120 km/h				
0% grade	160	21	210	900
3.67% grade	275	33	345	2450

$$P_{\text{prop}} = F_{t\text{dynamic}} \times v_{\max} \tag{10.4}$$

Based on the detail specifications of the Proton Saga (see Appendix A), with the adhesion coefficient value μ_r of 0.02, highest gradient $sin\theta_g = G$ equals to 3.67%, air density ρ_{air} equals to 1.164 kg/m^3 and maximum travelling speed $v_{\max}$ of the Proton Saga EV is expected to be 120 km/h. Discharge current drawn from the battery (I_b) by the motor to meet the power demands of the car in different road conditions has been estimated as follows:

$$I_b = \frac{P_{Prop}}{V_{m(r)}} \tag{10.5}$$

In this investigation, the motor rating voltage ($V_{m(r)}$) has been set at 100 V. Table 10.1 shows the maximum power required for the vehicle to go on a 3.67% grade at a top speed of 120 km/h. The vehicle can reach a maximum speed of 120 km/h thanks to the Proton electric motor (Model AC50-26.28.2, made by HPEVS, Inc.), which has a peak power rating of 52 kW at 3245 rpm. In the meantime, Table 9.1 in Chap. 9 provides specifics of the vehicle's traction parameters and associated heat generation at an intermediate speed of 60 km/h and a maximum speed of 120 km/h.

10.3 Simulation Study

10.3.1 Traction Torque and Current

The needed traction torque and discharge current change as the vehicle speed increases are depicted in Fig. 10.1. The profiles are predicated on a 0% gradient, or in order, at the level of the road. As the speed increases, the figure illustrates how much discharge current is needed to move the vehicle forward. It was anticipated that the vehicle could reach a maximum speed of 100 km/h in the first running mode and 120 km/h in the second, while an average speed of 100 km/h was anticipated in the second running mode. According to the figure, the battery pack must supply

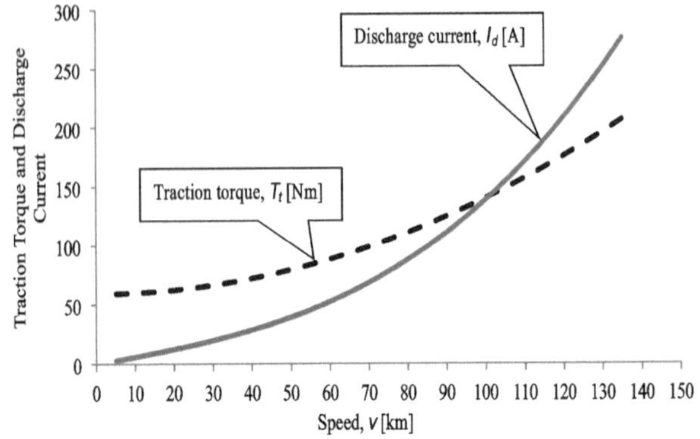

Fig. 10.1 Traction torque and discharge current corresponding to the vehicle speed at 0% gradient [4]

between 135 A for the first running mode and 90–210 A for the discharge current in the second running mode.

In contrast, the third running mode required the car to be driven as far as feasible in two hours or until the battery ran completely flat, with an average speed of between 50 and 60 km/h. The number indicates that the average discharge current would be between 40 and 55 A. Likewise, the amount of traction torque and discharge current needed to move the car would rise significantly if the Sepang International Circuit's (SIC) maximum gradient of 3.67 percent were taken into account. The discharge current during the second running mode was between 175 and 330 A, as illustrated in Fig. 10.2, and during the third running mode, it was between 90 and 120 A.

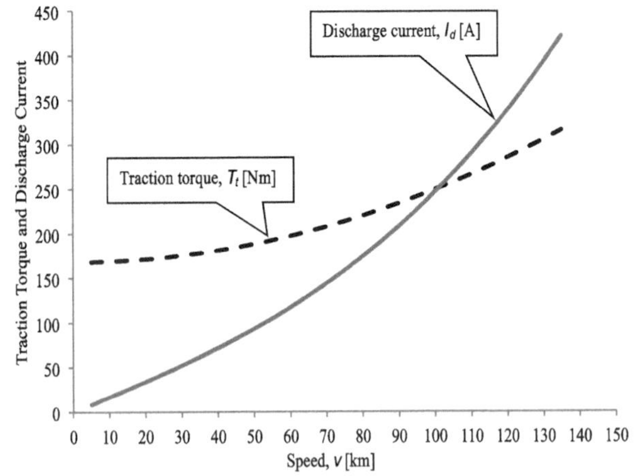

Fig. 10.2 Traction torque and discharge current corresponding to the vehicle speed at maximum 3.67% gradient [4]

10.3.2 Heat Generation

Fast charging and rapid draining cause the battery cells to generate heat. The battery cells may sustain damage or possibly catch fire if the temperature rises to 50 °C or higher. Therefore, the battery's cooling system is necessary to keep the battery's temperature within the ideal operating temperature range of 30–40 °C. The heat generation inside the battery modules is depicted in Fig. 10.3. According to simulation results, resistive heat (Joule effect) is the primary source of heat generation. Equation 9.5 demonstrates that the battery's heat generation is influenced by both the discharge current and the polarization resistance [7, 9, 10].

The discharge current needed to drive the car at various speeds on a 0% and 3.67% road gradient has been simulated in order to determine the amount of heat generation vs discharge current and cooling load versus heat generation into the battery. Three running modes were taken into consideration while calculating the simulation's parameter (discharge current): (i) acceleration of a quarter mile (402 m), (ii) shortest time and highest speed in 11.4 km of travel, and (iii) farthest distance in a single full charge battery of 100 V and 86 Ah. Figure 10.4 illustrates that as speed and discharge current increase, the battery pack's internal heat generation will climb. Table 10.2 summarizes the findings.

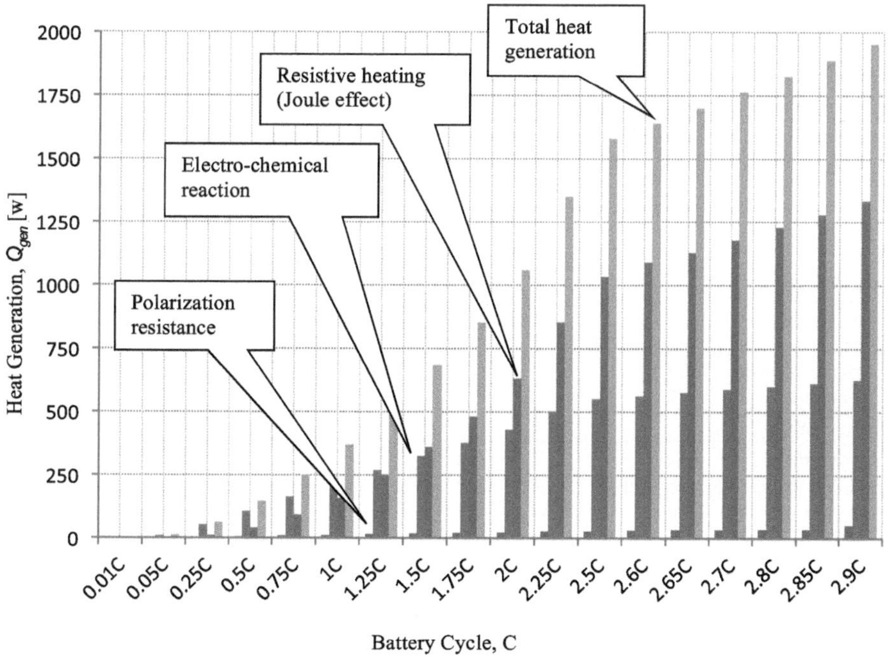

Fig. 10.3 Estimation of heat generation for each factor at different battery cycle [4]

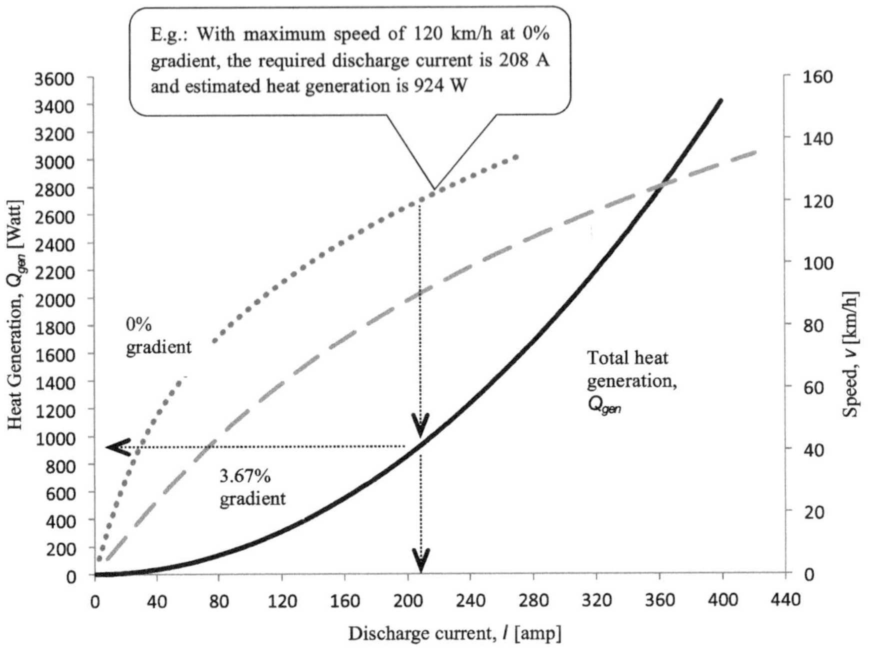

Fig. 10.4 Total heat generation within the battery pack [4, 7]

Table 10.2 Estimation of required vehicle input and output during PGMC 2012 [6, 8]

Running mode	Estimated vehicle speed, *v* [km/h]	Estimated operating time, *t*	Required discharge current, *Id* [A]	Estimated heat generation, Qgen [W]
1st) Quarter mile (402 m) acceleration	Max: 100	25 s	135	200–400 Average: 300
2nd) Shortest time and maximum speed in two laps (11.4 km)	80–120 Average: 100 Max.: 120	20 min	130–270 Average: 200	500–1600 Average: 1050
3rd) Farthest distance (travel by single charge)	50–60 Average: 55	90 min	65–90 Average: 80	50–200 Average: 125

Furthermore, as the gradient increases, the discharge current would likewise rise in proportion. As a result, more heat will be produced. For instance, if the car is traveling at 120 km/h on a level road with no gradient, 208 A of discharge current is needed, producing 924 W of heat. Similarly, if the car is driven at the same speed on a slope with a maximum gradient of 3.67%, the necessary discharge current is roughly 335 A, which will generate roughly 2500 W of heat.

10.3.3 Battery Temperature Analytical Modelling

Case 1: Without Cooling

Based on the anticipated discharge current, which ranges from 50 to 350 A, Fig. 10.5 displays the battery pack's temperature profiles when no cooling (no BCThMS) is applied. The simulation ran for ninety minutes at one-minute intervals. It was taken into account that the battery's initial temperature was supposed to be 30 °C, which was the same as the ambient temperature. The car may still be able to be driven with no BCThMS provided to the battery pack at low discharge current rates, such as below 100 A. However, at high discharge currents, the battery may explode because of the hyperheated consequence from the generated heat.

Furthermore, Fig. 10.6 illustrates how an increase in discharge current during a specific operating time causes the battery temperature to rise. Figures 10.5 and 10.6 demonstrate that as the battery discharge current rises, the battery temperature rises considerably, potentially resulting in damage, fire, or a shorter battery life. Additionally, when the Proton Saga EV is driven for a different operating time, such as 60 min and 90 min, a comparison is done to determine the safe zone (battery modules temperature not exceeding 40 °C) and critical zone (battery modules temperature exceeding 40 °C).

As illustrated in Fig. 10.7, a maximum discharge current of 90 A is advised if the vehicle is to be driven for 60 min. This area is known as the "safe zone," as shown by the A-B-C plot. The battery's temperature will increase sharply over the intended maximum temperature of 40 °C if the discharge current is pulled at a rate greater

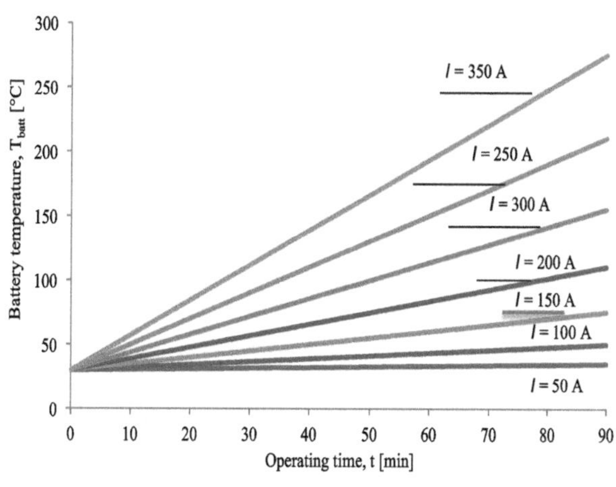

Fig. 10.5 Comparison of battery temperature rises at various discharge currents for 60 and 90 min of operating time [7]

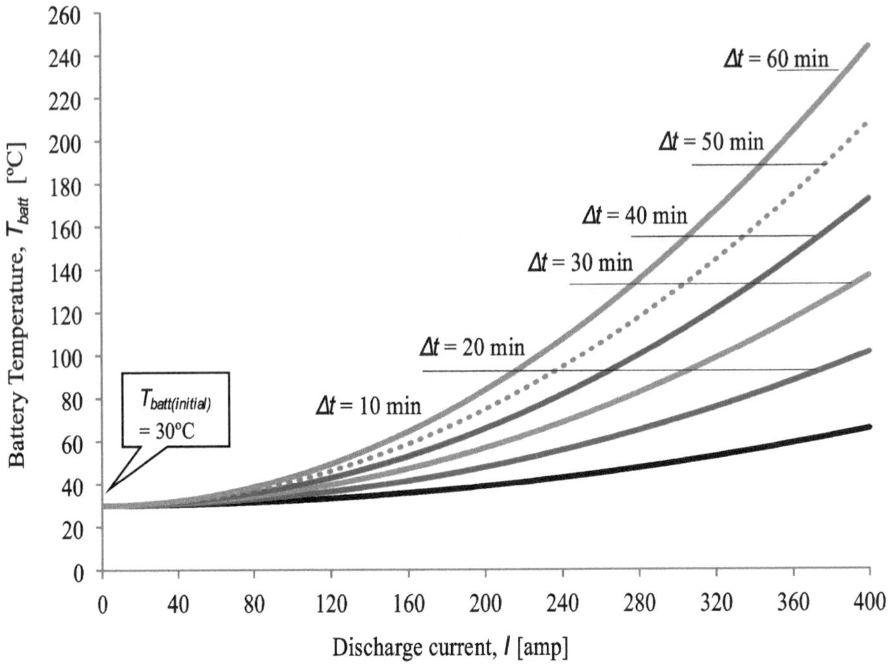

Fig. 10.6 Battery temperature rises with increasing of discharge current [6, 8]

than 90 A. We refer to this area as the critical zone (B-C-D-E). Likewise, a maximum discharge current of 80 A (within the safe zone; A'-B'-C') is advised if the car is to be driven for 90 min. According to the figures, a 12% decrease in discharge current and a safe zone are achieved by varying the running period by 30 min. Thus, the greater the discharge current extracted from the battery, the steeper the temperature increase.

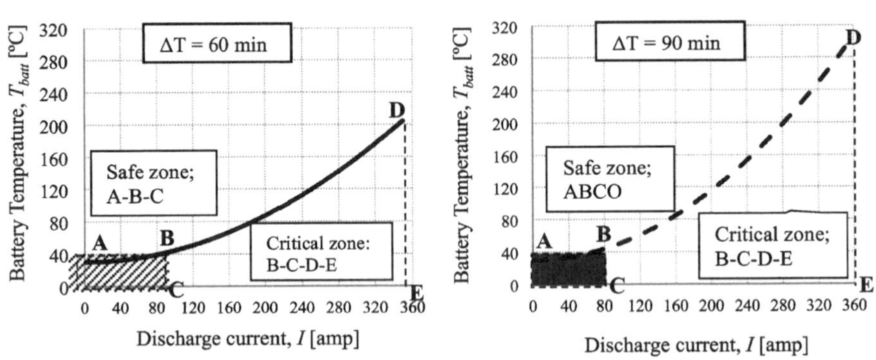

Fig. 10.7 Battery temperature rises with an increase of discharge current [4, 6, 8]

Case 2: With Cooling

As Fig. 10.6 illustrates, the battery temperature rises noticeably as the discharge current increases. For instance, the battery pack's temperature rises above 60 °C when the car is driven at 100 km/h because the motor must draw 160 A to meet the power demand. The battery temperature rises to 45 °C when the car reaches 60 km/h because the motor has to draw 100 A. The development of BCThMS aims to regulate the temperature between 20 and 40 °C.

Therefore, a simulation research on forced air cooling, liquid cooling, and evaporative cooling has been carried out in order to choose the best cooling system for the EV. For the simulation of each cooling system, the following assumptions have been made:

- Based on the maximum operating time in SIC during PGMC 2012, the vehicle's operating time is 90 min [7].
- The battery discharge current (during PGMC 2012) ranges from 50 to 350A, which the motor must draw in order to generate the vehicle's traction torque of 280 Nm and speed of 120 km/h [7].
- For air cooling systems, the heat transfer coefficient is 40 W/m^2K; for liquid cooling, it is 200 W/m^2K; and for evaporative cooling, it is 500 W/m^2K [2].

10.3.4 Force Air Cooling

The battery pack's temperature profiles as a function of running time are displayed in Fig. 10.8. According to the results, if cold air—which is defined as ambient air with a temperature of 30 °C—is blown over the battery surface while driving, the battery's temperature will rise for a while before stabilizing. When the battery's heat generation and dissipation are equal, the situation is said to be in a steady state. Additionally, it demonstrates that the battery's maximum temperature can reach between 31 and 37 °C when 100–150 amps of discharge current is extracted from it. However, the temperature rises over the intended working temperature of 40 °C at high discharge rates or above 200A.

The temperature rises considerably when the vehicle's operating speed is between 120 and 130 km/h and the discharge current is between 300 and 350 A. The battery's performance would drastically decline and its lifespan would be in jeopardy at this rate of discharge, which would boost the battery's temperature to 60 °C and higher. We therefore want the alternative cooling system.

The heat transfer coefficient for liquid cooling is 200 W/m^2K, which is 80% greater than that of air cooling. Figure 10.9 illustrates that the battery temperature reaches 29 °C, which is regarded as the ideal battery temperature, for discharge currents of 50–150 A. The maximum battery temperature increases to 35.5 °C with discharge currents between 200 and 350 A. The outcome shows that liquid cooling can maintain the battery modules' temperature between the intended range of 20 and 36 °C—the liquid's (water's) greater heat transfer coefficient. This outcome is only

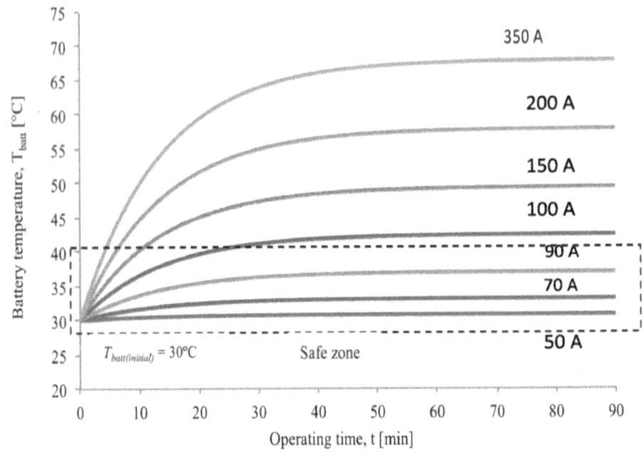

Fig. 10.8 Temperature profiles of battery pack using air cooling [7]. Liquid Cooling

possible, though, if the cooling process maintains a constant water flow rate to the battery modules.

Therefore, it is necessary to continuously supply electrical energy to the cooling components, such as the water pump, in order to maintain the high rate of heat removal. Consequently, this process adds to the battery's duty, which is a disadvantage of this cooling system approach. Additionally, there is a leak in the liquid cooling system, which could result in an electric short circuit or, in the worst situation, a battery power pack fire.

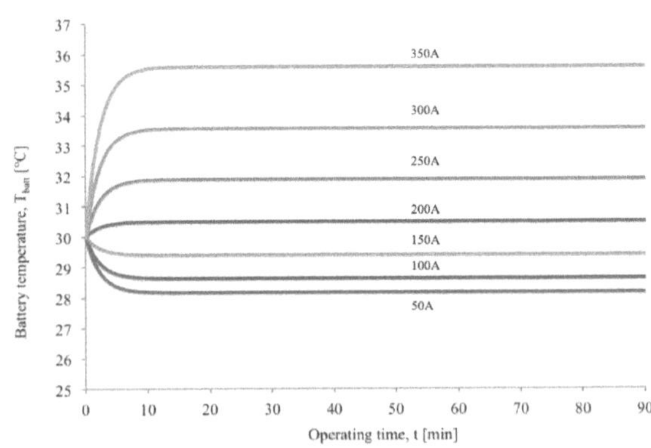

Fig. 10.9 Temperature profiles of battery modules using liquid (water) cooling [4, 7]

Solenoid valve of the cooling system is in default set, only the pump speed has been placed in variable speed,
$N_p = k\,\Delta T_b$, N_p is the compressor speed, rpm, k expansion valve setting coefficient, and ΔT_b battery temperature gradient.

10.3.5 Evaporative Cooling

As illustrated in Fig. 10.10, evaporative cooling using refrigerant R143a has a heat transfer coefficient of 500 W/m^{2}K, which is 150% greater than liquid cooling. Maintaining the mass flow rate of refrigerant while keeping the battery temperature between 20 and 30 °C is unquestionably beneficial. It displays the battery pack's temperature profiles over time in a range of discharge currents from 100 to 350 A. According to the results of the simulation, an evaporative cooling system may regulate the battery's temperature between 20 and 40 °C.

This system's benefits include using the battery while the EV is operating in passive mode, which might last for around ½ hour, and then switching to active mode for the evaporative cooling system after ½ hour. Nevertheless, compared to liquid and air cooling systems, this cooling technique uses a lot more power.

10.4 Selection Creteria of Cooling System

For battery-powered EVs, a battery cooling system is much better if it uses less electricity to keep the battery at the ideal temperature.

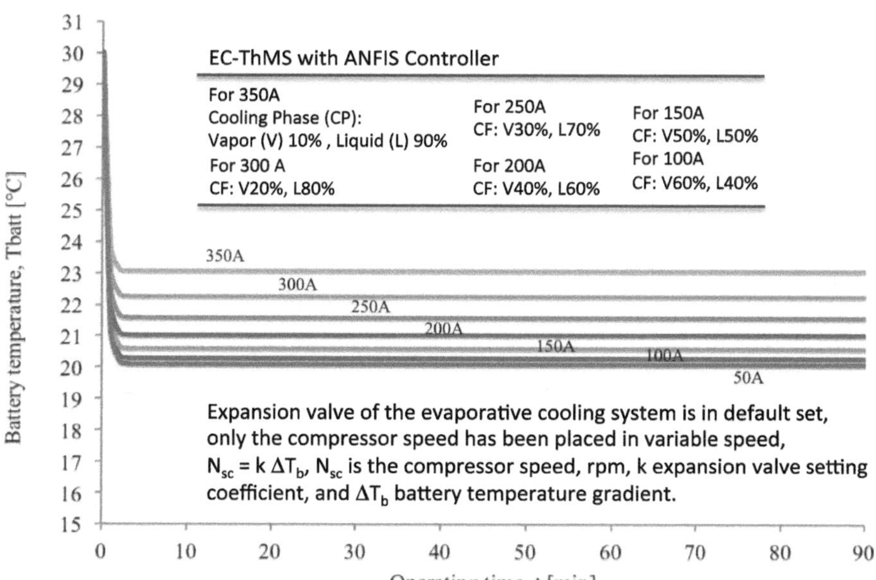

Fig. 10.10 Temperature profiles of battery modules using evaporative cooling [4, 6, 8]

10.4.1 Air Cooling System

The air-blower's capacity determines how much electricity the air-cooling system uses. Twenty fans with a 30W power consumption each have been utilized in the air-cooling system to disperse roughly 500 watts of heat. In order to run fans at 7000 rpm during discharging current, 600 Wh of total power must be delivered continually, which is quite high for an EV battery pack.

10.4.2 Liquid Cooling System

If the 1.0 kW electric pump continues to function during battery discharging, the liquid cooling system can keep the battery temperature below 40 °C by sustaining a liquid flow rate of 1.6 l/m down the 35.5 mm pipe to the hollow rectangular duct of battery modules. The pump requires a total of about 1 to 2 kW to run in order to maintain the battery temperature below 40 °C.

10.4.3 Evaporative Cooling System

If a 2.6 kW electro-compressor compresses the low pressure refrigerant R134a with a mass flow rate of 16.5 g/s to the evaporator ducts (hollow rectangular duct) with several mini-channels, the evaporative cooling system may also maintain the battery temperature between 20 and 40 °C. Figure 10.11 provides a summary of the power usage for evaporation cooling, liquid cooling, and air cooling.

10.5 Safety System

For EVs, system safety must be integrated within the battery pack to prevent short circuits, damage, or battery pack burning.

Air Cooling System

For discharge currents of 200 A and higher, the air-cooling battery thermal management system would not be able to sustain the battery's intended temperature. As a result, operating a vehicle at a higher speed while drawing more current and climbing a steeper slope while carrying a larger cargo battery could result in damage or fire.

Liquid Cooling System

Leakage is the liquid cooling system's other main drawback, as it can result in an electric short circuit.

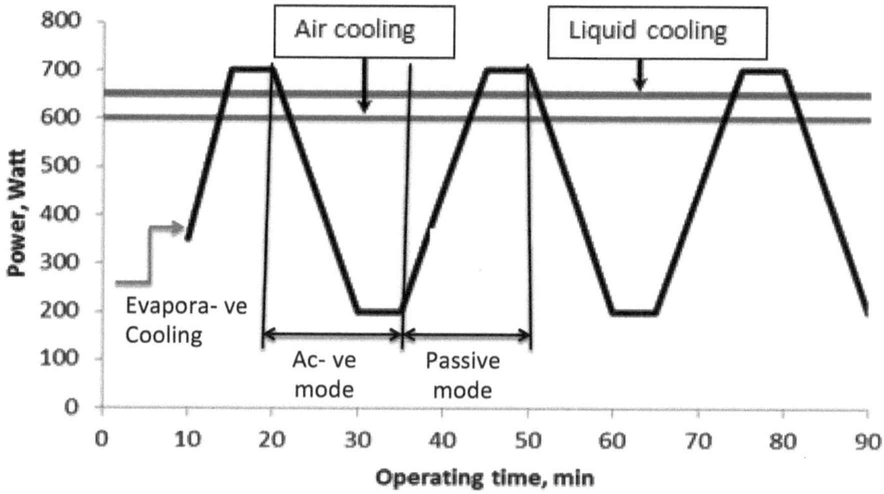

Fig. 10.11 Cooling system power consumption [4]

Evaporative Cooling System

In comparison to air-cooling and liquid-cooling systems, evaporative cooling is thought to be the optimum thermal management and battery safety technology. Additionally, because it can operate in both passive and active modes, it is an energy-efficient battery cooling solution.

10.6 Development of Battery Pack

The battery pack, which is made up of several battery modules, each of which has four cells, is the primary source of power for EVs. Two cells in series and two in parallel have been used to create a module. The vehicle is propelled and the battery thermal management system is powered by the electric discharge current that is extracted from the battery modules. Estimating the vehicle's total power, which is roughly 33 kW at 100 V and 86 Ah, is the first stage in the battery pack development process.

In order to generate a total output of 84.6 kW, 13 modules were electrically connected in series to create the battery pack used in this study for the Proton Saga EV. The modules, as depicted in Fig. 10.12, have a nominal voltage of 7.5 V, an impedance of 1.64 m'''', a normal capacity of 86 Ah @ 0.5 C, a power density of 1.813 W/kg @ 25 °C, a maximum continuous discharge current of 460 A @ 25 °C, a discharging cut-off voltage of 6.00 ± 0.05 V, a maximum charging current of 172 A and a charging cut-off voltage of $8.3^{n}0.05$ V, and a working temperature of −25 °C to 50 °C. Additionally, two (2) cells connected in series and two (2) connected in

Fig. 10.12 Detail dimension of the battery module [7]

parallel make up each module. As seen in Fig. 10.12, each cell has a nominal voltage of 3.75 V and a capacity of 43 Ah.

Aluminum plate with a density of 1636 kg/m^3, specific heat capacity (cp) of 1.377 kJ/kg·K, and thermal conductivity (k) of 0.427 W/m·K was used to make the battery pack shell. Each shell plate had layers of insulated rubber applied to prevent any potential electrical short circuit incidents. There were fourteen slots to accommodate fourteen evaporator ducting and thirteen battery units. 14 circular holes were cut in the top cover of the casing for the secondary inlet pipe, and 14 holes were made in the bottom base of the casing for the secondary inlet pipe (for refrigerant flow) to pass through.

To guarantee that copper plate connecters could be attached outside the battery casing, 26 square holes were arranged in tandem on the front cover for the terminals of the battery modules to pass through. To ensure the battery casing would be securely placed inside the trunk compartment, four (4) attachments were constructed and fastened to the vehicle's under rear panel. The battery housing with 13 modules is shown in Figs. 10.13 and 10.14. Table 8.1 lists the battery module's parameters.

Fig. 10.13 Top view at terminal position of battery modules in 13-series

Fig. 10.14 Battery casing with 13 units of battery modules [4]

10.7 Development of BThMS

Resistive heating in changing current and resistance needed for atom movement, as well as electrochemical processes in altering entropy, are the causes of the heat produced inside the cells during battery charging and discharging. High rates of discharge cause the battery's temperature to rise, which could cause damage or even fire. As a result, it's critical to create a cooling system to disperse the heat. To choose the best cooling method for the battery, a simulation research was carried out on three distinct cooling media: evaporative cooling, liquid (water) cooling, and force air cooling. The ideal cooling system is thought to be an effective cooling system that keeps the battery's temperature between 20 and 40 °C. The thermal management of evaporative cooling batteries was chosen because to its great cooling performance and energy efficiency.

10.8 EC Battery Thermal Management System

The Proton Saga EV is equipped with an evaporative cooling battery thermal management system (EC-BThMS) that uses the refrigerant cycle principle. During charging and discharging, the heat produced inside battery modules is collected by the refrigerant and released into the surrounding air through the heat exchanger. The system's schematic diagram is displayed in Fig. 10.15.

The compressor, condenser, expansion valve, and independent cooling ducts are all features of the EC-BThMS. The heat exchanger for the EC-BThMS is developed using a total of 14 cooling ducts.

The following assumptions are made for the development of the cooling ducts:

i. Flow is one-dimensional steady state, homogeneous and two-phase;
ii. Duct cross-sectional area is constant, and similar to all ducts; and
iii. Pressure of each inlet and outlet are the same.

Copper in the shape of a rectangular hollow was used to build the cooling ducts. Due to its high density of 8950 kg/m^3 and excellent thermal conductivity of 391 W/m·K, copper was selected as the duct material [9]. The dimensions of each cooling duct are 360 mm in length, 184 mm in height, 10 mm in inner space width, and 1 mm in wall thickness. The purpose of the cooling ducts was to evenly absorb the heat from the battery modules' two surfaces. In addition, the cooling ducts' width should not exceed 10 mm to precisely match the trunk compartment's dimensions. The cooling ducts' thickness should be reduced to improve convection heat transfer efficiency. For thermal contact, each cooling duct was placed in close proximity to the battery modules and positioned between them, side by side. The cooling modules and ducting were set inside an aluminum casing.

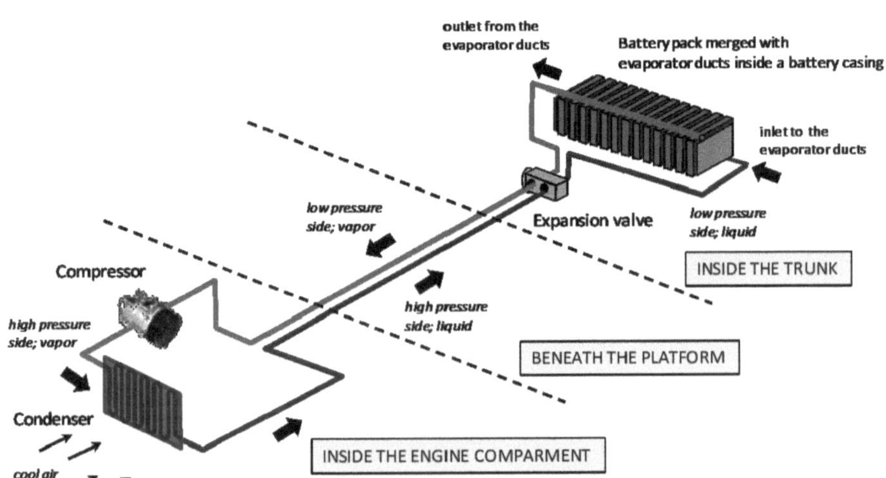

Fig. 10.15 Structure of evaporative cooling system [7, 10]

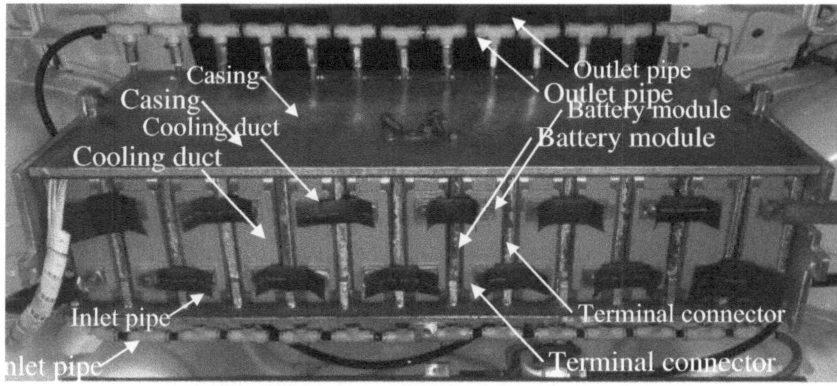

Fig. 10.16 The developed prototype model of eBThMS for the Proton Saga EV [7]

The cooling media utilized in the evaporative cooling system was R-134a. The refrigerant enters the cooling ducts through an inlet port (primary and secondary inlet pipes) at the bottom of each duct and exits through an outlet port (primary and secondary outlet pipes) at the top of the ducts. There were fourteen secondary input and output pipe units, each constructed from copper pipe with an internal diameter of 5 mm. To ensure that the refrigerant entered the cooling ducts at the same pressure, they were connected in parallel at the cooling duct.

As seen in Fig. 10.16, all of the secondary inlet and outlet pipes were then connected to a single primary inlet pipe and a primary outlet pipe, respectively, before being connected to the hoses. The original A/C system pipe or hosing was used to operate the remaining connections, which included the compressor, condenser, and expansion valve.

A discharge line was linked from cooling ducts to compressor to condenser level, and a suction line was connected from the condenser to expansion valve to cooling ducts, allowing the refrigerant to flow in a closed circuit. In order to circulate the refrigerant throughout the system, the compressor would suck low-pressure vapor and send it as high-pressure vapor to the condenser. The refrigerant transforms into a high-pressure liquid at the condenser from high-pressure vapor. When the high-pressure liquid went through the expansion valve, it would partially flash as it entered the cooling ducts. As a result of this close-loop procedure, the battery's intended operating temperature range can be reached. As previously stated, the battery's ideal operating temperature falls between 20 and 40 °C.

10.9 AI Control Strategy of the EC BThMS

The PID controller used in the development of the EC-BThMS activates it when the battery temperature rises beyond the threshold of 35 °C. It is more efficient if the system can only be turned on when the battery temperature is about to reach the peak ideal temperature of 40 °C, as seen in Fig. 3.20, in order to keep the battery pack's temperature within the intended range of 20–40 °C. The simulation result is displayed in Fig. 10.17 when three (3) distinct discharge current values—50 A, 250 A, and 350 A—are taken from the battery while the evaporative cooling system is automatically operating within the intended temperature range of 20 to 40 °C.

As seen in Fig. 10.18, the EC-BThMS Fuzzy Intelligent Control System (FICS) has been created. A group of thermocouples respond, activating the FICS. The temperature of each battery module is tracked by a set of thermocouples. In order to regulate the flow rate of refrigerant, a fuzzy logic controller (FLC) regulates the power supplied by the battery pack to the DC motor integrated within the compressor [5, 7, 11]. In order to produce the torque (T) using FLC, the battery power regulating mechanism regulates the variable power input (P) to the compressor motor. It is observed that the vehicle speed and battery state of discharge (SoD) affect the battery module temperature sensing inputs.

Figure 10.18 displays the throttle valve and fuzzy logic controller schematic diagram. This understanding of the behavior of the system enables the development of a set of general rules, which are explained [6, 8, 9]. TE (C), TPE (C/s), and MFR (kg/s) are the units of the factors that are employed. A fuzzy associated memory or decision (also known as a decision rule) is created as regulation rules for the two inputs and one output.

Large negative rate of error (LNRTE), small negative rate of error (SNRTE), zero rate of error (ZRTE), small rate of error (SPRTE), and large positive rate of error (LPRTE) are used for the rate of temperature error (RPE) in fuzzification, while the

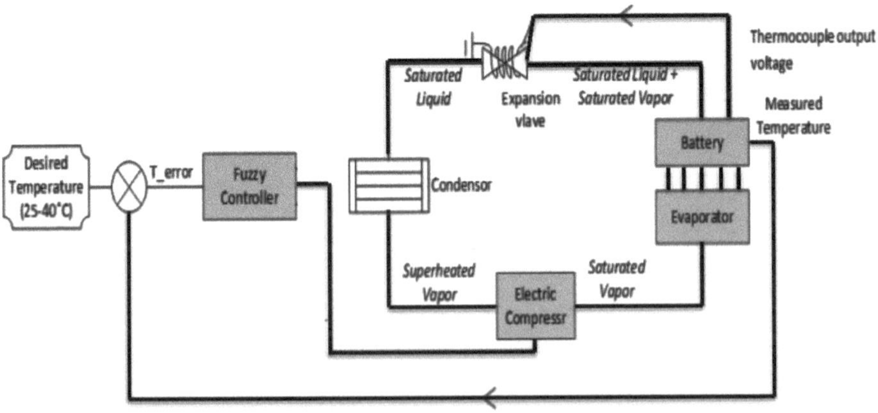

Fig. 10.17 Fuzzy Intelligent System for EC-BThMS [7]

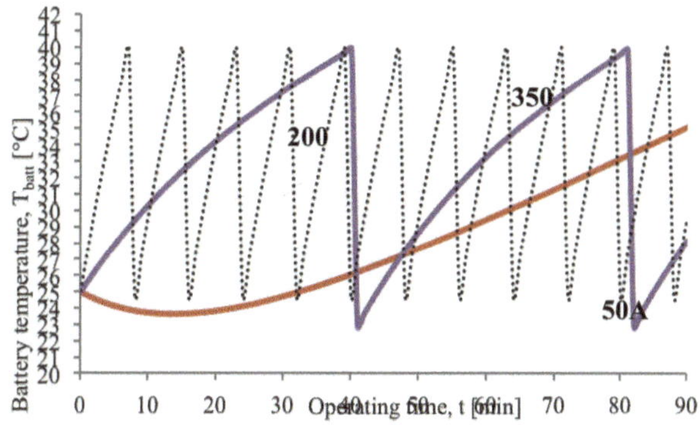

Fig. 10.18 Temperature profiles of battery pack of EC-BThMS embedded [3, 4]

linguistic variables large negative error (LNTE), small negative error (SNTE), zero error (ZTE), small positive error (SPE), and large positive error (LPTE) are used for the temperature error (TE). The ·flow rate-of-refrigerant (m) is also determined by the linguistic variables' very little mass flow rate of refrigerant (VL_m), little mass flow of refrigerant (Li_m), medium mass flow rate-of-refrigerant (M_m), and big mass flow rate-of-refrigerant (L_m).

As seen in Fig. 10.18, the FLC controller immediately activates the cooling system in one minute when the battery temperature hits 40 °C. Temperature rise profiles indicate that the battery's heat generation is absorbed by the refrigerant in a natural convection process (passive mode), whereas temperature drop profiles indicate that the battery's heat creation occurs under two different cooling conditions. The outcome shows that, with a fully charged battery, heat dissipation in passive mode enables the vehicle to run for a longer amount of time than heat dissipation in active mode.

While the C-BThMS absorbs the generated heat by using battery power when the compressor is triggered with FLC, the EC-BThMS may absorb the generated heat in passive mode without using any battery power because the system is inactivated. Thus, the energy-efficient BCThMS for EVs can be defined as the system with automated activation mode [5, 7, 10].

10.9.1 EV_EC-BThMS Field Testing

The Proton Saga EV now has the developed EC-BThMS fitted. Experimental studies of its performance have been conducted on Sepang International Circuit (SIC) and the real road. Three (3) distinct running modes served as the foundation for the experiment's parameters, which are explained below:

10.9.2 First Operating Mode (Acceleration for a Quarter Mile)

From a standstill, the car was accelerated to the end of the quarter mile, or roughly 402 m. The battery's temperature was recorded at 30 °C at the conclusion of the test.

10.9.3 Second Method of Operation (Maximum Speed and Shortest Time)

The car ran two laps (11.4 km) at the track in the quickest amount of time. The vehicle's maximum recorded speed was 127 km/h, and the battery's temperature was 31 °C.

10.9.4 Third Running Mode (Farthest Distance

Until it stopped, the car was powered by a fully charged 86-amp, 100-V battery. The battery's highest temperature was measured at 37 °C, and the maximum distance traveled was 69 km [1, 3, 6].

10.10 Conclusion

In comparison to air-cooling and liquid-cooling systems, evaporative cooling is thought to be the optimum thermal management and battery safety technology. Additionally, because it can operate in both passive and active modes, it is an energy-efficient battery cooling solution.

Acknowledgements The project was financed by PROTON, an automotive manufacturer in Malaysia, for which the authors are thankful. The International Islamic University Malaysia's Research Management System (RMC) provided financing for the creation of the intelligent system, and Eastern Michigan University let the writers to work on this project for additional modification.

References

1. Ataur, R., Rafia, A., & Jafar, A. H. (2019). Green transportation system (1st ed). IIUM Press, International Islamic University Malaysia, Kuala Lumpur, Malaysia. ISBN 978-967-491-021-1.

2. Soldo, V., Grozdek, M., & Curko, T. (2006). Two-phase heat transfer during refrigerant evaporation in a solar collector tube. In *61th ATI National Congress-International Season "Solar Heating and Cooling"* (pp. 2–7).
3. Heckenberger, T. (2009). Li-ion battery cooling: More than just another cooling task. *Technical Press Day, 2009*, 4–9.
4. Helmi, A. (2014). Development of evaporative cooling battery thermal management system for electric vehicles. MSc Thesis, International Islamic University Malaysia.
5. Rahman, A., Jaffar, A. H., & Ihsan, S. I. (2022). IoT-based electromagnetic actuator for CVT: Basic design and prototyping'. *International Journal of Powertrains, 11*(1), 38–61.
6. Rahman, A., Afroz, R., & Alam, Z. (2014). Development of electric vehicle: Public perception and attitude, the Malaysian approach. *World Review of Intermodal Transportation Research, 5*(2), 149–167.
7. Rahman, A., Farhana, S., & Hawlader, M. N. A. (2015). Fuzzy intelligent controlled two phase fuzzy controlled evaporative thermal management system for EVs/HEVs. *International Journal of Automotive Technology, 28*(4), 297–304.
8. Rahman, A., Farhan, S., Ahmed, H., & Hawlader, M. N. A. (2014). Fuzzy controlled evaporative battery cooling system for EVs/HEVs. *International Journal of Electric and Hybrid Vehicle System, Inderscience, 8*(2), 233–240.
9. Li, M., & Zinkle, S. J. (2012). Physical and mechanical properties of copper and copper alloys. *Comprehensive Nuclear Materials, Elsevier, 4*, 667–690.
10. Haines, R. W., & Wilson, C. L. (2003). *HVAC systems design handbook* (4th ed.). McGraw-Hill.
11. Kim, G., Gonder, J., Lustbader, J., & Pesaran, A. A. (2008). Thermal management of batteries in advanced vehicles using phase-change materials. *The World Electric Vehicle Journal, 2*(2), 46–59.

Chapter 11
Performance of Evaporative Battery Thermal Management System

Abstract Battery Thermal Management System (BThMS) controls the safety and performance of electric cars. Studies have been done on the effectiveness of different BThMS charging with current and discharging in both traction and cruising modes. The performance of the EC BThMS is 25.35% better than that of the AC-BThMS and 21.12% better than that of the LC-BThMS. Energy savings are also 23% greater with AC-BThMS and 57.5% greater with LC-BThMS. Thus, the invention in intelligent evaporative BThMS for contemporary electric vehicles is highlighted by this difficulty. By keeping an eye on the battery's temperature between 30 and 40 °C in both the state of charge (SoC) and the state of discharge (SoD), the EC BThMS's integrated intelligent system maximizes battery performance. If the battery temperature rises beyond 40 °C, particularly when the car is in driving mode and there is cell damage, the smart system alerts the user. Thus, by lowering the frequency of battery recharging, intelligent EC BThMS can increase battery life, guarantee battery safety, and conserve energy.

Keywords Battery thermal management system · Intelligent evaporative cooling · State of charge · State of discharge

11.1 Introduction

The motor of an electric car uses battery power to move the vehicle along various road profiles. Warming of the battery's cells is caused by high discharge currents. Temperatures exceeding 40 °C shorten battery life and can potentially start a fire. In order to regulate the battery temperature within the range of 20–40 °C, the suggested system, EC-BThMS, was created by evaluating the total cooling loads [1–3]. With the EC-BThMS, a fuzzy regulating system has been introduced to regulate the expansion valve and electro-compressor in response to battery temperature sensors [4–6].

EVs have been driven at 60 km/h on a 0% grade and 40 km/h on a 5% gradient on the IIUM campus, and 130 km/h on a 0% gradient and 50 km/h on a 3.67% gradient using an 8.6 kWh battery pack. The performance of the air conditioning battery

A. Rahman et al., *Autonomous Driveline System for Electric Vehicle*, Green Energy and Technology, https://doi.org/10.1007/978-981-96-8418-2_11

thermal management system (AC-BThMS) and EC-BThMS has been compared using the same vehicle. According to the results, EC-BThMS can save 17.69% more energy than AC-BThM 1 and 23% more than AC-BThM 2.

Keeping the battery temperature at a suitable level is one of the primary challenges facing the development of EVs. This is crucial for preserving the battery's lifespan and safety in addition to its performance and capacity. For traction and fast speed, electrical automobiles must have a powerful battery pack. Strong electrical currents are produced during the brief peak load times, and because of internal resistance, the Li-ion cells warm up significantly. The ideal operating temperature range for lithium-ion batteries is between 25 and 35 °C., which could lead to balance utilization of the active material and potentially to improve the performance and span the battery life. Charging and discharging mode of a battery produces heat. Particularly during the high rate discharges, a tremendous amount of heat is produced. In the event that this heat cannot be released right away, the battery temperature quickly increases [7, 8]. Onda et al. examined the heat behavior of Li-ion batteries under cycles of rapid charging and discharging. They found that rapid cycles of charging and discharging generate a lot of heat, which causes the battery's temperature to climb sharply. One of the most crucial problems with lithium-ion batteries at high charge/discharge rates is heat control and management, which can shorten battery life or possibly start a fire, according to reports [9–11].

The performance analysis of the ECThMS for three distinct modes of operation, both theoretically and empirically, is presented in this chapter. The EC-BThMS was the subject of an experimental investigation, and the results were compared to those of an air-cooling system. An intelligent system based on heat generation and battery discharge current input has maintained the mass flow rate of refrigerant. Thermo-couples installed in each cell have been used to measure the amount of heat generated.

11.1.1 Theoritical Performance of EC-BThMS

Matlab Simulink has been used to conduct the simulation result. Three vehicle operating modes have been examined: maximum travel distance with a fully charged battery, high speed running mode for 11.4 km of travel, and ¼ mile acceleration.

The necessary compressor power and refrigerant mass flow rate for each operating mode are shown in Figs. 11.1 and 11.2. The maximum battery discharge has been determined to obtain the vehicle's maximum acceleration speed in the ¼ mile acceleration mode. The vehicle motor has drawn a maximum current of 120 A in this running mode (quarter mile acceleration) in order to accelerate the car to 120 km/h in 5 s. Despite the battery drain being 1.4 °C, the measured temperature being 30 °C. A maximum of 5.8 g/s is discovered for the refrigerant mass flow rate, and the compressor requires roughly 1100 W@550 of battery power to run.

The vehicle's operational speed was measured at 135 km/h in the second running mode, with the shortest time of 11.4 km. The average speed was 60 km/h. In 17 s,

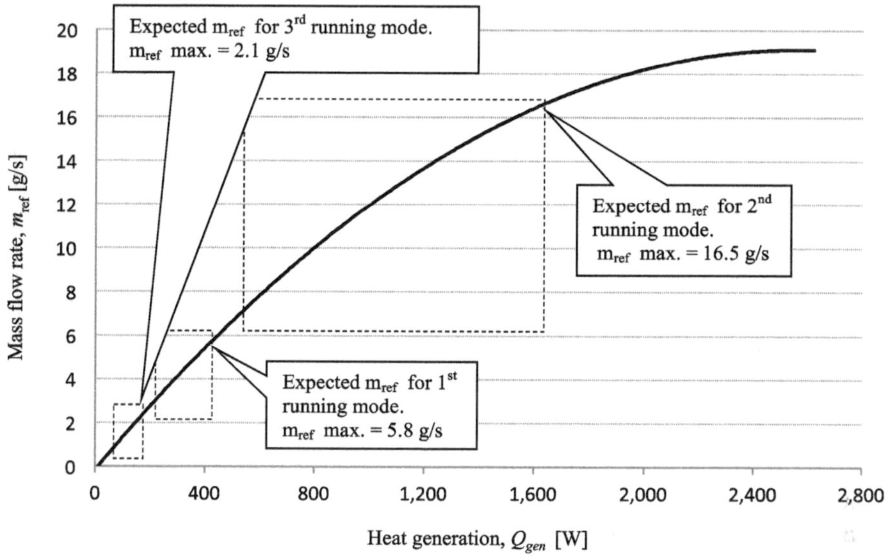

Fig. 11.1 Refrigerant mass flow rate corresponding with heat generation [4, 11]

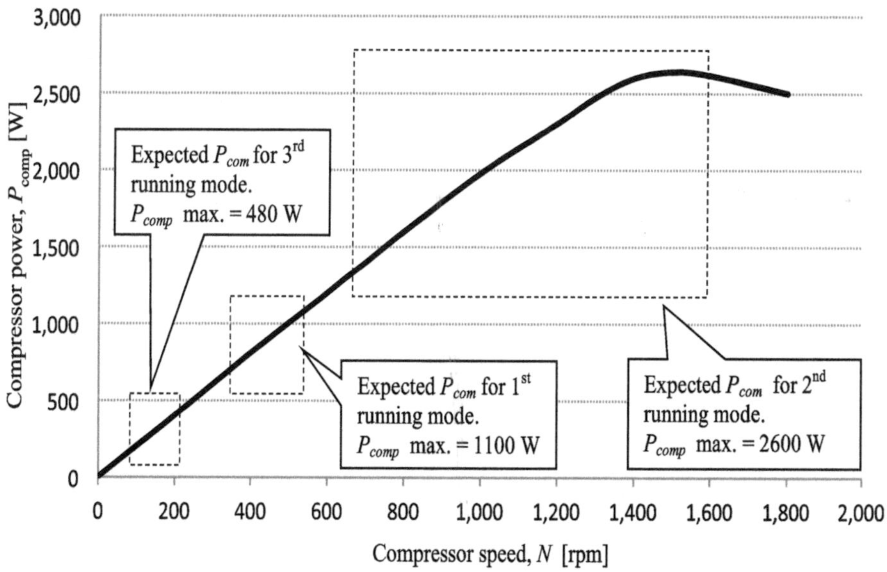

Fig. 11.2 Compressor power corresponding with compressor speed [5, 12]

the car covered 11.4 km. After traveling 11.4 km, the battery discharge current was measured at 100 A and averaged at 70 A. The temperature was recorded at 36 °C, and the maximum refrigerant mass flow rate was 16.5 g/s. The compressor required approximately 2600 W@1600 of battery power to operate [3, 13, 14]. With a single fully charged battery, the car has covered 69 km in the third running mode.

The temperature was measured at 35 °C, the battery discharge current was 70 A, and the average was 60 A. The refrigerant mass flow rate was at its maximum of 2.1 g/s, requiring around 480 W@550 of battery power to keep the compressor running. Considering the heat transfer coefficients of 40 W/m K for air cooling, 200 W/m K for liquid cooling, and 500 W/m K for evaporative cooling, three distinct battery cooling systems—air, liquid, and evaporative—were tested for a vehicle traveling 130 km/h and having a 200 A discharge current.

The low heat transfer coefficient of the battery means that the air cooling system can only regulate the battery's desired temperature if the current drawn from it is less than 150 A. Thus, the air-cooling battery thermal management system may result in fire or damage to the battery in order to increase the vehicle's speed and torque. Furthermore, the air-cooling system uses a lot of electricity because it requires a lot of fans to run continuously in order to cool the batteries [15–17]. The liquid cooling system's increased heat transfer coefficient and fast rate of heat removal make it highly promising for controlling the battery's target temperature, even when 400 A of current is extracted from it [18, 19].

The longer the battery is left operating without a cooling system and the higher the discharge current, the higher the battery temperature. The simulation study if discharge liquid is kept evenly distributed across the battery surfaces by the 1.0 kW electric pump, the liquid cooling system can keep the battery temperature below 40 °C. The evaporative cooling system uses a 2.6 kW electro compressor to compress R134a, which keeps the battery temperature between 20 °C and 40 °C. Because their temperatures fall between 20 °C and 40 °C, evaporative systems are far safer than other types.

11.2 Cooling System Development

This study designed an EC-BThMS (evaporative cooling battery thermal management system) for the Proton Saga EV's battery pack. This cooling system was created using the "Refrigerant Cycle" concept. Figure 11.3 illustrates the construction of the EC-BThMS, which includes a set of thermocouples, a compressor, condenser, expansion valve, and independent cooling ducts. For the cooling ducts, the following assumptions have been made: (i) the flow is homogeneous, two-phase (saturated liquid and vapor), one-dimensional steady state; (ii) the duct cross-sectional area is constant and comparable to all ducts; and (iii) the pressure at each inlet and outlet is the same.

The EC-BThMS operates on the premise that heat produced within battery modules is absorbed by the refrigerant before dissipating into the ambient air. Because

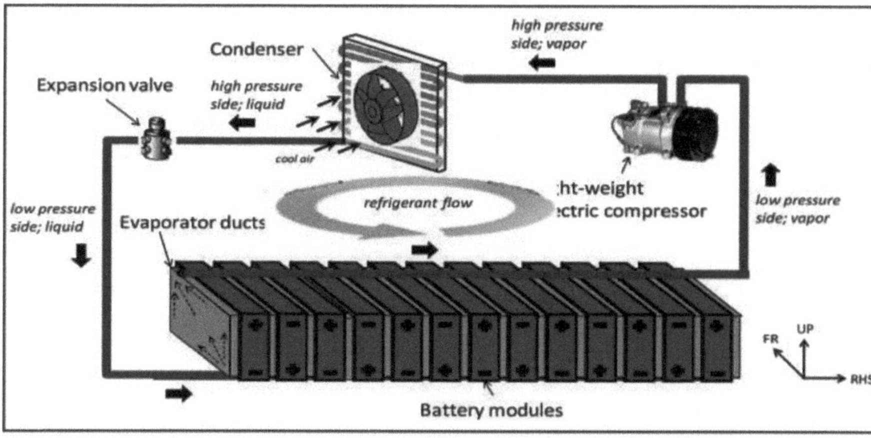

(a)

(b)

Fig. 11.3 Evaporative battery thermal management system [4, 11]

of its high heat conductivity (386 W/m K), copper was selected as the material. The battery modules were sandwiched between each cooling duct, which was positioned extremely close to the modules (for thermal contact). Additionally, every evaporative duct includes an inlet port that permits low pressure refrigerant (80% saturated liquid and 20% saturated vapor) to enter the ducts and exit as low pressure refrigerant (roughly vaporized) via discharge pipes. The cooling media that has been employed is R-134a.

Table 11.1 Battery pack and cooling system

Battery modules (13 modules in series (each module has 4 cells which two cells in series and two in parallel)		
Nominal voltage (V)	Volt	96
Normal capacity (Q)	Ah	86
Energy capacity (E)	kWh	8.6
Power density (P/W)	W/kg	1.813
Internal resistance (R)	mΩ	1.64
Total surface area (A)	m	1.73
Weight (W)	kg	62.4
Internal resistance (R)	mΩ	21.32
Overall module heat capacity (c)	J/kg·K	12025
Total entropy change per module (ΔS)	J/mol·K	12.00
Battery pack with 13 modules and 14 evaporative ducts Battery pack with 13 modules, 13 ducts and casing)		
Duck size = 184 × 36 × 10 mm, Battery pack size with aluminum casing Size = 930 × 360 × 184 mm Each duct has one intake hole and one exhaust hole		
Cooling medium of the battery, Refrigerant, R-134a is compressed by 24 V electric motor compressor		
Heat transfer coefficient considered for R-134a (h)	W/m·k	500

In order to control the battery temperature between 25 and 40 °C, the EC-BThMS requires 1.6 kW of power to run the electro compressor at varying speeds depending on the expansion valve's opening. This makes the system energy heavy. At the same time, the expansion valve is set according to the thermocouple output signal. Every battery module now has a thermocouple installed in it. Five to thirty percent of the expansion valve opens, allowing the refrigerant to enter the evaporator as saturated liquid and vapor. The battery pack specifications with an evaporator are displayed in Table 11.1.

11.2.1 EXPERIMENTAL PERFORMANCE of EC BThMS

A laboratory experiment was carried out, and Fig. 11.4 displays the outcome. In order to meet the vehicle's acceleration load requirements, the controller developed a motor torque of 120 N.m at 3,000 rpm by drawing the maximum current of 550 A. Due to the excessive heat development into the module's cells, the battery can only last a few minutes at a discharging current of 550 A (LG battery manufacturer). This could lead to a reduction in battery life, deterioration in battery performance, and even fire.

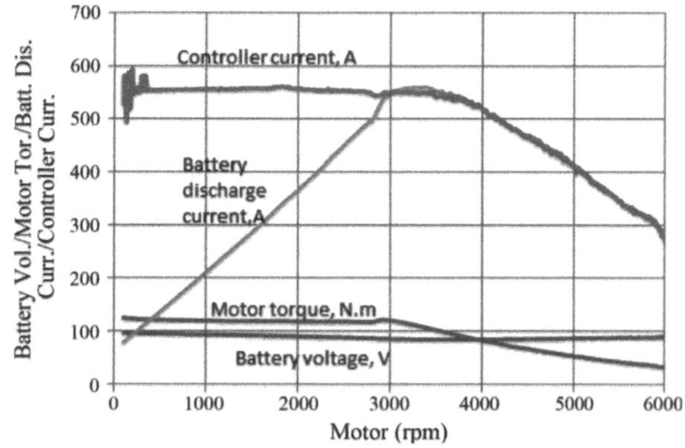

Fig. 11.4 Battery discharge current [20]

For instance, the battery pack's temperature rises by roughly 70 °C while the car is driven at 100 km/h because the motor must draw 160 A to meet the power demand. However, the battery temperature rises to 45 °C if the motor draws 100 A at 60 km/h. Three distinct cooling systems have been used to study battery temperature in order to determine the best cooling method for maintaining battery temperatures between 20 °C and 40 °C.

11.2.2 EC-BThMS: Field Experiment

On the campus of International Islamic University Malaysia, a study on battery discharging currents was carried out using varying vehicle speeds. A digital display mounted on the dashboard has been used to track the battery's temperature for each SOD and SOC, and data has been automatically saved. Using a digital tachometer, the motor RPM has been tracked for the battery temperature that corresponds to the SOD and SOC. The car was driven at a speed of 45–60 km/h as part of a demonstration test to see how well all the parts of the evaporative cooling system worked.

There are 2.3 km of roads encircling the IIUM campus, with a maximum road gradient of 3%. The EC-BThMS was used for the study's refrigerant loading and unloading. A digital clip-on multimeter was used to record the maximum discharge current of 130 A in 10 s at vehicle speeds between 0 and 35 km/h [4, 8, 9].

The battery's temperature and discharge current at various running times are displayed in Fig. 11.5. For a battery temperature of 50 °C, Fig. 11.5a records a discharge current of 130 A, which is lower than the subsequent discharge current and operating time. This is mostly because the cells were internally cooled when the car first started up. The battery's highest recorded temperature was 63 °C. Despite the nonlinear vehicle traction, rising trends were observed when the battery temperature was increased to 60 °C. Even with the increased discharge current of the battery, which is mostly caused by the heat absorbed by the refrigerant, Fig. 11.5b demonstrates that the battery temperature was observed in the range of 27–38 °C. It's intriguing that while the battery pack's temperature was below 30 °C, the EC-BThMS did not activate during the vehicle's 14-min total operating period.

The comparison of the battery temperature profile between loading and unloading refrigerant into the EC-BThMS during the first, second, third, and fourth testing is displayed in Fig. 11.6, respectively. It can be argued that the battery was configured to function without a cooling system when refrigerant was not loaded into the system. As can be seen in Fig. 11.6a, the battery's starting temperature was nearly equal to the ambient temperature before the refrigerant was loaded. The battery's temperature climbed slowly when the car was driven at a moderate pace of 35–55 km/h without adding refrigerant to the EC-BThMS. After eight rounds, or roughly 40 min of running, and a total distance of 21.6 km, the battery achieved its maximum temperature of 40 °C.

In contrast, the battery's starting temperature dropped by roughly 1–3°C from the ambient temperature when the experiment was repeated by injecting the refrigerant into the EC-BThMS. When the refrigerant runs into the cooling ducts, it absorbs

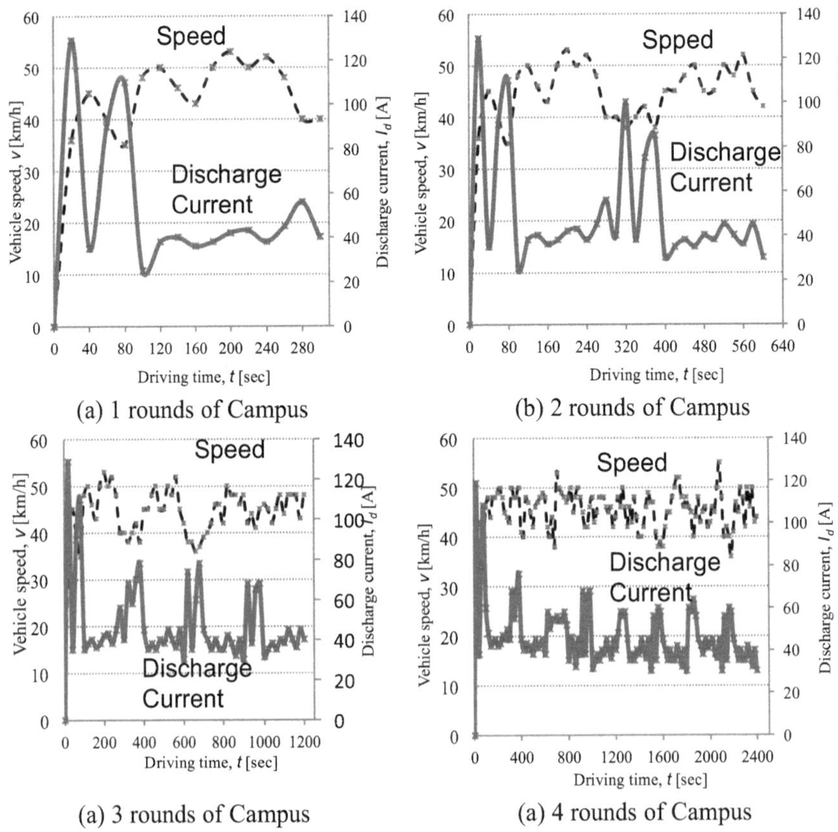

Fig. 11.5 Vehicle performance with EC-BThMS [21]

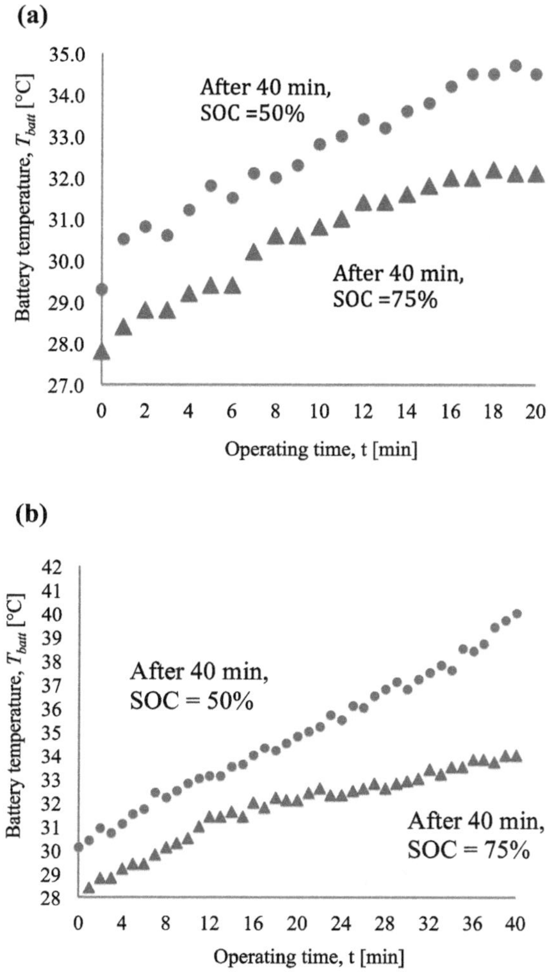

Fig. 11.6 Battery temperature profile **a** without loading R143a **b** with loading R143a based cooling system [3, 12]

heat, which is thought to be the cause of the initial slight drop in battery module temperature. As seen in Fig. 11.6b, the battery's maximum temperature after eight rounds of running through the campus road was 34.1 °C after the test was repeated four times with an extended test duration. This showed that loading and unloading refrigerant to the cooling system reduced the highest temperature by 40% between tests.

Table 11.2 Performance of EC-BThMS over the AC-BThMS

BThMS	Performance (measured)				
	Time for ¼ mile acceleration (Sec)	Travelling 11.5 km		Travelling distance @ V_{cut} (km)	EC-BThMS performance energy saving (%)
		Time (min)	Max velocity (km/h)		
EC-BThMS	22	17.35	119	68.62	
LC-BThMS	51	26.51	119	59.94	17.69
AC-BThMS	25	26.19	91	57.75	23

11.3 Performance Comparison

Two different EVs with air cooling battery thermal management systems (AC-BThMS 1 and AC-BThMS 2) have been used to compare the performance of EC-BThMS. Whereas AC-BThMS had 12 fans, AC-BThMS1 was an air-cooling battery thermal management system with 8 fans at the battery pack. Cold air is extracted from the ambient air and blown to the battery modules in both systems. During vehicle operation, all of the fans for these AC-BThMS were maintained running in order to blow air onto the surfaces of the battery modules.

The EV with EC-BThMS can save 17.69% more energy than the EV with AC-BThM 1 and 23% more than the EV with AC-BThMS 2, according to Table 11.2. This is due to the potential of the EC-BThMS, which can keep the battery pack's temperature between 20 and 40 °C [4, 10].

As seen in Fig. 11.7, the temperature increase inside the modules had also been examined. Since the figure showed very little fluctuation, it can be said that the temperature increase is consistent throughout all 13 modules, which may be because of the clever EC-BThMS's efficacy.

11.4 Performance Interpretation

The efficiency of the battery cooling system employing the evaporative cooling approach was one of the primary reasons why the IIUM was able to win the overall champion title in the PROTON GREEN MOBILITY CHALLANGES (PGMC) 2012 [3, 4, 10]. The air-cooling battery thermal management system (AC-BThMS 1 and AC-BThMS 2) has been used to compare the energy efficiency of the developed EC-BThMS with two other EVs. This is due to the EC-BThMS's capability of keeping the battery pack's temperature between 20 and 40°C.

Table 11.3 demonstrates that the EV with EC-BThMS is 25% more efficient than the EV with AC-BThM 1 and 29.2% more efficient than the EV with AC-BThMS 2 in controlling the battery temperature rise. It is also 21.12% more efficient than the EV with AC-BThM 1 and 25.35% more efficient than the EV with AC-BThMS 2 for the

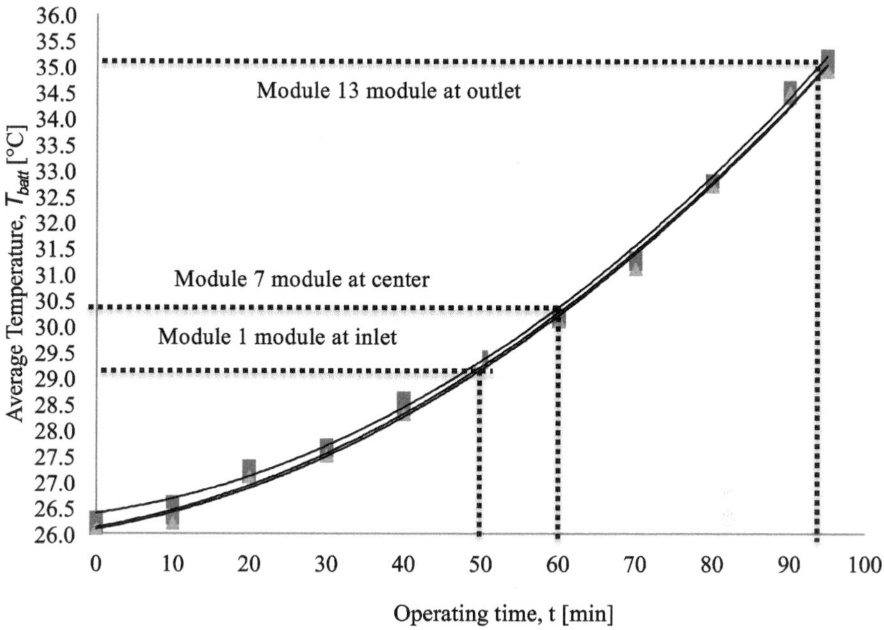

Fig. 11.7 Temperature profile within the battery modules for a single full charge [3, 12]

Table 11.3 Performance analysis of EC-BThMS over two others AC-BThMS (II)

Testing mode	Battery Thermal Management System (BThMS)				
	Temperature (°C)			Performance of EC-BThMS (%)	
	EC-BThMS	LC-BThMS	AC-BThMS	LC-BThMS	AC-BThMS
¼ mile acceleration	36	45	46.5	25	29.2
Time in 11.5 km travelling	47	56.5	57	20.21	21.275
Farthest distance @ V_{cut}	35.5	43	44.5	21.12	25.35

farthest distance, which allowed the EVs with EC-BThMS to travel a significantly greater distance than the other two EVs.

11.4.1 EC-BThMS Validation

Figure 11.8 shows the relationships between the experimentally measured and simulation-predicted temperature profile values. The goodness of fit, or correlation coefficient, for the temperature profile at discharge current 80 A is 0.9730. It shows that there is a closed agreement between the projected and measured data, proving

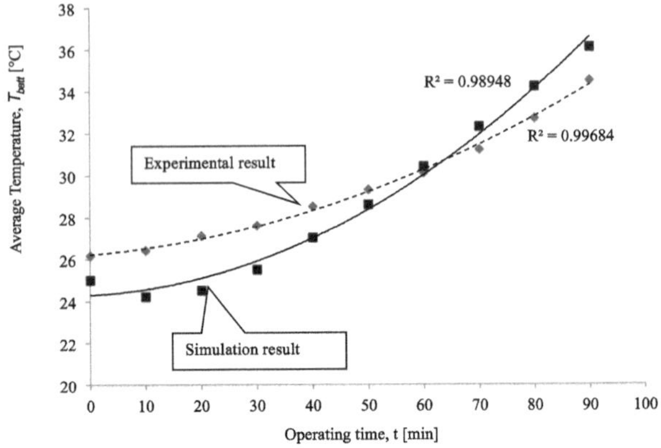

Fig. 11.8 Validation of battery temperature profile between experimental and simulation [4, 11, 21]

the correctness of the simulation's outcome. On the other hand, 9.027% is the mean relative error between the measured and anticipated values from the temperature profile mathematical model. The impact of the external environment is the reason for the relative error of simulation values over experimental data.

11.5 Conclusion

A prototype of an embedded battery management system (BMS) called BThMS has been developed for 30 kWh (150 V and 200 Ah) in order to track its temperature, voltage level, BSOH(t), and SOC(t). It is a good method with 99 ± 1% accuracy. When the Proton EV is driven at 120 km/h, the maximum temperature is anticipated to be 38 °C. The automated eBThMS can regulate this 8.6 kWh battery by keeping it from rising above 40 °C. Future temperature control technology for electric car batteries is anticipated to be BMS_EBThMS. Graduate students and industry R&D will find this study helpful in their research on improving the design of an electric vehicle BThMS.

Acknowledgements The authors are grateful to PROTON, a Malaysian car manufacturer, for funding the experiment. The authors would like to thank the Research Management System (RMC) of International Islamic University Malaysia for funding the creation of this project's intelligent system and Eastern Michigan University for allowing them to continue working on it.

References

1. Kim, G., Gonder, J., Lustbader, J., & Pesaran, A. A. (2008). Thermal management of batteries in advanced vehicles using phase-change materials. *The World Electric Vehicle Journal, 2*(2), 46–59.

2. Kim, G., & Pesaran, A. (2007). Battery Thermal management design modeling. *The World Electric Vehicle Association Journal, 1*, 126–133.
3. Ahmad, H. (2014). Development of evaporative cooling battery thermal management system for electric vehicles. MSc Thesis, International Islamic University Malaysia.
4. Rahman, A., Farhan, S., Ahmed, H., & Hawlader, M. N. A. (2014). Fuzzy controlled evaporative battery cooling system for EVs/HEVs. *International Journal of Electric and Hybrid Vehicle System, Inderscience, 8*(2), 233–240.
5. Anderson, D. L. (2009). An evaluation of current and future costs for lithium-ion batteries for use in electrified vehicle powertrains. Master Dissertation, Nicholas School of the Environment of Duke University.
6. Campbell, J. B., Tolbert, L. M., Member, S., Ayers, C. W., Ozpineci, B., & Lowe, K. T. (2007). Two-phase cooling method using R134a refrigerant to cool power electronic devices. *Industry Applications, IEEE Transactions, 43*(3), 648–656.
7. Rahman, A., Helmi, A., & Hawlader, M. N. A. (2017). Two-phase evaporative cooling battery thermal management technology for electric vehicle. *International Journal of Automotive Technology, 18*(5), 875–882.
8. Duan, X., & Naterer, G. F. (2010). Heat transfer in phase change materials for thermal management of electric vehicle battery modules. *International Journal of Heat and Mass Transfer, 53*(23–24), 5176–5182.
9. Heckenberger, T. (2009). Li-ion Battery Cooling: More Than Just Another Cooling Task. *Technical Press Day, 2009*, 4–9.
10. Ataur, R., Rafia, A., & Hassan, A. J. (2019). Green transportation system (1st edn). IIUM Press, International Islamic University Malaysia, Kuala Lumpur, Malaysia. ISBN 978-967-491-021-1.
11. Rahman, A., Afroz, R., & Alam, Z. (2014). Development of electric vehicle: Public perception and attitude, the Malaysian approach. *World Review of Intermodal Transportation Research, 5*(2), 149–167.
12. Helmi, A. (2014). Development of evaporative cooling battery thermal management system for electric vehicles. MSc Thesis, International Islamic University Malaysia, 2014.
13. Soldo, V., Grozdek, M., & Curko, T. (2006). Two-Phase Heat Transfer During Refrigerant Evaporation in a Solar Collector Tube. In 61th ATI National Congress-International Season "Solar Heating and Cooling", September 14, 2006, pp. 2–7.
14. Nelson, P., Dees, D., Amine, K., & Henriksen, G. (2002). Modeling thermal management of lithium-ion PNGV batteries. *Journal of Power Sources, 110*, 349–356.
15. Kizilel, R., Sabbah, R., Selman, J. R., & Al-Hallaj, S. (2009). An alternative cooling system to enhance the safety of Li-ion battery packs. *Journal of Power Sources, 194*(2), 1105–1112.
16. Mousavi, M., Hoque, S., Rahnamayan, S., Dincer, I., & Naterer, G. F. (2011). Optimal design of an air-cooling system for a Li-ion battery pack in electric vehicles with a genetic algorithm. In *IEEE Congress on Evolutionary Computation (CEC)* (pp. 1848–1855).
17. Nellis, G., & Klein, S. (2009). *Heat transfer*. Cambridge University Press.
18. Ramotar, L. (2010). Second law analysis of a liquid cooled battery thermal management system for hybrid and electric vehicles. Master dissertation, University of Ontario Institute of Technology.
19. Wu, M., Liu, K. H., Wang, Y., & Wan, C. (2002). Heat dissipation design for lithium-ion batteries. *Journal of Power Sources, 109*, 160–166.
20. https://www.hpevs.com/hpevs-ac-electric-motors-wiring-schematics-programming-instructions-troubleshootinginformation.html. Retrieved January 17, 2012
21. Rahman, A., Farhana, S., & Hawlader, M. N. A. (2015). Two phase fuzzy controlled evaporative thermal management system for EVs/HEVs. *International Journal of Automotive Technology, 28*(4), 297–304.

Part IV
Autonomous Two-Speed Transmission for EV

Chapter 12
Autonomous Transmission

Abstract EVs with a single speed and two speed transmissions are available on the market. Although a multi-speed transmission can increase an electric vehicle's traction performance, it also adds weight, expense, and efficiency. But when the EV is driven on a mountainous road, the two-speed gearbox that is available on the market with two DC motors causes slower gear shifts, which interrupts the power supply. This study introduces a servo-electric gear shift system for a self-contained two-speed gearbox. The market is anticipated to expand at a CAGR of 5–7% between 2021 and 2031. With a bold long-term growth perspective, the market will progressively recover in the near future. The market has grown as a result of rising global demand for electric vehicles and growing worries about carbon emissions.

Keywords Electric vehicle · Autonomous transmission · Market demand · Emission

12.1 Introduction

Electric vehicles, as opposed to gasoline engines, power electric vehicles. Like an accelerator pedal, the controller provides power to the electric motor. Rechargeable batteries are used for storage after being charged by household energy. The opposite of a car with an internal combustion engine is an electric vehicle. Because of these factors, automakers have started to enhance the functionality of each part of their vehicles. However, the 1970s saw a sharp increase in interest in electric vehicles as people's concerns about protecting the environment and conserving non-renewable energy sources grew [1].

Efficiency is largely determined by how well an electric vehicle's power train performs. A power train's primary parts are an electric motor and transmission. Vehicle performance is determined by the performance of the transmission system. It is more advantageous to increase the transmission system's efficiency than the electric motor's efficiency in order to increase the transmission's efficiency [2]. The

A. Rahman et al., *Autonomous Driveline System for Electric Vehicle*, Green Energy and Technology, https://doi.org/10.1007/978-981-96-8418-2_12

1830s saw the invention of the first electric vehicles for demonstration, and the late nineteenth century saw the introduction of commercial electric vehicles.

Commercially available electric automobiles are already in their third century and have proven to be more successful than many other technical innovations. However, internal combustion engine (IC) vehicles often have a longer range and are relatively easy to refuel when necessary, making them more successful than electric vehicles. Together with recent advancements in battery and fuel cell technology, environmental concerns—particularly those related to noise and exhaust emissions—have the potential to tip the scales in favor of electric vehicles in the present day. Therefore, it is important that the principles underlying the design of electric vehicles technology and environmental issues related can be fully understood [3, 4].

One way to conceptualize electric vehicles is as a conglomeration of several subsystems. In order for an electric car to function, each of these systems must communicate with the others. The subsystems can be controlled by a variety of technologies. The main components of these subsystems and how they contribute to the system as a whole are depicted in Fig. 12.1. While, some of these components require minimal interaction, others must actively collaborate with some of the others. In any case, all of these systems work together to make the electric automobile function [5].

Effective power is produced by internal combustion engines (ICEs) at particular rpm ranges for efficient cruising or quick acceleration. Shifting gears in the appropriate RPM range is therefore necessary to ensure proper power distribution. Usually, while driving from a stop, the lowest speeds provide the most torque. Gradually, the

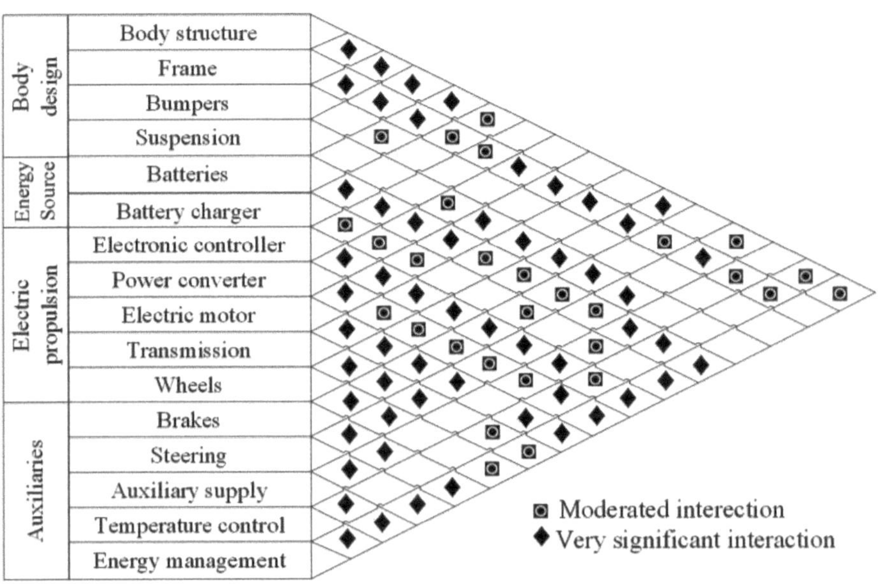

Fig. 12.1 Major EV subsystems and their relations [5, 6]

torque drops as the vehicle accelerates. Because of this, manufacturers meticulously analyze and calibrate gear ratios to match engine power in order to optimize effective power in each gear. With electric automobiles, this isn't true. An electric motor eliminates the need for a multi-speed transmission in an electric vehicle.

Electric motors generate constant torque at any speed within a certain range, but internal combustion engines need several gears with varying gear ratios to produce power. Unlike internal combustion engines, which require a torque build-up procedure, electric motors provide power instantly. To optimize electric motor efficiency without shift gear, automakers employ precisely measured gear ratios.

Furthermore, the majority of electric motors can easily operate at speeds above 10,000 rpm. Since many internal combustion engines only run at 6,000 rpm, electric motors can provide consistent torque across a much wider range of rpms. EVs with multi-speed gearboxes will only result in inefficiencies like increased weight and manufacturing expenses.

There will be a greater need for electricity as the number of electric vehicles rises. Matching solar and wind power to battery charging will become increasingly crucial as the proportion of renewable energy sources increases. Therefore, in order to meet the objective of reducing greenhouse gas (GHG) emissions overall, the energy sector's structure needs to be adjusted to the new electricity user. To meet the higher energy demand brought on by an 80% share of electric vehicles in the EU by 2050, more electricity generation will be needed. By 2030, electric vehicles will account for around 4–5% of Europe' total electricity usage, up from 0.03% in 2014, and by 2050, they will account for 9.5% [7]. The development of an electromagnetic autonomous transmission for an electric vehicle's two-speed gearbox is the primary focus of this project.

12.1.1 Research Gap of EV's Transmission

Leading automakers used a two-wheel drive system with a centrally positioned electric motor that is coupled to the wheels via a single-speed gearbox, differential, and axle shafts to create the first contemporary all-electric vehicles [8, 9]. Although it is a cost-effective option, the basic single-speed drivetrain is fully optimized for the electric car, which is the future standard, and does not imply a performance limitation [10]. According to customer reports, a compact transmission that is less than one-third the size and weight of a multi-speed transmission can replace it.

It makes it possible to install the motor in the transmission tunnel, which improves weight distribution and makes room underneath the hood for the controller and battery charger connectors. Efficiency ranges from 25 to 85% for users switching from ICE to EV [11, 12]. Although an electric vehicle's traction capability can be enhanced by a multi-speed motor, its weight, cost, and efficiency are all increased. For rapid gear changes, electric cars do not need a clutch, nor do they need a complicated gear shifter with moveable rods. To enable various installation places, it is preferable to use cable or electrical shifting.

It would require a lot of amps to run the gear with doubled power without a transmission. An EV can reach a speed of approximately 48 km/h with a motor torque of about 136 Nm at 2000 rpm and a battery current of roughly 650 A. A third gear requires a much more complex shifter because it isn't only backwards and forwards; it also requires sideways shifter movement. The best solution for the EV/HEV is a 2-speed small gearbox, which can be mounted anywhere and may even be fastened directly to the motor as a two-cog reduction box with an additional gear [13].

The Exagon Furtiv e-GT shifting mechanism was created by the French business Exagon Motor. Each of the two motors in its design has a two-speed gearbox. When the car is in start mode, both motors are operating in first gear. One motor goes into second gear when the vehicle accelerates. The other, which will be an idler, will stay in the first gear. This method is as complicated as a manual transmission, radical, and costly [14]. The single motor mode epicycle transmissions seen in the Volt, Ampera, Toyota Prius, Auris, Fusion, and C-MAX have two or three clutches and run at a single speed.

There are numerous moving elements throughout the entire transmission. When a driver shifts the gear lever to the gear position on a Toyota PRIUS, they may not be aware of whether the transmission is truly moving into gear or the gear they are in. Since the second motor works to modify the driving ratio of the torque of the main drive motor at a higher speed, cars with plugins, like the GM-Volt, already have a variable speed gearbox.

The addition of a multi-speed transmission to the electric axle has a considerable benefit, and the addition of a new four-speed system has an extra benefit that varies from minor to major based on the driving cycle and vehicle application. In addition to optimizing gear ratios and engine size, state selection and torque distribution algorithms have a significant impact on the energy results of a twin-engine four-speed transmission [15]. These algorithms will need to be further developed and analyzed.

For electric vehicles, a two-speed transmission with one speed for driving and another for highway speed is most likely appropriate. With good efficiency, it can travel between 75 and 120 km/h, and the rpm efficiency range is sufficiently broad [16]. For increased acceleration and range, the Porsche Taycan's rear axle is equipped with a two-speed system. The Porsche Taycan is the first EV to have a two-speed automatic transmission at the rear axle. At higher speeds, this multispeed gearbox seeks to increase efficiency and acceleration.

A conventional multi-speed transmission's implementation for electric cars presents additional difficulties for a number of reasons. In addition to covering maximum engine rotational rates up to 14,000 revolutions per minute or more for specific engine types, electric motors can transmit high torque at low speeds and maximum power across a broad range of engine speeds [16, 17]. Because multi-speed transmissions reduce motor torque, power, and efficiency and because they are big, heavy, and take up a lot of room in electric vehicles, they are therefore not appropriate for use in these vehicles.

12.1.2 Focusing Scope of EV's Transmission

This study's scope has been restricted to the construction of a control system and electromagnetic two-speed gearbox. A synchronizer, a wheel speed sensor (WSS), a motor speed sensor (MSS), an FLC controller, a 24 V DC battery pack, a set of electromagnet actuators, and a set of gears with gear ratios of 3.5 (1st gear) and 1.3 (2nd gear) have all been integrated into the design and development of the autonomous two speed gearbox. Based on the axial force required to move the gear from first to second gear and vice versa, an electromagnetic actuator has been created.

The synchronizer is pushed to the appropriate gear by the wheel speed sensor and motor sensor, which energize the EMAs with the controller to move the gear from first to second and second to first. Gear shifting has been synchronized by the controller in the EMAs' switching system. This provides a more precise time for the EMA to shift the gearbox's gear. To follow the EV mode, a 24 V DC motor is employed in conjunction with the gearbox. A 24 V DC lead-get battery powers the EMA and motor.

The gearbox's gear ratio is optimized using an analytical model that takes the EV into account both when cruising and when in traction mode. The maximum traction torque needed for the EV on a 20% incline and the maximum speed for the EV when cruising have been determined through vehicle dynamics analysis. The design of the electromagnetic actuator, which is the main focus of this study in accordance with the automation system, is also optimized using an analytical model.

EMAs may negatively affect the environment, passengers, and the growth of the gear shifting force due to flux leakage. Polypropylene has been used to stop the flux leak. Controlling the power supply to the EMAs has avoided heat generation in the EMAs. According to a prior study on the creation of EMA for the CVT, if the current supplied exceeds 17 A, the EMA's magnetic force substantially drops. However, according to Original Equipment Manufacturer (OEM) guidelines, the EMA coils' current supply should be greater than 17 A [18].

By using more effective design solutions, the new design also lessens component friction. This should increase the smoothness of the transmission by lowering wear between the parts. For smoother shifting, the synchronizer takes the place of the shift lever. The synchronizer's minor elements and components allow it to operate normally. Because more precise spacing is preferred, gears now engage correctly with very little clearance. The new prototype is three times larger than the previous one, which was half the size of the real gearbox component but nearly the same. The transmission produces less noise when a lubrication system is in place because of bearing rotation and component friction.

The construction of an electromagnetic actuator (EMA) to actuate the gear and synchronizer in the current 2SGB system is the focus of this work. The present limitations of the current 2SGB, such as the lack of an automated system and inadequate alignment that results in insufficient grip between contract units, prompted this study to consider a novel and creative actuation method for 2SGB. This study strongly recommends the availability of a fast-responding electromagnetic actuator.

An EMA is made up of a coil of wire with a magnetic flux return channel and a plunger that is powered by electricity and passes through the center of the coil. EMA is developed by adapting consecutive objectives. First, determine the gear ratio and torque that the car needs to be propelled. With the aid of the intended motor specification, the torque is calculated based on the gear specification. With the use of the shift fork, a shifting force is required to engage the synchronizer to the gear cone.

Following the specification of the shifting force, the electromagnetic force is calculated [19] to determine the maximum electromagnetic force required to complete the actuation. In MATLAB (FLC), the maximal electromagnetic force serves as the platform for program iteration to choose the appropriate parameter value required for solenoid development. The development of actuators and the improvement of their performance depend heavily on the values of factors like wire diameter, number of windings, wire length, supply current, etc. EMA performance may be further enhanced by an intelligent actuation system.

12.2 Overview of the EV's Transmission

The development of clean, efficient, and ecologically friendly cars for urban transportation is strongly encouraged by both economic and environmental considerations. Although cars play a significant role in our daily lives, internal combustion engine (IC) vehicle exhaust emissions are thought to be the primary source of greenhouse gas emissions that cause global warming. The pollution issue is only made worse by the yearly rise in the number of automobiles on the road.

The low energy conversion efficiency of internal combustion engines has an economic component as well. Even though there aren't many more alternative electric cars, it makes a difference because efficiency is determined by how much crude oil is converted to traction at the wheels. Alternative energy sources power highly efficient electric motors and controllers, and electric cars offer the means to establish a green, efficient, and clean urban transportation system. Since electric cars produce no emissions, they can effectively address the issue of pollution in the environment. As a result, the only vehicles that can achieve zero emissions are electric ones.

The term "transmission" encompasses the whole system that converts engine crankshaft power to the rear wheels and, if required, adjusts engine torque to the support wheels. An apparatus that delivers various gear ratios or gear ratios between an automobile vehicle's engines and drive wheels is called a transmission system. The requirement for a transmission arises from the fact that a vehicle's tractive effort and resistance to movement fluctuate with varying speeds. Below are a few examples of the transfer functions:

- Provides high starting torque, hill climbing, acceleration and lifting when high tractive effort is required.
- Variable torque with a set of gears.

- Vehicle speed can be varied while keeping the EV's motor speed within a certain limit.
- The transmission also provides a means to support the vehicle by changing the direction of rotation of the drive.

Energy-saving technologies like neutral-idle control and a mechanical torque converter lock-up, which separate the engine from the transmission parts when the car is stationary, are also included in the newest transmissions.

The efficiency of electric motors still varies at different speeds and variable efficiency, even though the majority of developers feel that only one speed is required because most electric engines attain their peak torque at 0 rpm. Therefore, the engine may be operating at 70% or even 60% of capacity at lower speeds, even though it may be peaking at 90% efficiency. Thus, at various speeds, a multi-speed transmission can maximize engine efficiency. Transmission systems come in a variety of forms, each having unique uses and underlying theories.

These include electromagnetic transmission, continuously variable transmission (CVT), dual clutch transmission, automatic transmission, and manual transmission. Every transmission system has advantages and disadvantages, but selecting the right kind of gearbox for electric cars is crucial. The following subsections provide a brief description of each of these transmission technologies.

12.2.1 Automatic Manual Transmission

When you have a manual transmission, the gear ratio of your car is not automatically altered. The motorist must shift gears on their own. The "clutch" is an extra pedal seen in cars with manual transmissions. Pressing this pedal shifts gears. The applied load determines the efficiency of this transmission, which is the most efficient one on the market. Its efficiency can get up to 97% during a normal driving cycle. Manual transmissions have an average acceleration time of 16.4 s from 0 to 100 km/h. Compared to anything other than the newest automatic gearboxes, it is typically more cost-effective, efficient, and inexpensive to produce.

The straightforward mechanical construction of this transmission, as seen in Fig. 12.2, requires less driver effort than a manual transmission. The output shaft's and the parallel lay shaft's gears always mesh in a manual transmission. The chosen gear is then moved to the output shaft by means of a synchronizer.

Because of its straightforward engine, reduced complexity, and lighter weight, the manual transmission is a cost-effective vehicle. An automatic gearbox only has one gear that can be used in any situation, whereas a manual transmission allows the driver to better control the vehicle. The driver must possess some knowledge of the gear-changing procedure, which is a drawback. An individual cannot accomplish this without the necessary information because they are not familiar with gear shifting. The driver needs to be more attentive and utilize both hands when operating a vehicle.

Fig. 12.2 Manual transmission components [20]

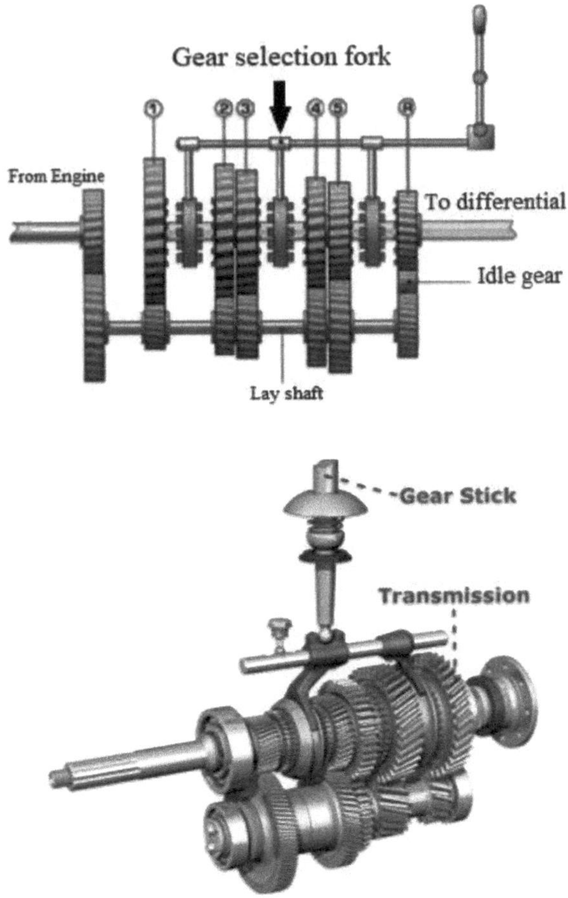

This can be explained by the fact that he must simultaneously control the steering wheel and the gear. Every gear change requires the driver to engage the clutch, and occasionally these shifts are not smooth, making the driver uncomfortable and possibly causing muscle soreness if they spend a lot of time in the automobile. Because of its size, weight, difficulty shifting gears, and maintenance requirements, it is not a good fit for an EV transmission. One fundamental explanation for why EVs can avoid having a multispeed transmission is the long operational range of electric motors.

12.2.2 Single Speed Transmission

The primary benefit of an electric motor is its torque characteristic, which allows it to deliver the most torque possible from startup to relatively low speeds and then

Fig. 12.3 A single speed transmission for EV [5, 6]

proportionately to the maximum power as the motor speed increases. The majority of commercially available EVs use powertrains that are directly connected to the driving wheels via a single reduction ratio in order to take advantage of the previously noted benefit. A standard single-speed gearbox (transmission) is depicted in Fig. 12.3. One reduction wheel torque electrical motor with a 10,000 rpm or even double capacity is a feature of the electric car that is now on the market. It can also be made accessible as needed.

These benefits, however, do not imply that the vehicle can start with the full motor torque for traction and rest when climbing a slope. Additionally, when the vehicle climbs a slope, the motor is unable to produce traction torque due to a single reduction at high speeds, which results in the vehicle using more energy than is necessary.

12.2.3 2-Speed Automatic Transmission

According to the report's author [21], a multi-speed gearbox's continuous chain transmission enables "energy to flow continuously between the engine and the wheels, even throughout the change", distinguishing it from competing products. Transmissions for electric vehicles could be altered in the future using this technique. As seen in Fig. 12.4, a multi-speed gearbox can be used as a solution to satisfy the design specifications for electric vehicle acceleration and maximum speed. Therefore, it is possible to install a 2-speed transmission to boost the available wheel torque, shorten acceleration times, and raise the feasible road grade while encouraging a more efficient power supply. A two-speed transmission adopted for an electric drivetrain has energy consumption benefits over a single-speed equivalent.

The two automakers who presently sell electric cars using a dc motor actuated 2-speed transmission (dcMA2ST) made by ZF gear manufacturing firm are Porsche (Porsche Tycan) and Audi (E-Tron GT). According to both automakers, dcMA2ST

Fig. 12.4 A 2-speed DC Motor actuated transmission for EV [22]

saves roughly 5% of battery energy and is appropriate for use in electric vehicles. They did note, however, that the dcMA2ST's shifting time is shorter than that of the manual transmission, which results in the car losing traction power when climbing a slope.

An acceleration gain of 15% and a range improvement of 7–15% are claimed by Inmotive, a Canadian company that has introduced its own two-speed transmission for EVs. Porsche claimed to have included a two-speed system in the Taycan because of these two benefits. The only drawbacks are the additional weight, expense, and complexity of the shifting gear, which results in a delayed response in traction during slope climbing.

12.2.4 3-Speed Transmission

Two e-Car motors, each producing 148 kW, power a France Furtive-eGT electric car. They are located in the center and connected to a cutting-edge three-speed semi-automatic gearbox called ZTI (Zero Torque Interruption). Furtive-eGT accelerates incredibly well, reaching 100 kph in 3.5 s. When the e-GT is in beginning mode, both motors are in first gear. One motor shifts into second gear when the vehicle accelerates. The second motor went into second gear as the driver continued to accelerate. There have been reports of noises coming from an automobile's transmission, including humming or clunking, whining, burning aromas, and gears slipping.

An engineering firm situated in the UK has made the decision to replace the transmission and examine an alternative system on the electric car. As seen in Fig. 12.5, they have developed a three-speed gearbox especially for electric cars, which are intended to improve energy efficiency. There is no mention of the actuation mechanism anywhere.

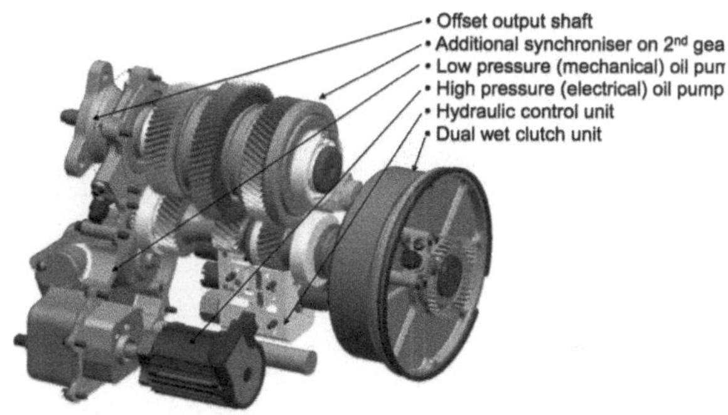

Fig. 12.5 Antonov a 3-speed transmission for EV [7, 23]

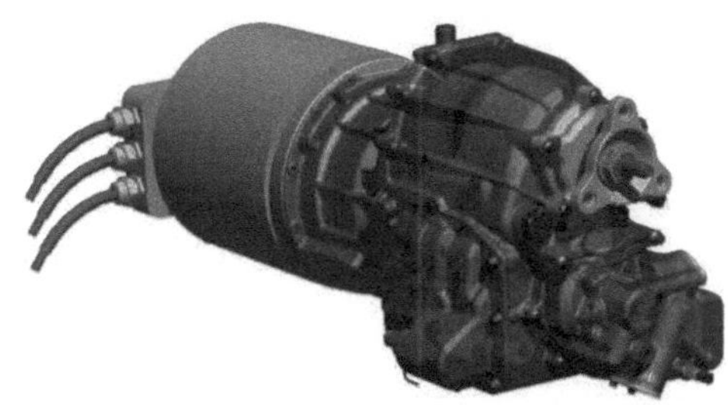

12.2.5 4-Speed Automatic Transmission

The Eaton Power Systems category includes the EV transmissions as one of its product offerings. As seen in Fig. 12.6, Eaton provides a range of EV transmissions for buses and commercial vehicles. Eaton EV transmissions provide the performance and efficiency needed for the most demanding applications, having been in safe, dependable operation for more than 15 years and 2 billion miles.

12.2.6 6-Speed Sequential EV Transmission

A six-speed automatic transmission, which is typical in today's market, is shown in Fig. 12.7. A six-speed planetary gearbox typically consists of an oil pump, clutches, a torque converter, a planetary gear set, and a valve body. The transmission control unit activates the solenoid valve. The 6-speed automatic transmission uses six different driving ratios to provide the best power and fuel efficiency, according to the manufacturer. The automatic transmission knows when to shift ratios and does it for you, so you just need to operate the vehicle. Automatic transmissions are easier to use than manual transmissions, which need you to swap gears by hand and give you little control over how your car accelerates and decelerates.

Fig. 12.6 4-Speed transmission for EV

Fig. 12.7 6-Speed sequential transmission [7]

12.2.7 8-Speed Automatic Transmission

When it was fitted to the Chrysler 300 in 2011, the 8-speed transmission made its debut in the US. The consumer benefits greatly from the 8-speed transmission's exceptional efficiency. Fuel efficiency is the first and most requested benefit. This can increase fuel efficiency by up to 14% over contemporary 5-speed gearboxes and by 11% over the outdated 6-speed gearbox. The torque converter's incredibly quick lockup, which leads to quicker engagement and less slide, is an additional advantage. The clutches inside the block will last longer if the torque converter engages more quickly and slippage is reduced. Another advantage of this upgraded drivetrain is that it allows for incredibly smooth gear changes that happen so quickly that the driver frequently cannot see.

12.2.8 9-Speed Automatic Transmission

The transmission allows the car to move at the specified speed by transferring power from the motor to the wheels. Because the transmission is responsible for converting the appropriate amount of power into speed, even a little transmission problem can have a significant impact on an automobile's performance. Such issues are easily identified. The world only recently learned about the nine-speed automatic transmission. An automobile having a front engine for passengers. Figure 12.8 shows that it has six shift elements, four basic gear sets, and a nine-speed planetary gear. To save room, gear sets are nested. Cam lock-up clutches are used in passenger car power shift transmissions for the first time in this system, and ZF's well-proven control technology contributes to the exceptional shifting performance renowned for 8 HP.

The classic benefits of all future ZF automatic transmissions—great economy and dynamics with low fuel consumption, agility paired with driving comfort, and unmatched driving pleasure—are renewed by the new, innovative 9-hp automatic transmission for cars with front transverse engines. When combined with the 2.0 L Turbo engine, the new nine-speed transmission helps the Malibu achieve 33 mpg on the interstate, which is 3% better than the 2016 model with the six-speed automated transmission. Additionally, the transmission has stop/start technology, which enables the engine to shut down under specific stop-and-go driving circumstances, potentially saving gasoline. Additionally, there is an accumulator to store energy for silent engine restarts.

The new 9T50 nine-speed automatic features GM's first use of a selectable one-way clutch and an on-axis design, meaning all of the gears line up with the crankshaft. By reducing the package size, both characteristics aim to make it around the same size as a six-speed transmission. The hydraulic system of the 9T50 is electronically managed to provide the volume and pressure needed for different working temperatures for maximum effectiveness. Additionally, it is filled with DEXRON-VI fluid, which, under typical driving circumstances, doesn't require changing.

Compared to freewheeling versions, the automatic transmission uses modern torque converter technology to help provide smooth takeoff performance, and five planetary gear sets with four stationary clutches and three rotating clutches save space. Automatic grade braking, which chooses a lower gear to help maintain the target vehicle speed when coasting or stopping on a downgrade, is one of the shift and active oil management settings.

Shifting delays, leaping or grinding when accelerating, the vehicle trembling on the road, or whistling sounds and a burning smell coming from below the hood are

Fig. 12.8 Chevrolet Blazer with 9-speed transmission

all signs of a Chevrolet Blazer transmission issue. Fuel economy could be lost due to transmission issues with the Chevrolet Blazer.

12.2.9 10-Speed Automatic Transmission

This transmission is a component of a joint venture between General Motors and Ford Motor Company, which aims to divide the development of two transmissions—a transverse 9-speed and a longitudinal 10-speed—and give each company the freedom to alter and manufacture it to suit their own requirements (Ford, SAE International). Compared to its cast iron predecessors, Ford's 10-speed transmission is substantially stronger, lighter, and more compact due to its steel and aluminum construction. You can easily handle the Ford 10-speed transmission as soon as you step behind the wheel.

12.2.10 Dual Clutch Transmission

As illustrated in Fig. 12.9, the quick dual-clutch gearbox (DCT) is another excellent option for automatic transmissions. Although they are frequently criticized, DCTs, which are frequently found in high-performance cars, also make their way into more conventional vehicles. Since its introduction in 2011, Ford's dual-clutch transmission, which powers the Fiesta and Focus, has been the subject of several owner complaints regarding roll, stalling in town, and lack of throttle response when driving.

Electric motors are used to drive the clutches and shift fork drives in place of hydraulic systems, which lowers these losses. Even though the DCT's losses are already far lower than those of a traditional automatic transmission, the electrically powered DCT is anticipated to save an additional 2% or more on fuel usage.

Honda announced in 2015 that it was developing an eight-speed DCT that would use a torque-to-DCT converter for the first time due to the DCT's unsatisfactory start. This could result in more efficient driving characteristics at low speeds compared to a typical automatic transmission. Additionally, the improved idle acceleration is a result of the increased torque converter torque. AMT is inferior than a gearbox with a double clutch.

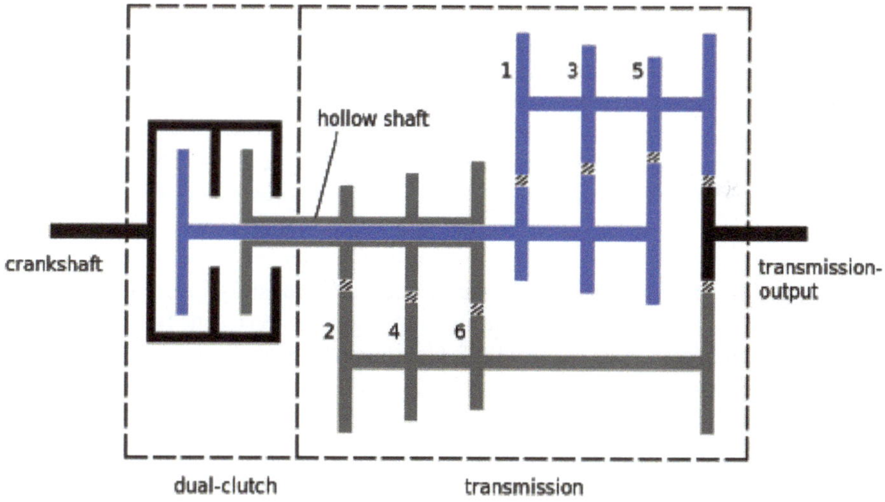

Fig. 12.9 Diagram of dual clutch transmission

12.2.11 Gear Shift Strategy

The two main issues with electric vehicle (EV) multi-speed transmissions (MSTs) gear shifting are speed and smoothness. Automatic transmissions (ATs), dual-clutch transmissions (DCT), continuously variable transmissions (CVT), and automated manual transmissions (AMTs) are examples of multi-speed transmissions. Choosing a different gear causes a stepped change in the transmission ratio, which abruptly alters the electric motor operating point. An energy-deficient operation state and/or low driveability can arise from a gearshift-induced change in the motor's operating point.

The PID controllers for a seamless and quick gearshift have been researched by the authors. They stated that even in the event of an unidentified external disturbance, a PID-controlled method shows great promise for achieving the intended gear-shifting. According to the authors' (Mashadi et al., 2007) fuzzy shifting method, an AT can eliminate needless shifts that happen when the driver's purpose is ambiguous or missed. Because of this, a multi-speed transmission's ideal gearshift technique is crucial to a vehicle's excellent performance.

12.2.12 Performance of Automatic Manual Transmission

The automatic transmission, or AMT, is an improved version of the manual transmission. Automobiles equipped with an automatic manual transmission (AMT) for shifting control offer numerous advantages, including lower fuel consumption, increased driving effectiveness, and enhanced shifting performance (Table 12.1). The advantages of AMT over conventional automated transmissions are its smaller weight, increased efficiency, and generally lower production costs.

A manual transmission with a zero shift automatic transmission may shift gears in zero seconds. In comparison to an alternative automatic transmission, it is inexpensive and simple to construct. Compared to manual transmissions, AMTs offer superior driving comfort and shifting quality [20]. Compared to manual and torque converter automatic transmissions, this one uses less gasoline. The primary drawback is a single clutch AMT, which causes a painful acceleration delay for drivers. AMT is deemed unacceptable for the US market due to this issue.

Table 12.1 Fuel economy and acceleration time of AMT over MT and AT [20]

Popular car in India	City fuel economy (km/l)	Highway fuel economy (km/l)	0–100 (km/h) acceleration time (s)
Manual	12.1	16.2	16.4
Automatic	9.0	13.7	21.2
Zeroshift (simulated)	12.1	16.2	15.4

12.2.13 Continuous Variable Transmission

Continuously Variable Transmissions (CVTs) are gaining popularity because of their straightforward mechanical construction and potential fuel efficiency. In 2014, the EPA predicts that the U.S. CVT market share would rise to 19.3%. This type of gearbox has the ability to shift gradually through an endless number of effective gear ratios between maximum and minimum. An endless number of gear ratios are available with the CVT, which can lower emissions of pollutants and increase fuel efficiency. There aren't any gears in this design at all. Rather, as seen in Fig. 12.10, it features two pulleys joined by a belt. The engine drives one pulley while the other is connected to the rest of the transmission with wheels. The magic of a CVT is that the size of the pulleys can vary from small to large, and there can be an infinite number of sizes in between, basically, shifting the gears all time.

The purpose of the CVT control system is to optimize the electric motor operating points' efficiency. Even while engine losses are reduced, the CVT's generally worse efficiency characteristics do not translate into lower energy consumption, and a two-speed transmission has a substantial advantage over a single-speed transmission. Although pulley resizing can be accomplished in a variety of methods, the fundamental idea is always the same. As a result, the first engine pulley will be small, while the second, like the first gear, will be enormous. However, the engine pulley gradually grows larger as the vehicle accelerates, much like another pulley's size diminishes. For the driver, this means that the engine can go straight to the best RPM range for a given situation. The result is smoother, more efficient and faster travel than a conventional automatic machine.

The CVT system offers several benefits, including consistent step-less acceleration and engine speed that remains constant regardless of vehicle speed. It responds quickly to changes in throttle and speed and loses less power. When compared to a manual transmission, it takes 75% less time to accelerate to 100 km/h. Lower manufacturing costs and the engine's capacity to operate at its most efficient speed across the output speed range can make up for the belt-driven design's roughly 88% efficiency, which is less than that of a manual transmission.

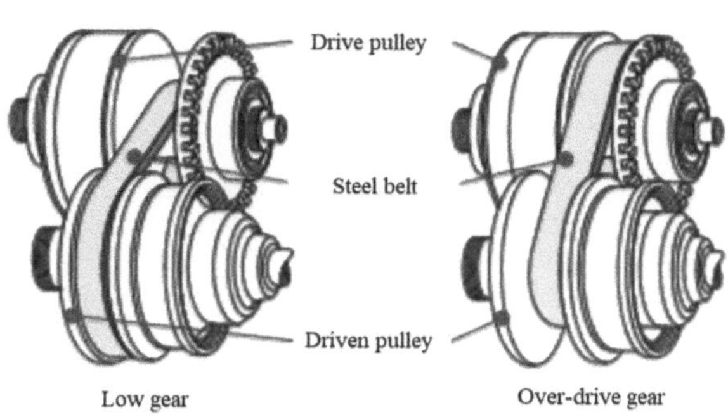

Fig. 12.10 CVT components [24]

However, the CVT has certain drawbacks as well. For instance, it can only be used in compact cars because it employs steel belts, which aren't suitable for heavy vehicles because they can't transmit as much torque. Furthermore, because of the friction between the belt and the pulley, it wears out easily. Long-distance driving in temperatures above 30 °C results in rattling sounds from a change in oil viscosity and poor grip when accelerating on uphill roads.

12.2.14 Bosch CVT

Today, the only car with a geared transmission is the Porsche Taycan (and its sister model, the Audi ETron GT), which includes a two-speed gearbox for the rear motor. Figure 12.11 shows how the Bosch CVT Transmission uses the well-known torque of electric motors to provide a mix of increased speed cruising ability, efficiency, and acceleration. Neither changeable ratios nor gear shifts are visible. The defunct Coda Sedan also tried to optimize its electric motor by altering a CVT (continuously variable gearbox), but unfortunately, no one is around to see how it would have been done.

Bosch CVT's EV-specific CVT has been promoting as increasing an electric vehicle's range. With this transmission, battery packs can be a little bit smaller and electric motors don't need to be as strong as they are now. Additionally, CVT4EV should boost off-road vehicle performance and supply additional torque for towing tasks. Bosch Transmission Technology, a Dutch division of the business, is now testing the transmission in a Volkswagen e-Golf; however, no commercial use has been disclosed as of yet. Bosch Transmission Technology, a Dutch division of the business, is now testing the transmission in a Volkswagen e-Golf; however, no commercial use has been disclosed as of yet.

Although transmissions with set gears persisted in part because they were easier to hybridize, it seemed many years ago that the majority of internal combustion vehicles were going for a CVT future. Honda boasted that it has effectively eliminated the transmission from its following two-motor hybrid attempts because the engine, CVT, and electric motor combo didn't work so well in earlier iterations of the Civic Hybrid and Insight.

12.2.15 Electromagnetic Actuated CVT

An EMA, shown in Fig. 12.12, is made up of 3000 wire turns on a steel bobbin with a solenoid length of 220 mm. It can generate an electromagnetic force of 200–300 N when 5.16–6.23 A of current is given. Steel was used in the construction of the hollow bobbin to provide a stronger electromagnetic force. The secondary pulley's movable sheave is activated, the copper coil is held in place, and the iron disk is drawn in by its powerful electromagnetic field.

Fig. 12.11 BOSCH CVT for EV

Fig. 12.12 EMA with Toyota Wish CVT in testing

The EMA-CVT test-rig result concludes that the full-scale EMA-CVT guarantees gear ratios for reduction, cruising, and reverse without producing rattling noise. Additionally, the EMA-CVT guarantees consistent power transfer during slope ascending and accelerates the vehicle in 5–7 s by applying the payload (Payload, WP = $W\sin^{i}$). It makes the car more energy efficient by lowering power transmission losses by roughly 75% and saving engine power by 20%. To validate the test-rig result, a road test is necessary. The prototype's centre manual switching system (CMSS) ensures the EMA operation with actuating the CVT system in case of IOT-based system failure.

12.2.16 EMA-CVT Test-Bed

Toyota wish CVT (model k311) was originally made with a spring at primary pulley of stiffness 117 n/cm and a hydraulic pumping system attached with the secondary pulley to control the gear ratio of the CVTs required. The hydraulic pumping is powered by engine. The CVT has been dismantled from the original set up. However, its primary pulley spring has remained same. Only the secondary pulley's hydraulic actuation system has taken away to attach with the proposed EMA.

EMA-CVT driving torque for the primary pulley has been developed with an electrical motor as shown in Fig. 12.13. This solenoid actuator allows the push mechanism by energizing the coil in order to engage exact gear ratio. A load sensor sensed the load weight that mimics the traction torque at the propeller shaft. The associated load signal then sent to the current or power level estimator to estimate a suitable electrical current to be supplied to the EMA to create enough force for every situation. Besides, there two rpm sensor (tachometer) that continuously measure the rotational speed of the motor and the propeller shaft and estimated the actual gear ratio of the vehicle.

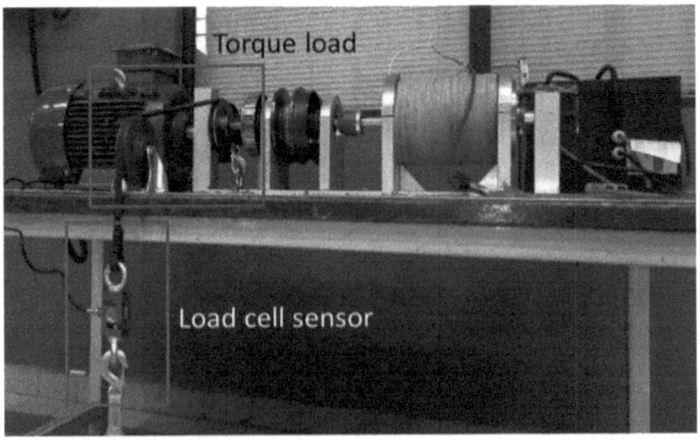

Fig. 12.13 Test bed for EMA laboratory testing [24]

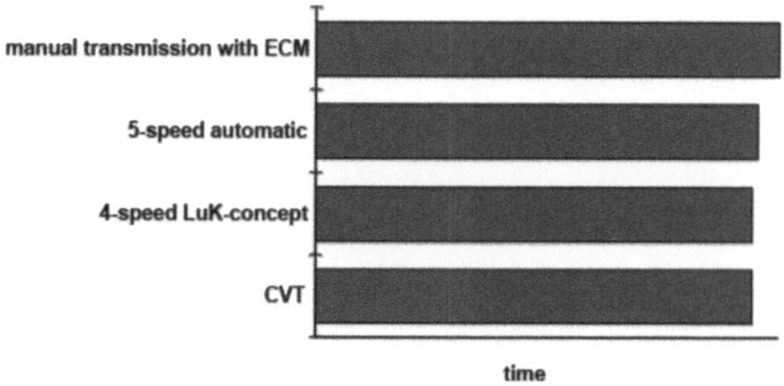

Fig. 12.14 Comparison of acceleration

12.3 Transmission Comparison

12.3.1 Acceleration

In comparison to the automatic transmission, the manual transmission offers about 8% acceleration from 0 to 48 km/h and 4–5% acceleration from 0 to 97 km/h and 72–105 km/h. A car with a CVT will only be roughly 3% faster than a car with a 5-speed automatic transmission when it comes to accelerating from 0 to 97 km/h [25]. Therefore, the Perodua MyVi 1.3 AT's acceleration time from 0 to 100 km/h with a curb weight of 1450 kg is utilized as a benchmark. It can be presumed that there is minimal change in the percentage increase in acceleration time between 0–97 km/h and 0–100 km/h. The acceleration time can then be calculated for other transmission types as shown in Fig. 12.14.

12.3.2 Transmission Losses

Transmission losses are not a set amount and are determined by the transmission's design, which varies greatly throughout automakers. Because manual transmissions have fewer gear losses and parasitic loads, they are the most efficient. Additionally, there is no need for forced cooling or an oil pump to lubricate the gears, which lowers the transmission's efficiency. In contrast to a traditional automatic transmission, which loses 15–20% of engine power, these characteristics enable the transmission to change torque with a power loss of only 5–15%.

The inefficient torque converter during unlocking in idle and operating conditions at low speeds, forced lubrication, braking with an open clutch, parasitic losses from constant pump operation for the pressure and cooling clutch, and micro-slip losses all contribute to the relatively low efficiency of the automatic transmission. According to the research, CVT transmissions often have higher losses than automatic gearboxes. The causes of losses include hydraulic pump mechanisms, pulleys, and belts.

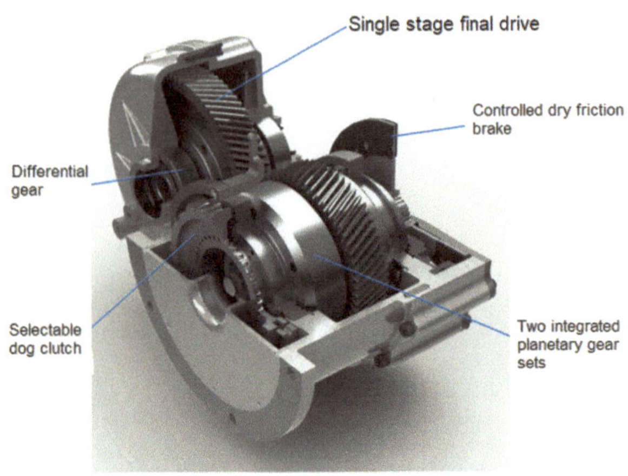

Fig. 12.15 Novel twin speed transmission

12.4 Overview of 2-Speed Automated Transmission

Compared to vehicles powered by internal combustion engines (ICE), electric vehicles have a number of special features. The electric motor can cover a broad engine speed range up to 14,000 rpm and provide very high torque at low rpm (torque at rest). Peak power can also be transmitted across a broad speed range, unlike ICE. These characteristics point to the creation of a novel transmission system that might take the place of traditional multi-speed gearboxes, which are ineffective, too big and heavy, costly, and cumbersome for electric cars. One possible option for electric vehicles is the two-speed transmission depicted in Fig. 12.15.

Numerous advancements have been achieved in order to create a working prototype. If this technology is developed successfully, the size of the transmission can be reduced while the stop torque needs are further reduced. Increasing the range can also maximize efficiency. To guarantee ideal acceleration, the top speed can also be raised. The cost of assembly should be kept to a minimum by using fewer parts and reducing weight. In order to guarantee seamless gear shifting without torque interruption during gear changes, the transmission's efficiency should therefore be increased. Since electric vehicles can achieve these gains while conventional ICE drivetrains cannot, these changes must be implemented. The latest technology in research is single-speed, two-speed, or two-speed transmissions for electric vehicles.

12.5 Two Speed Transmission for EV's

Performance and energy economy are two major benefits of multi-speed transmissions over traditional single-speed gearboxes for electric powertrains with a central electric motor. Particularly, the two-speed gearbox offers the ideal balance between the ease of use of a small, light transmission and the benefits of a multi-speed transmission. For instance, switching from a single-speed to a seamless multi-speed transmission can improve overall electric transmission efficiency and boost vehicle performance. This is because electric transmission systems need to be designed to be efficient, small, easy to use, and inexpensive to produce.

The efficiency of electricity consumption in each driving mode is increased by using a two-speed transmission rather than a single-speed transmission since the electric vehicle has a basic retarder with a fixed deceleration rate that varies depending on the driving mode. The distribution features of the working area for various driving modes can be diminished by variations in energy efficiency brought on by changing the average speed and putting 2-STM into practice. "Although electric motors can operate at a fairly wide range, this does not imply that they are equally effective at all speeds. The torque curve of a typical traction motor, as seen in Figs. 12.16 and 12.17, is well suited for vehicle propulsion, with a wide constant power zone and maximum torque from zero speed.

However, there is a 'sweet spot', typically at medium speed and medium to high loads, where the delivery of power is most efficient. A choice of gear ratios allows the motor to be kept in this operating region during more of the drive cycle".

It is widely acknowledged that the powertrain of an electric vehicle (EV) is a single-speed drivetrain, depending on the electric motor's torque distribution to guarantee optimal performance under all operating circumstances. Although this configuration is easy to execute, it does not allow the vehicle to operate at its full efficiency since the traction motor is frequently operating at loads and speeds that do not yield the best results. Engine efficiency usually decreases to 60–70% at low and high vehicle speeds and when there is little load. Due to their limited speed range and lack of torque at 0 rpm, combustion engines in traditional cars require multi-speed gearboxes. Higher speed is preferable over 8-speed, 9-speed, etc., if at all possible. Because the electric motor can span a wide range of speeds and produces strong torque at 0 RPM, a two-speed transmission is preferable to a 4-speed or higher. The advantages of lower fuel usage are anticipated to diminish because automatic and dual-clutch gearboxes have a tendency to increase gear ratio and ratio spread. In comparison to the 2-speed gearbox, the other transmission is inefficient, too big, and too heavy. Furthermore, compared to electric cars, it is more costly and inconvenient.

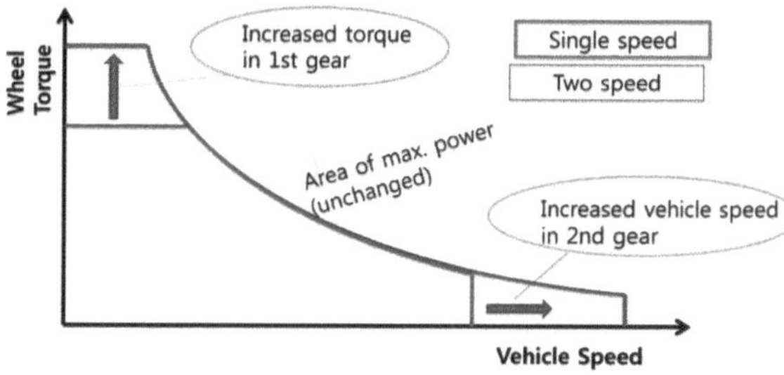

Fig. 12.16 Two speed transmission speed and torque range over single speed transmission

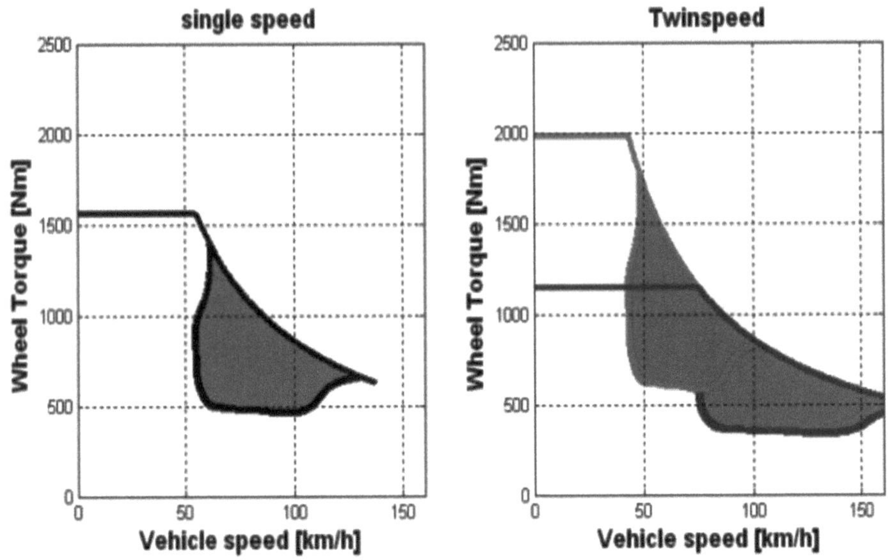

Fig. 12.17 Effective area of single speed vs double speed [26]

12.5.1 *Gear Shifting Force*

The capacity to continuously match intended subjective performance against objective measurements that are measurable, reproducible, and trustworthy is one of the issues involved in creating a competent gear shifting mechanism in a vehicle. Since they are the most efficient means of comprehending and developing client wants, traditional subjective design procedures utilizing professional evaluators can be successful. The physical effort required by the driver to choose and engage the correct gear by using the lever is a critical component of the majority of objective approaches for evaluating the quality of gear changing. Each transmission's synchronization time or the overall transmission time is used as a quality indicator.

Gear shift quality (GSQ) parameters, including shifting/gear selection forces, perception of various shifts/shift phases, gate detection, and lever travel, have been measured using both dynamic and static testing methods. This has aided in the development of objective metrics that can be used to define GSQ in the vehicle segment and an objective methodology for evaluating it.

One crucial factor that is solely dependent on the synchronizer is the quality of a manual transmission's dynamic gear changing. The dynamic shift parameter's behavior during gear changes is extremely non-linear and unpredictable, making it difficult to achieve satisfactory shift quality. The most intricate aspects of a manual transmission are the synchronizers' dynamic behavior and design. Calculations anticipate the behavior of most dynamic and static characteristics, including breakout force, synchronizing force, clamping force, and end stop force, which are dependent on the gearbox architecture and synchronizer design. A vehicle's shift performance was measured using Ricardo's Gear Shifting Quality Assessment (GSQA) toolset. Initially, basic GSQA data was collected, then one by one the gearboxes were upgraded with variable parameters and GSQA data was collected again.

Position sensors track the position of the shift fork (sliding clutch) in automated manual gearboxes (AMT) and dual clutch transmissions (DCT). The picture below illustrates how the sliding sleeve's location varies when the gears are changed. The shift fork, also known as the sliding sleeve, begins in the center position and moves in the direction of the synchronizer ring in the synchronizer approach. According to Fig. 12.18, the shifting force is normally between 60 and 120 N during this phase. The synchronization point begins at 3–6 mm, while the shift fork's overall travel can range from 8 to 12 mm.

Vehicle transmissions, such as dual clutch transmissions (DCT), automated manual transmissions (AMT), and manual transmissions (MT), frequently require synchronizers. The synchromesh's primary purpose is to minimize aberrant effects during gear changes, ensure correct transmission operation, and improve shifting comfort.

The shift force applied on the shift fork is seen in Fig. 12.19. The shift force is minimal during the initial free flight. The shifting force is increased to push the sleeve when it makes contact with the plunger. The shift force is quickly decreased until it approaches a minor constant when the sleeve interacts with the spline on the outer ring at the end of synchronization [3].

12.6 Automotive Actuators

The automotive industry typically provides a wide range of driving solutions designed to increase the members of the vehicle's comfort and safety. Although they are arguably the least well-known components, automotive actuators have had the biggest influence on the regulation of novel mechanisms. A power source and a control signal are necessary for the actuator. The control signal can be either electrical voltage or current and has a comparatively low energy content. Pneumatic pressure, hydraulic fluid pressure, or electric current can all serve as its primary power source. The actuator responds by transforming the energy from the source into mechanical motion when it receives a control signal.

12.6.1 Hydraulic Actuators

The hydraulic actuator, depicted in Fig. 12.20, uses hydraulic energy to control the movement of mechanical devices through a hydraulic cylinder with a sliding piston. Linear, rotational, or oscillatory movement are the outcomes of mechanical

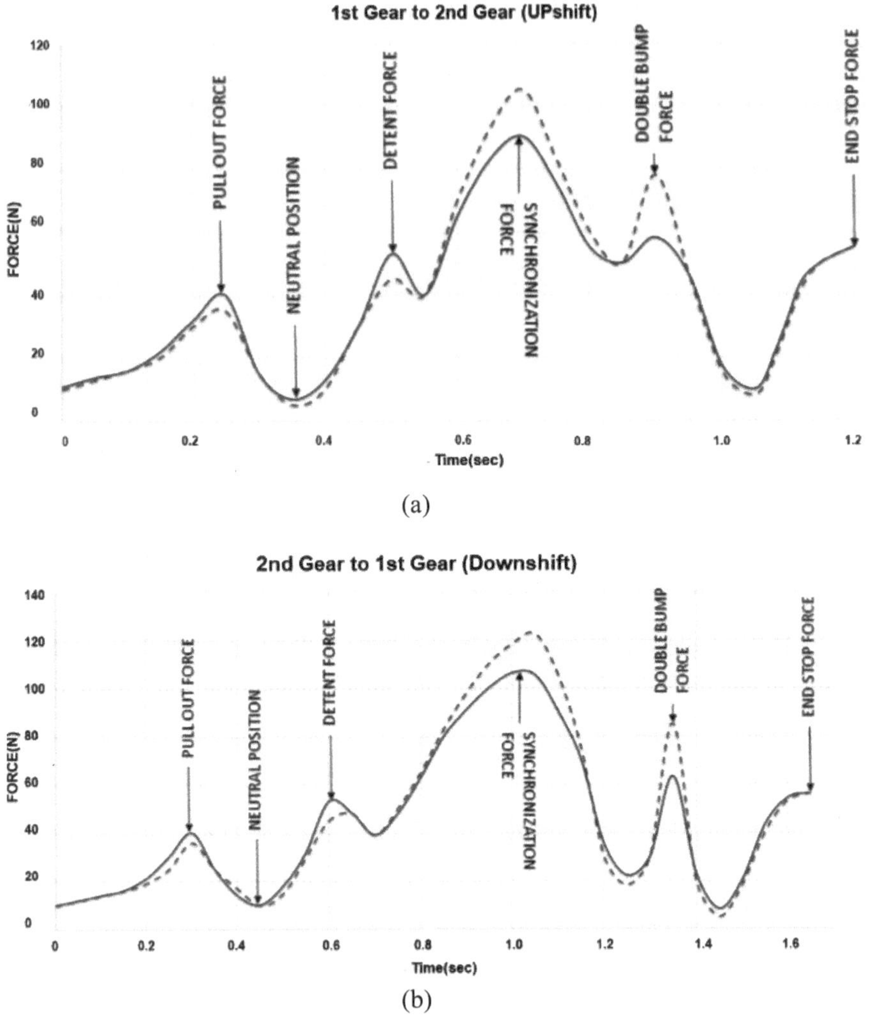

Fig. 12.18 **a** Upshifting force for 1st gear to 2nd gear engagement. **b** Downshifting force for 2nd gear to 1st gear engagement [26]

movement. Because fluids are incompressible, hydraulic actuators are able to provide high output forces. Single-acting and double-acting hydraulic actuators are the two varieties available. A single-acting hydraulic actuator uses a spring to return the piston to its starting position while applying pressure to only one side of the piston to move it in the desired direction. The hydraulic double-acting actuator, on the other hand, uses differential pressure on both sides of the piston to move it in a certain direction.

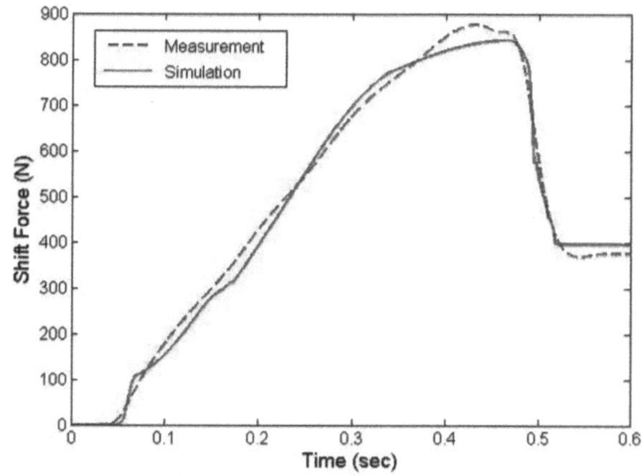

Fig. 12.19 Shift force acting on the shift fork [27]

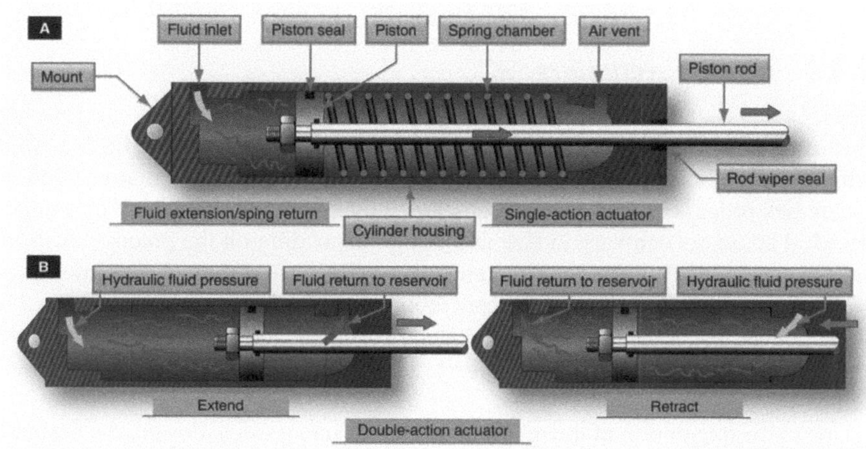

Fig. 12.20 Hydraulic-pneumatic linear actuator

The hydraulic actuators are robust and appropriate for situations requiring a lot of effort. Compared to pneumatic cylinders of the same size, they are capable of producing 25 times as much force. Additionally, they function at pressures of up to 4000 psi. Because the fluids are incompressible, the hydraulic actuator can maintain a constant force and torque without adding extra fluid or pressure. With little power loss, hydraulic pumps and motors can be placed far apart. There will be fluid leaks in the hydraulic system. Numerous more components are needed for hydraulic actuators, such as a fluid reservoir, motor, pump, exhaust valves, heat exchangers, and noise-cancelling devices. Because of this, linear systems are big and challenging to locate.

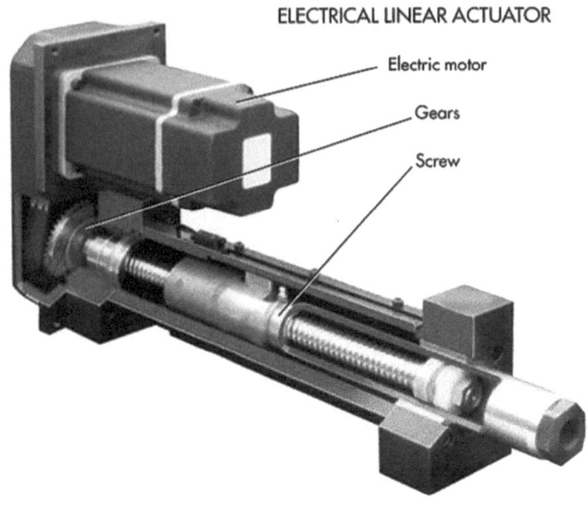

Fig. 12.21 Electrical linear actuator

12.6.2 *Electric Actuators*

As illustrated in Fig. 12.21, an electric actuator is a mechanical apparatus that transforms electrical energy into kinetic energy in a single linear or rotating motion. It increases process complexity and efficiency by automating the valve or damper. Electrical actuators can vary in size and design depending on the precise functions they are decided to do during their intended path. Because automation technology is moving toward a large rise in decentralization, electric drives are currently used in more applications. Now that the newest electric drive technology is available for the new process controllers, it will be simpler to adhere to the recently revised automation regulations.

Due to the integration of the motor and actuator into a single design, the design is quite straightforward. It is also capable of being reprogrammed and linked to several devices. It has controls for output power, torque, position, and speed and gives instant feedback. An electric drive's high cost and the gearbox and engine breaking from heat over extended use are drawbacks. Too big of a motor can make it difficult to install the engine requirements, which must be changed for various uses.

12.6.3 *Electromagnetic Actuator*

As seen in Fig. 12.22, an electromagnetic actuator uses the electromagnetic–mechanical principle to convert mechanical and electrical energy into one another. Circuits that are electrical and magnetic make up its two primary parts. A fixed frame and a freely moving shaft make up the magnetic circuit, while cylindrical coils are attached to the frame to provide the electrical circuit. The purpose of a nonmagnetic sliding

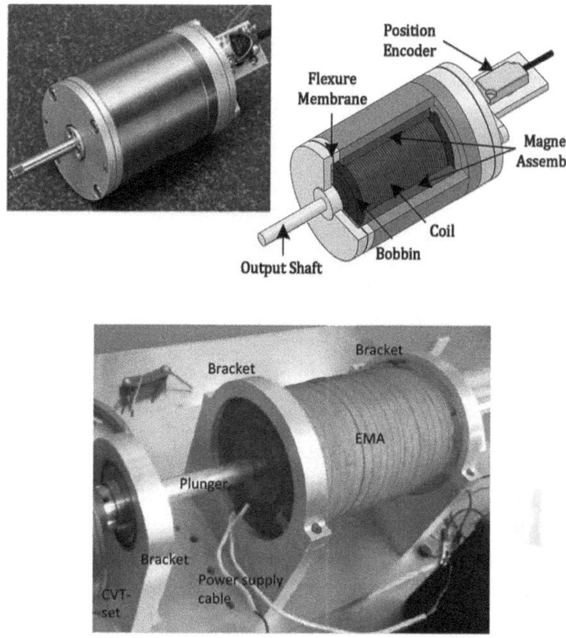

Fig. 12.22 Electromagnetic actuator

shell is to lower friction on the moving shaft and avoid collisions. A magnetic field created by the coil's current flow would either attract or repel a freely moving shaft, depending on the output current flow.

12.7 Control System for Aem System

The automotive industries of today are developing extremely quickly, utilizing a wide range of innovative technology and intelligent systems to meet the expanding needs of the public and private sectors. Engineers and researchers are frequently compelled by the need for such systems to push several tasks that are typically performed by humans into technology, such intelligent vehicles. A system is said to be intelligent if its behavior is modified in response to demands, circumstances, and prior knowledge. By incorporating concepts from experts or intelligent biological systems, intelligent systems provide alternatives to traditional methods. To mimic the operation of an intelligent biological system, researchers are looking for intelligent systems and control strategies (e.g., the human) to solve the technological problem.

Fuzzy logic is currently being employed in conjunction with vehicle expertise for a variety of applications, including modeling the interaction between the vehicle and the ground, estimating velocity in automobiles, controlling energy flow in wheel-electric vehicles, and more [28]. Because it can handle imprecise data, fuzzy logic control, as illustrated in Fig. 12.23, has been widely used in other engineering

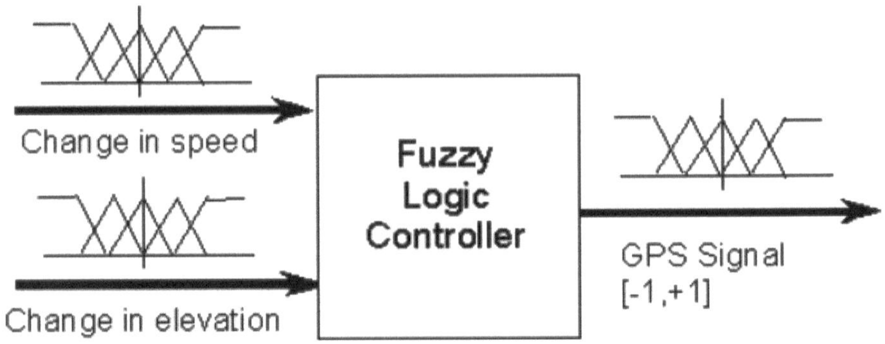

Fig. 12.23 Predictive controller inputs and output [30]

systems. Using velocity, front and rear wheel acceleration, and undercarriage velocity as input data [29], employed a fuzzy logic controller to control the active suspension system of a half-car model. The controller's outputs were active forces that improved the vehicle's handling, driving, and ride comfort.

In addition to describing the corresponding inference systems based on fuzzy rules, the Fuzzy Logic Expert System (FLES) makes it possible to express nebulous human conceptions using fuzzy sets. The fuzzy controller decides whether to carry out control actions based on the surroundings and the motion of the vehicle. According to reports, fuzzy logic systems' strong performance is attributed to their capacity to quickly provide an operational estimation of intricate, frequently unknown system input–output rules as well as to enumerate linguistic input variables. Additionally, due to its ease of implementation, fuzzy controllers have reportedly been employed extensively in the industry [19, 30].

The fuzzy inference system can be used to enhance the EMA system's performance. Fuzzy control systems were employed by several researchers to increase controlling precision and parameter self-adjustments. For a hydraulically operated CVT system, a simulation-based model was developed. To increase the accuracy of the system, two fuzzy logic controllers were created and the parameter self-tune technique was used. Exciting simulation and experimental data have been used to assess this controller's performance. Using an auto-tuning fuzzy control technique is another way that intelligent control systems are used to improve performance. To follow the necessary control variables, PID controllers are combined with an integrated power screw controller.

To regulate a DC servomotor so that the axial-pulley position meets the required point, they suggest using a fuzzy-PID controller. To address the PID system's tuning issue, fuzzy logic was used to determine an appropriate weight value for modifying the proportional gain in relation to the system error's current value and its error rate change. The authors suggest that additional research may be required to place the controller on an actual vehicle and assess its functionality.

12.8 Autonomous Transmission Market

The market for electric vehicle transmissions was valued at USD 7,256.7 million in 2021 and is projected to grow to USD 43,689.0 million by 2030. A compound annual growth rate (CAGR) of 22.4% is anticipated for the market over the forecast period. Additionally, in 2021, the market had a volume of 4,076.7 thousand units.

Many nations' preference for electric vehicles has helped to alleviate a number of problems related to the economy, power, ecology, and other areas. The development of automobiles with advanced technology and newer features that are affordable for all target groups in multiple economies is the main goal of many businesses in the electric vehicle transmission market and its parent sector. In order to maintain their position in the market, established businesses have a great grasp of demand. The market is expanding because to growing worries about carbon emissions. However, the demand for electric vehicles is closely correlated with the number of charging stations. The expansion of the electric car market is hampered by the absence of these conditions in many economies.

12.9 Growth Influences: Rising Adoption of Electric Vehicles

The global increase in the number of electric car purchases is fueling the EV transmission industry's rapid expansion. In the first half of 2021, for example, the Volkswagen Group's worldwide deliveries of fully electric vehicles more than doubled, largely due to demand in Europe. With 128,078 EV sales, up 156%, Europe was the automaker's leading EV market. Consequently, the global market for electric vehicles is expanding at a significantly faster rate due to this strong demand.

Due to the current revolution over the depletion of fossil resources, many nations are giving serious consideration to creating policies that will improve the environment and, consequently, society and the community as a whole. Every year, 995 metric tons of carbon emissions are produced in India. The adoption of EV tactics by people as a result of these concerning figures has indirectly increased demand for the EV transmission sector. The market for electric vehicle transmissions is divided into several segments, including distribution channel, vehicle type (design/model), vehicle type (power), transmission type, and transmission system.

12.10 Two Speed Gearbox Market Forecast

According to the most recent Fact.MR report, the two-speed gearbox market is expected to develop at a compound annual growth rate (CAGR) of 5–7% between 2021 and 2031. The market will gradually rebound in the near future, with a positive outlook for long-term growth. The main factor propelling the market's growth has been the desire for electric vehicles worldwide.

12.11 Conclusion

The majority of electric vehicles employ two-speed gearboxes. Strict government emission laws and regulations, along with the growing demand for high-performance, low-emission, fuel-efficient automobiles, are driving the market's expansion. The introduction of alternative fuel vehicles into the market has been made possible by rising fuel prices and the environmental impact of conventional vehicles. It is anticipated that the consumer desire for electric and hybrid vehicles would drive demand. Asia Pacific is anticipated to develop at the quickest rate during the projection period, followed by North America and Europe.

Acknowledgements The authors are grateful to the Ministry of Higher Education Malaysia to finance this project under the Prototype Research Grant Scheme (PRGS) 2022. The authors are also grateful to the laboratory technician helping to develop this project.

References

1. Kim, G., Gonder, J., Lustbader, J., & Pesaran, A. A. (2008). Thermal management of batteries in advanced vehicles using phase-change materials. *The World Electric Vehicle Journal, 2*(2), 46–59.
2. Kim, G., & Pesaran, A. (2007). Battery thermal management design modeling. *The World Electric Vehicle Association Journal, 1*, 126–133.
3. Ahmad, H. (2014). *Development of evaporative cooling battery thermal management system for electric vehicles*. MSc Thesis, International Islamic University Malaysia.
4. Rahman, A., Farhan, S., Ahmed, H., & Hawlader, M. N. A. (2014). Fuzzy controlled evaporative battery cooling system for EVs/HEVs. *International Journal of Electric and Hybrid Vehicle System, Inderscience, 8*(2), 233–240.
5. Anderson, D. L. (2009). An *evaluation of current and future costs for lithium-ion batteries for use in electrified vehicle powertrains*. Master Dissertation, Nicholas School of the Environment of Duke University.
6. Un-Noor, F., Padmanaban, S., Mihet-Popa, L., Mollah, M. N., & Hossain, E. (2017). A comprehensive study of key electric vehicle (EV) components, technologies, challenges, impacts, and future direction of development. *Energies, 10*(8), 1217.
7. Campbell, J. B., Tolbert, L. M., Member, S., Ayers, C. W., Ozpineci, B., & Lowe, K. T. (2007). Two-phase cooling method using R134a refrigerant to cool power electronic devices. *Industry Applications, IEEE Transactions, 43*(3), 648–656.
8. Rahman, A., Helmi, A., & Hawlader, M. N. A. (2017). Two-phase evaporative cooling battery thermal management technology for electric vehicle. *International Journal of Automotive Technology, 18*(5), 875–882.
9. Duan, X., & Naterer, G. F. (2010). Heat transfer in phase change materials for thermal management of electric vehicle battery modules. *International Journal of Heat and Mass Transfer, 53*(23–24), 5176–5182.

10. Heckenberger, T. (2009). Li-ion battery cooling: More than just another cooling task. *Technical Press Day, 2009*, 4–9.
11. Ataur, R., Rafia, A., & Hassan, A. J. (2019). *Green transportation system*, 1st edn. ISBN 978-967-491-021-1. IIUM Press, International Islamic University.
12. Rahman, A., Afroz, R., & Alam, Z. (2014). Development of electric vehicle: Public perception and attitude, the Malaysian approach. *World Review of Intermodal Transportation Research, 5*(2), 149–167.
13. Kizilel, R., Sabbah, R., Selman, J. R., & Al-Hallaj, S. (2009). An alternative cooling system to enhance the safety of Li-ion battery packs. *Journal of Power Sources, 194*(2), 1105–1112.
14. Soldo, V., Grozdek, M., & Curko, T. (2006). Two-phase heat transfer during refrigerant evaporation in a solar collector tube. In *61th ATI national congress-international season "solar heating and cooling"*, September 14, pp. 2–7.
15. Mousavi, M., Hoque, S., Rahnamayan, S., Dincer, I., & Naterer, G. F. (2011). Optimal design of an air-cooling system for a Li-Ion battery pack in electric vehicles with a genetic algorithm. *IEEE congress on evolutionary computation (CEC)*, pp. 1848–1855.
16. Nellis, G., & Klein, S. (2009). *Heat transfer*. Cambridge University Press.
17. Nelson, P., Dees, D., Amine, K., & Henriksen, G. (2002). Modeling thermal management of lithium-ion PNGV batteries. *Journal of Power Sources, 110*, 349–356.
18. Ramotar, L. (2010). *Second law analysis of a liquid cooled battery thermal management system for hybrid and electric vehicles*. Master dissertation, University of Ontario Institute of Technology.
19. Wu, M., Liu, K. H., Wang, Y., & Wan, C. (2002). Heat dissipation design for lithium-ion batteries. *Journal of Power Sources, 109*, 160–166.
20. Heath, R. P. G., & Child, A. J. (2007). *Zeroshift automated manual transmission (AMT)*. SAE Paper No. 2007-26-061, pp. 693–696.
21. Keynes, M. (2007). *Zeroshift automated manual transmission (AMT)*. Symp. Int. Automot. Technol., pp. 693–696.
22. Faid, S. (2015). *A highly efficient two speed transmission for electric vehicles*. EVS28, KINTEX, Korea, May 3-6, 2015.
23. Sorniotti, A., Holdstock, T., Everitt, M., Fracchia, M., Viotto, F., Cavallino, C., & Bertolotto, S. (2013). A novel clutchless multiple-speed transmission for electric axles. *International Journal of Powertrains, 2*(2–3), 103–131.
24. Rahman, A., Sharif, S. B., Hossain, A., Mohiuddin, A. K. M., & Zahirul Alam A. H. M. (2012). Kinematics and nonlinear control of an electromagnetic actuated CVT system for passenger vehicle. *Journal of Mechanical Science and Technology 26*(7), 2189–2196.
25. Heywood, J., & MacKenzie, D. (2012). Acceleration performance trends and evolving relationship between power, weight, and acceleration in us light-duty vehicles: Linear regression analysis. *Transportation Research Record, 2287*(1), 122–131.
26. Song, C. & Lee, S. (2015). *Design of a solenoid actuator with a magnetic plunger for miniaturized segment robots* (pp. 595–607).
27. Gonzalez, C. (2015). What's the difference between pneumatic, hydraulic, and electrical actuators. *Machine Design, 16*(04).
28. Economou, J. T., & Colyer, R. E. (2000). Modelling of skid steering and fuzzy logic vehicle ground interaction. In *Proceedings of the 2000 American Control Conference. ACC* (IEEE Cat. No. 00CH36334) (Vol. 1, No. 6, pp. 100–104). IEEE
29. Stříbrský, A., Hyniová, K., Honců, J., & Kruczek, A. (2003). Using fuzzy logic to control active suspension system of one-half-car model. *Acta Montanistica Slovaca, Ročník*, 8.
30. Rajagopalan, A., Washington, G., Rizzoni, G., & Guezennec, Y. (2003). Development of fuzzy logic and neural network control and advanced emissions modeling for parallel hybrid vehicles (No. NREL/SR-540-32919).

Chapter 13
Development of Autonomous Transmission

Abstract Development of a compact and autonomous energy-efficient transmission is a key technological issue for Next-Generation Vehicle (NxGV). The aim of this study is to fabricate a prototype of a compact and lightweight autonomous electromagnetic two-speed gearbox (AEM2SGB) for EV. The first-gear is for acceleration in uphill, which provides high torque, and the second-gear is for higher speed. A compact and lightweight laboratory-scale AEM2SGB has been developed. Furthermore, an autonomous fuzzy controlled system has been developed with associating the wheel speed sensor and motor sensor to control the power flow to the electromagnetic actuator (EMA). The test result of AEM2SGB's shift times in the first gear to be 1.5 s with a higher torque of 3000 Nm and 1.8 s in the second gear for 110 km/h with supplying current in the range of 12–16 A. Thus, the upscale AEM2SGB reduces the NxGV's transmission-weight about 40–50% and improves the NxGV performance by 30% and improves the battery life by 15%.

Keywords Electromagnetic gearbox · Fuzzy logic controller · Energy efficient · Compact and low cost transmission · Gear shifting time

13.1 Introduction

The research that went into the development of AEM2ST involved gathering knowledge about transmission systems, particularly two-speed transmissions and all associated mechanics, from reliable sources such as literature reviews, papers, and journal articles. The next step is to conceptually design the 2-speed transmission based on all the information the data has been obtained and calculated. For gear design, the forward and reverse gear ratios are calculated, and the size and specifications of the gears are determined, including the number of teeth, helix angle, pressure angle, tooth width, and center distance. Shaft and bearing dimensions are determined according to gearbox specifications.

The next step is to create and visualize the design concept in a 3D model using Solid works 2019 software. After creating the model, the design and compatibility of

A. Rahman et al., *Autonomous Driveline System for Electric Vehicle*, Green Energy and Technology, https://doi.org/10.1007/978-981-96-8418-2_13

each gear, shafts and bearings is analyzed. Simulating gear movement performance is carried out to check the output torque and speed of the gearbox and compared with the old prototype. When the result is satisfactory, the next step is to manufacture or manufacture the entire structure in a manufacturing company or store. The completed 2-speed transmission undergoes a test run with an electric motor and is equipped with an electromagnetic actuator to monitor its performance in real time and as appropriate. The flow of research work has presented in the Fig. 13.1. This transmission is proposed for the operation of vehicle of weight 3.0 kN around the IIUM campus. However, the AEM2SGB (AEM2ST) only tested in the laboratory due to the time constrain and the bad impact of COVID 19 pandemic.

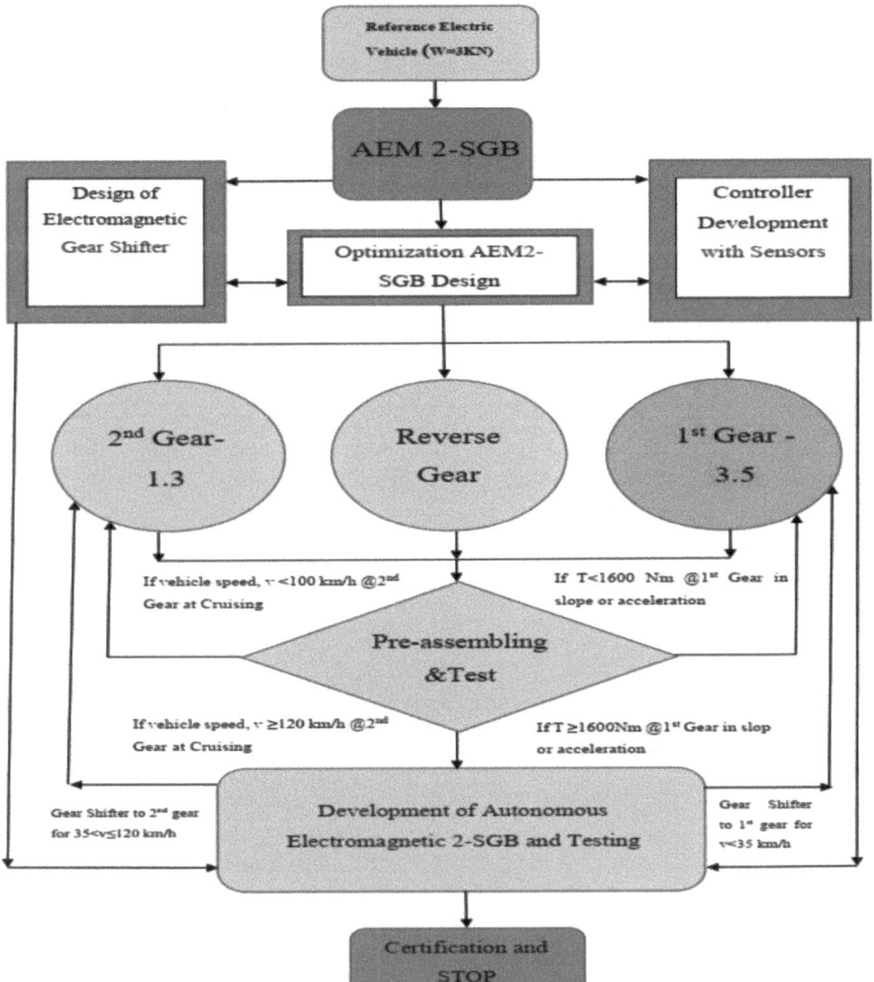

Fig. 13.1 Flow chart of the research work [1]

13.2 Development of an Analytical Design Concept

The transmission is the main element of the vehicle to make the power flow from the engine/electrical motor powered by battery to the driving wheel. The basic of this is the power flow to make vehicle in dynamics with maintaining traction and cruising. In general, the higher gear ratio (>3.0) is required for the vehicle in traction and lower ($\leq$1.3) for cruising control. The synchronizer mechanism should be with the gearbox to provide smooth, quiet and vibration-free gear shifting. An electromagnetic two-speed gearbox (AEM2SGB) for electric vehicles as shown in Fig. 13.2, has been designed from this study to improve dynamic of gear shifting performance and reduce the shifting time with the hassle of driver assistance.

The design concept of the proposed gearbox for the vehicle weight of 3.0 kN has been made for the dynamic in forward with two gear ratio. First gear of gear ratio 3.5 is made to ensure the traction force of the vehicle in hilly track of slopes of grade 20% and second gear ration 1.3 has been optimized for the cruising. While the third gear set has been made for the reverse. The reverse gear has been made in such a way that It will be controlled by the driver manually. A sets of electromagnetic actuator has been design to actuate the gearshift from 1st gear to 2nd gear and vice versa. The development of EMA has been made to develop the electromagnetic force equivalent to the axial force of the gear-shifting. The length of the electromagnet, number of turn of the coils and supply current to the coils has been optimized based on the development of the required electromagnetic force. However, the supply current has been limited to 17 A to the coils based on the OEM. The synchronizer mechanism has been made in such a way that it will allow the gear-engagement smoothly.

Autonomous system as shown in Fig. 13.3, has been developed using the fuzzy logic controller. It is made with RPM and slope sensor, which is placed at the motor shaft with a sprocket. The fuzzy logic controller controls the current flow to the EMA based on the signal of the RPM and slope sensor. The motor used in the AEM2SGB is a 24 V DC motor and the maximum speed was recorded at 350 rpm. This motor speed is 20 times lower than the 150 V AC induction motor that is available with the electric vehicle (EV), IIUM. The 24 V DC motor is classified into two modes: traction and high speed or cruising. The speed of the 24 V DC motor speed is considered below 250 RPM for vehicle traction on slopes 20% grade with maximum speed of 40 km/h and 250 RPM and above for higher speed. The controller codding has been made based on the output of the Mamdani Fuzzy Model's rule viewer.

13.3 Two Speed Gearbox Model

The electromagnetic 2-speed gearbox is made up of major components as shown in Fig. 13.4. When the driver pushes the acceleration pedal, the motor starts running and produced torque. The torque from the motor is then transferred through the sets of gears to produce the desired motion of the vehicle through the output shaft. The

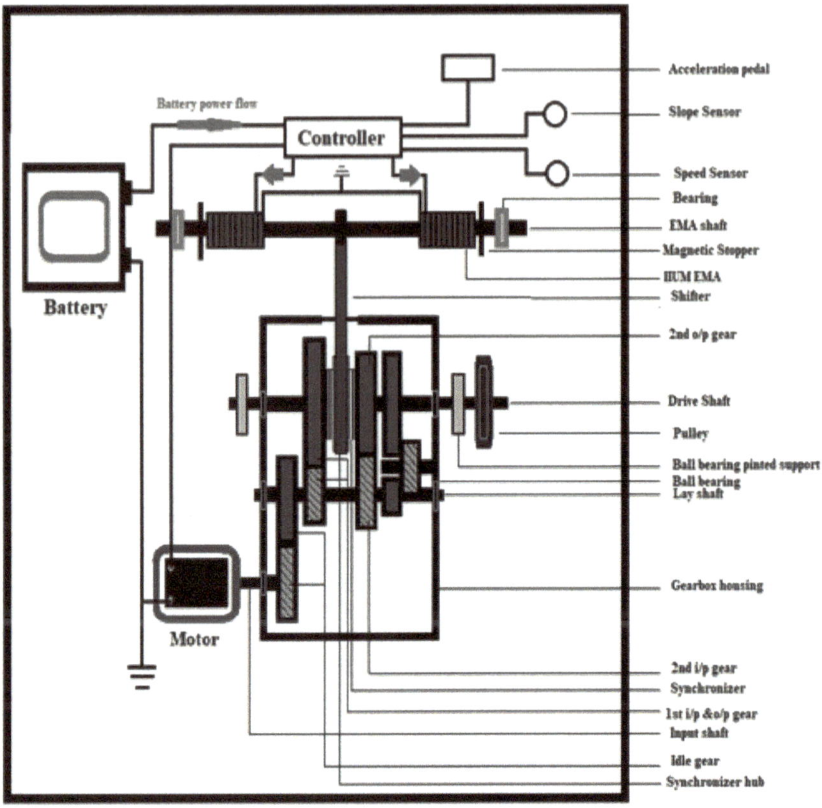

(a)

(b)

Fig. 13.2 Conceptual design of AEM2SGB **a** 2-D schematics diagram, **b** 3-D model of the gears [1]

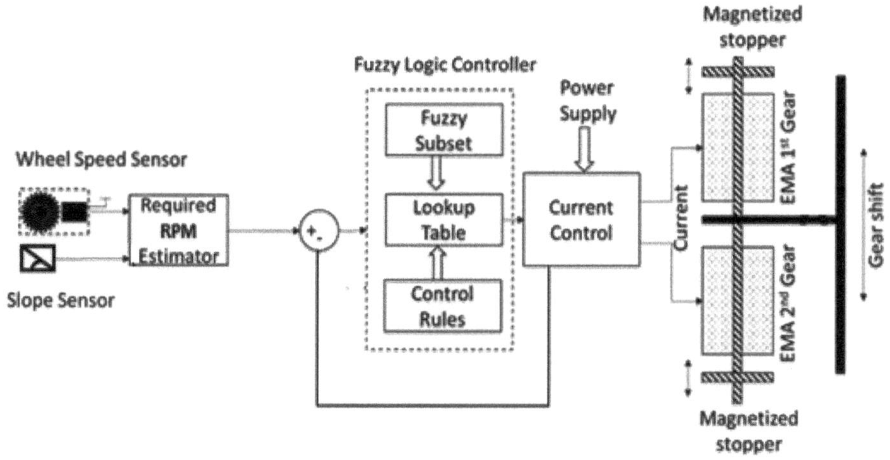

Fig. 13.3 Fuzzy logic control diagram of AEM2SGB [1]

transmission system will automatically shift the gear ratio depending on the condition of motion of the vehicle. The transmission is equipped with an electromagnetic actuator (IIUM) to operate the shifting of gears through the controller. The sensors will send signals to the controller and the controller will control the flow of current into the solenoid of the electromagnetic actuator. The current that passes through the solenoid will produce electromagnetic field that attracts the metal stopper besides the solenoid and simultaneously shifts the synchronizer's sliding sleeve to the left or right to engage with the gears. The motion of the vehicle is being detected by the sensors, which are the wheel speed sensor and slope sensor.

First Gear Engagement

When the vehicle starts to move from rest, the electromagnetic actuator will engage the first gear from neutral position to produce the torque to propel the vehicle forward. When the vehicle travels on a hilly track, the slope sensor will send signals to the controller for the transmission to shift into the first gear to produce more torque for the climbing. The transmission will also engage the first gear when the vehicle speed drops to a certain level to regain acceleration and torque through wheel speed sensor as shown in Fig. 13.5. The motor will transfer the torque and rotational speed through gear z1, z2, z3, and z4 to the output shaft.

Second Gear Engagement

Wheel speed sensor will send signals to the controller for the transmission to shift into the second gear when the vehicle starts to enter the cruising phase where the linear velocity of the vehicle increases until it reaches a certain point. The motor will transfer the torque and rotational speed through gear z1, z2, z5, and z6 to the output shaft as shown in Fig. 13.6.

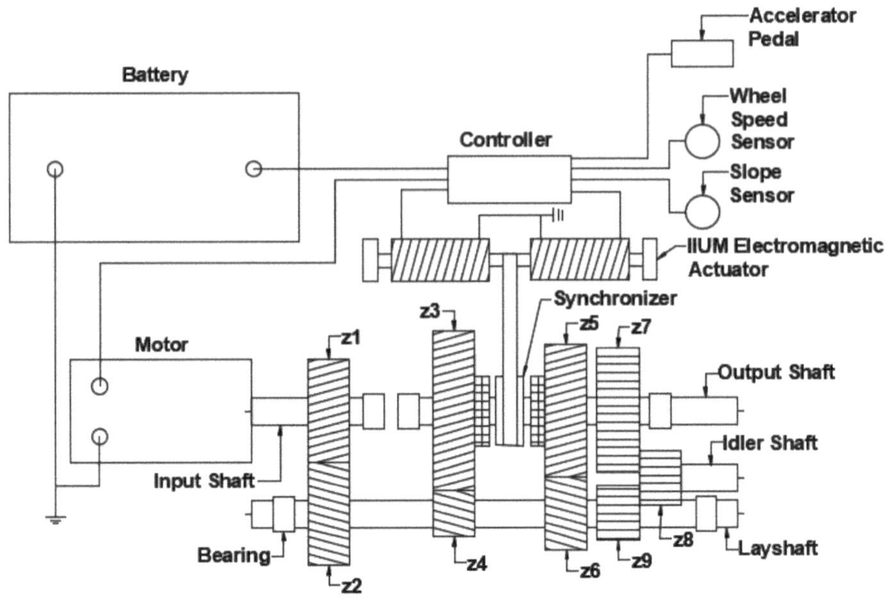

Fig. 13.4 Architecture of the AEM2SGB (or AEM2ST) [1]

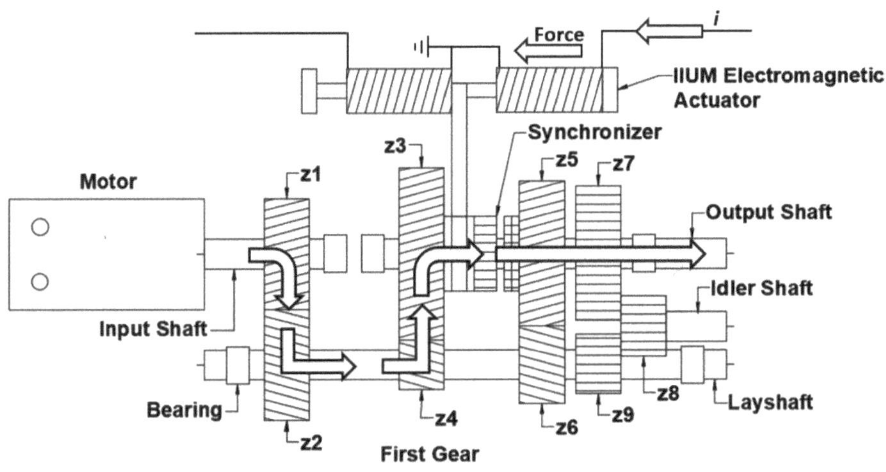

Fig. 13.5 First gear engagement for vehicle traction [1]

Reverse Gear Engagement

Figure 13.7 shows the reverse gear operation when the vehicle stops, the actuator should reset the position of the sliding sleeve to achieve neutral. The reverse gear can only operate when the vehicle is in full stop. However, for the reverse gear, the driver has to shift the gearbox manually. The manual shift will push the idler gear, z8

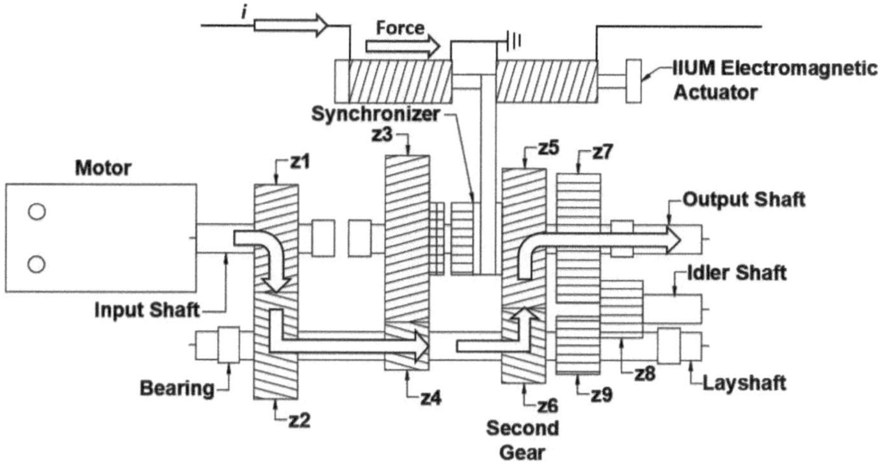

Fig. 13.6 Second gear engagement in vehicle speed up [1]

into meshing with gear z8 and z9. The rotation of the output shaft will be reversed to propel the vehicle backwards. The torque and rotational speed from the motor will be transferred through gear z1, z2, z7, z8, and z9 to the output shaft.

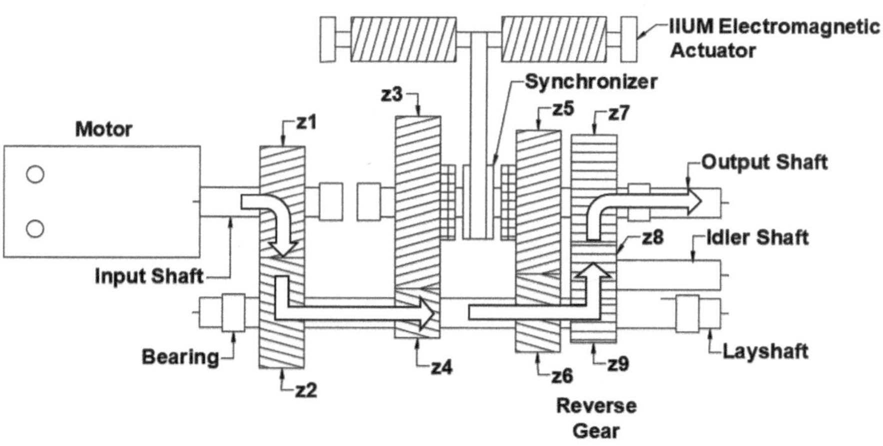

Fig. 13.7 Reverse gear operation mode [1]

13.4 Mathmatical Model

Transmission system function is to increase the torque or engine speed together with the differential. A higher gear ratio will increase torque while reducing the engine speed to the wheels. On the other hand, a lower gear ratio will increase the rotational speed while reducing the engine torque at the wheels. The final torque and wheel speed will be determined by the gear ratio of the differential as shown in Fig. 13.8.

The lowest and highest gear ratio of transmissions have the same formula which can be expressed by

$$T_w = T_m * GR * n_d * \eta_t \tag{13.1}$$

where, T_w is the wheel torque, T_m is the engine torque, η_d is the drive ratio, and η_m is the mechanical efficiency.

Vehicle Torque Requirement

In general, the torque requirement for the vehicle, T_w (or M_x) based on Fig. 13.8:

$$T_w = F_t r_w \tag{13.2}$$

where, r_w is the wheel radius. F_t (or F_x) is the force acting on the wheel and for accelerating vehicle, in which the Total resistive force, $\sum F_{\mathrm{R}}$ acts on the opposite of traction force as shown in Fig. 13.9, and the expression of F_t can be presented as:

$$F_t = ma + \sum F_R$$
$$\text{with} \quad \sum F_R = F_{rl} + F_g + F_a \tag{13.3}$$

$$F_{rl} = f_r W \mathrm{Cos}\theta$$

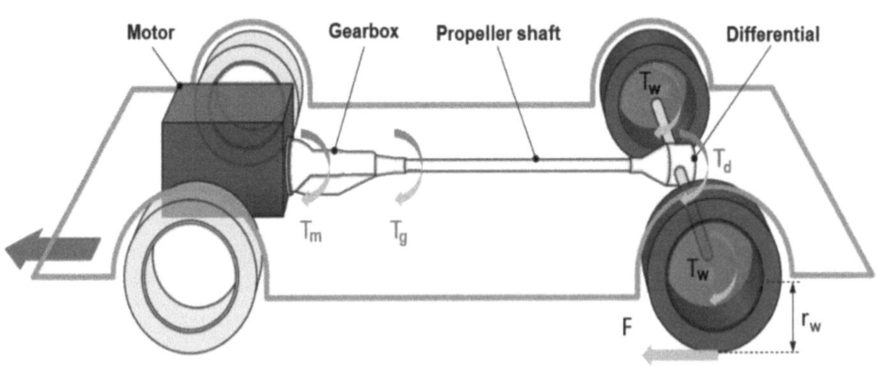

Fig. 13.8 Power train of 2-speed transmission [1]

$$F_g = W\mathrm{Sin}\theta$$

$$F_a = \rho_A C_D A_f \frac{v^2}{2} \quad (13.4)$$

Different types of resistive force are expressed as above. Where, F_{rl} is the rolling resistance force, where f_R is the rolling resistance coefficient, W is the mass of the vehicle, and θ is the angle of inclination. F_a is the aerodynamic force as shown in Fig. 13.10.

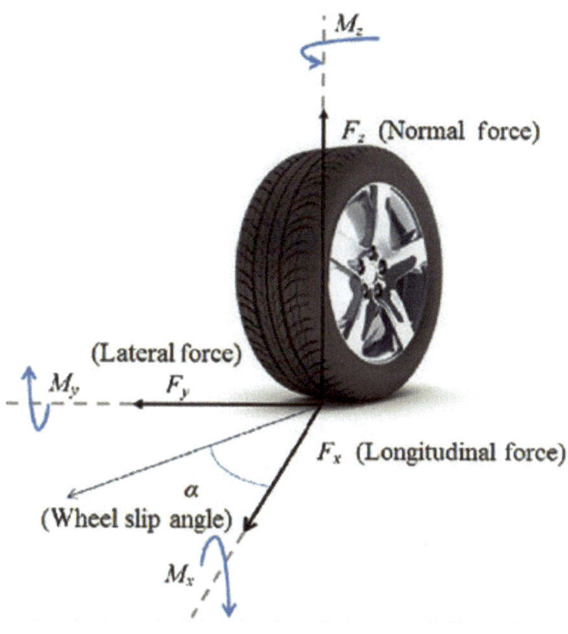

Fig. 13.9 Tractive force for various condition [6]

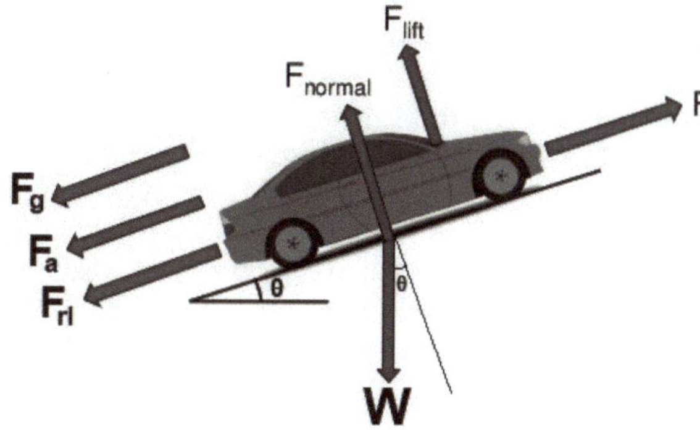

Fig. 13.10 Force acting on the vehicle

13.5 Gearbox Design Method

In designing the two-speed gearbox, many factors are needed to be considered. The gear ratios are the most important parameters that determine the effectiveness of the transmission on road, followed by the design of the gears, shafts, synchronizer, bearings as well as the choice of materials. The transmission or gear ratios is very crucial to convert the motor torque into the required wheel torque and wheel speed to produce the desired motion of a vehicle. The motor used for this project is HPEVS AC Electric-Motor Drive Systems (AC-50/AC-51). It is capable of producing maximum power of 53 kW and maximum torque of 163 Nm.

There are a few guidelines on designing an effective 2-speed transmission system.

1. The traction force produced by the wheels must be greater than the resistive force acting on the vehicle itself for it to propel.
2. The vehicle must be able to perform well on average slopes and have sufficient gradability.
3. The vehicle must have the maximum top speed of 120 km/h for the design at the cruising phase.

Gearbox Design

The transmission requires two forward gears that continuously mesh, so the two-speed manual transmission with reverse gear was chosen. The gearbox will have layshaft transmission, which is made up of three shafts. The input shaft and the output shaft are in the same line, which is connected by a double row cylindrical roller bearing, while the other shaft is a layshaft. Supposedly, if the transmission uses direct output (1:1 gear ratio) using layshaft transmission, the efficiency can reach up to 90% while having a very compact size. The first and second gear is a helical gear type, which operates greatly on high-speed operation, gives lower noise and wear as well as longer lifespan. The reverse gear is a spur gear type, which is used for gear sliding where the idler gear slides into two other gears. The gear shifting will be operated by a synchronizer system to change between the first and the second gear. Synchronizer is currently used in most manual transmission system due to its ability to remove shift shock while providing smoother shifting and better acceleration. The shifting operation will be conducted by an electromagnetic actuator (IIUM) that operates based on a controller and sensors.

Gear Selection

The forward gears (z1 to z6) will use helical gear type while the reverse gear (z7 to z9) will use spur gear type. Gear z1, z3 and z5 will be left hand helical gears while gears z2, z4 and z6 will be right hand helical as shown in Fig. 13.11. Gear teeth having a helix angle will spread the load and stresses to other teeth due to the engagement of many gear teeth in one time.

For this design, the input shaft does not use direct output transmission (1:1 gear ratio), so the input shaft and the output shaft will be treated separately.

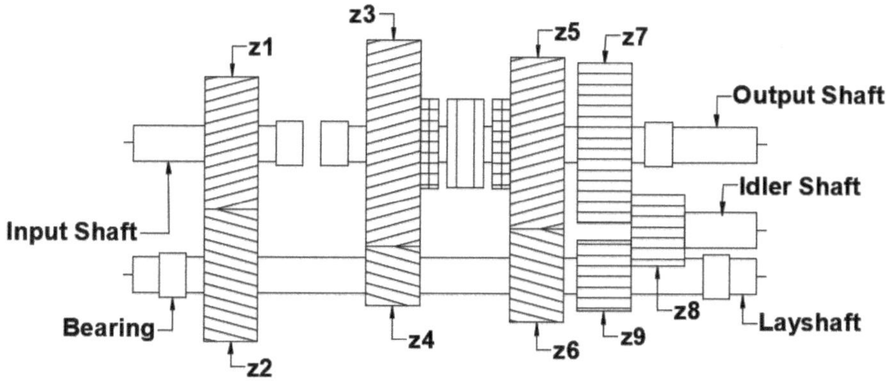

Fig. 13.11 Sketch of gear sets for the AEM2T for EV

Shaft Selection

The transmission uses layshaft transmission, which consists of three, rotating main shaft, which are input shaft, layshaft and output shaft. It also has a non-rotating idler shaft. For layshaft transmission, the input and output shaft are perfectly aligned. By having layshaft transmission, the size will always be compact and the efficiency will be higher. The conventional manual transmission uses layshaft transmission to create direct output (1:1 gear ratio) where the input shaft is connected directly with the output shaft for higher vehicle speed operation. However, for the prototype, the forward gear ratios are only two and the design will follow accordingly to the calculation for both gear ratio without considering the direct output operation. The prototype will also separate the input and output shaft to avoid confusion and to show clearly how it operates.

Gear Specification

The type of gear used for gearbox transmission are helical gear and spur gear. We choose the gear modulus according to national standards which is $mn = 3$ for helical gear and for spur gear, the $mn = 4$. However, for this model we decided to choose a smaller module to create a smaller gearbox model. With a smaller module, the range of number of teeth can be increased further without increasing the overall size of gearbox. With higher range of number of teeth, there is a higher chance to get multiple gear teeth combination for a specific gear ratio. So, the suitable module to be used are $mn = 2$ for helical gear and $mn = 2$ for spur gear.

Pressure Angle

For the transmission gear with low pressure angle, it has high contact ratio and usually gives a smoother transmission as well as lower noise during operation. For the gear with high pressure angle, the advantages are having a better bending and surface contact strength. Most gearboxes used on passenger cars have the pressure angle of 15°, 20°, and 25°. For the design, we choose the pressure angle of 20°.

Helix Angle

Helical gear is used this transmission. For a higher helix angle value, the overlap coefficient of gear meshing is also high as well as having smoother transmission and lower noise. The optimum helix angle for maximum bending and compressive strength is $\beta = 15^{\circ}$ [2].

Gear Design

For the design of the gear, the key parameters include the centre distance, gear modulus, pressure angle, helix angle, tooth width, gear teeth, and gear size. The parameters requiring calculations are shown below:

Centre Distance:

$$A = K_A \sqrt[3]{T_{1(\max)}}$$

where, K_A—centre distance factor, $K_A = 8.9$~9.3

$$T_{1(max)} = T_{m(\max)} * GR_1 * \eta_t$$

$T_{m(\max)}$—Maximum engine torque, Nm, GR_1—Gear ratio of the first gear, η_t—efficiency of automobile transmission, $\eta_t = 0.9$.

Tooth Width:

$$b = K_c m_n$$

where: K_c—coefficient of tooth width, for helical gear $K_c = 7.0 \sim 8.6$, m_n—normal modulus

Helical Gear Formula:

Pitch diameter: $d = \frac{m_n Z}{Cos\beta}$.

Addendum: $h_a = f_o m_n$.

Dedendum: $h_d = (f_o + c)m_n$.

Tip diameter: $d_a = d + 2h_a$.

where, Z is the number of teeth, m_n is the modulus of the tooth surface of the helical gear, f_o is an addendum coefficient, $f_o = 1.0$, β is the angle of inclination of the spiral, c is the radial gap coefficient, $c = 0.25$.

Spur Gear Formula:

Pitch diameter: $d = m_n Z$

Addendum: $h_a = f_o m_n$

Dedendum: $h_f = (f_o + c)m_n$

Tip diameter: $d_a = d_n + 2h$

where, Z is the number of teeth, m_n is the modulus of the tooth surface of the spur gear, f_o is an addendum coefficient, $f_o = 1.0$, c is the radial gap coefficient, $c = 0.25$.

13.6 Development of Electromagnetic Actuator

Shifting Force Development

Electromagnetic actuators have several advantages over other types of actuators. In particular, due to the availability of electricity in electric vehicles, the use of solenoid actuators to shifting gears in electric vehicles seems reasonable. Therefore, I propose to use solenoid actuators in electric vehicles for seamless transmission [3]. Actuating force or axial force of the synchronizer can be calculated by the following equation,

$$F_{ax} = \frac{2sin\alpha * J * \Delta\omega}{\mu * n_c * d_m * t} \tag{13.5}$$

where F_a, α, J, $\Delta\omega$, n_c, μ, d_m and t are the axial force, the cone angle for engagement, the angular speed differences of the synchronizing ring and gear cone, the number of cone, the friction coefficient of the cone, the mean cone diameter and the shifting time. Synchronization time depends on the coefficient of friction, moment of inertia, angular velocity and applied force on the fork. A smaller cone angle creates much more torque for a given coefficient of friction, but has a greater tendency to jam. The larger the cone angle, the lower the torque, but the likelihood of jamming is less and it is more resistant to changes. Synchronization of angular velocity is usually obtained by conical friction clutches, while the power transmission is usually realized by spline coupling. Depending on transmitted torque, one, two or three conical surfaces are applied. To reduce the shift force the following measures can be followed: increase diameter d_m, increase the number of cones n_c, increase c.o.f. (friction lining, oil), reduce cone angle α to avoid self-locking. In order to select a suitable solenoid actuator among low profile actuator with plunger, it is very important to know the performance observation and operating limits.

Electromagnetic Force (EMF)

The electromagnetic actuator has been designed to develop maximum electromagnetic force to overcome maximum axial force. The mathematical representation of the EMA is designed taking into account the dynamic behavior of magnetic flux, density, force, electromagnetic force and energy in accordance with Faraday's law, Ampere's law and Lenz's law, Maxwell's dynamic condition and modified equations [4]. In this study, the electromagnetic force should greater than the axial force (EMF $\geq F_{shift}$). The total electromagnetic force at the magnetic field of the current conducting solenoid is given by the general equation of Fleming's (LH) rule:

$$F_{\text{emf}} = \text{BI}1_w \tag{13.6}$$

where, F_{emf} is the electromagnetic force for the total length of the solenoid coil l_w, B is the magnetic flux density for the entire solenoid, I is the current passing through the conducting wire around which a magnetic field is created. It should be noted that the properties depend on the EMA: the geometry of the magnetic circuit, the size of the air gap in the magnetic core, the core material and the solenoid temperature.

The basic principle of EMA operation is related to the DC power supply. When power is applied to the EMA, the EMA actuators to the electromagnet and pulls the metal disc to the EMA and the forklift which is attached to the synchroniser plunger, pushes the synchroniser hub to the synchroniser ring so that it is attached with a gear cone and the gear ratio instantly as needed. The energy density EMA for single coil for instantaneous actuation can be defined as,

$$E_d = B^2/2\mu \tag{13.7}$$

where, B is the magnetic field and μ is the magnetic permeability. A magnetic permeability μ about 0.006 H/m. The total energy store in the EMA coils due to the supply current can be modelled as,

$$\begin{aligned} E_{\text{act}} &= E_d \\ \text{with } E_d &= \frac{1}{2\mu} B^2 V_{EMS} \end{aligned} \tag{13.8}$$

where, E_{act} is the energy actuation of the EMA system, E_d energy develops at the EMA due to supply current I and $V_{\text{EMA}} = \pi r^2 l_c$ is the volume of the solenoid., by simplifying the Eqs. 16.3–16.4, the energy developed by the EMA actuation system can be expressed as,

$$\begin{aligned} E_{\text{act}} &= \frac{\mu n^2 I^2 V_{\text{EMA}}}{2} \\ \text{with } n &= \frac{N}{L_{\text{sol}}} \end{aligned} \tag{13.9}$$

EMA2-SGB testing has been found to generate electromagnetic radiation, which is measured based on the temperature change around the free surface of the EMA, just as we feel hot under sunlight due to solar radiation. Electromagnetic waves have electric and magnetic fields that travel through empty space at the speed of light c. If its oscillation frequency is f, then it produces an electromagnetic wave with frequency f. The wavelength ($\lambda = c/f$) is set by electromagnetic waves that carry energy through free space.

The EMA power loss due to the current $I_0{}^2R_{\text{rad}}$ on its coils is associated with the radiation resistance, while the $I_0{}^2R_c$ power loss associated with the real resistance is associated with the ohmic heating of the EMA. R_{rad} may be determined by the equation,

$$R_{\text{rad}} = \frac{2\pi}{3\varepsilon_o c}\left(\frac{l_w}{\lambda}\right)^2 \tag{13.10}$$

Total power loss of the EMA can be defined as,

$$\begin{aligned} P_{\text{EMA(loss)}} &= I^2(R_c + R_{\text{rad}}) \\ &= I^2\left[R_c + \frac{2\pi}{3\varepsilon_o c}\left(\frac{l_w}{\lambda}\right)^2\right] \end{aligned} \tag{13.11}$$

The effective energy of the EMA to do the actuation of the sleeve to the gear cone can be expressed as,

$$\begin{aligned} E_{\text{effective(EMA)}} &= E_{\text{act(EMA)}} - E_{L(\text{ohomic})} - E_{L(\text{rad})} \\ &= \frac{\mu n^2 I^2 V_{EMA}}{2} - I^2\left[R_c + \frac{2\pi}{3\varepsilon_o c}\left(\frac{l_w}{\lambda}\right)^2\right]\int_0^t dt \end{aligned} \tag{13.12}$$

The Electromagnetic Actuator (EMA) has been designed with a copper coil winding based on the requirements of the electromagnetic force (axial displacement force). Magneto motive force (MDF) is the equivalent of the electromotive force (EMF or voltage) in an electrical circuit. Units of measurement are ampere-turns (A-t). A coil carrying an N-turn I-amplifier generates MMF in NI ampere-turns. An electromagnetic drive is used to maintain the gear ratio or gear ratio of the vehicle to adjust the gear ratio as needed. The DC power supply from the battery to the EMA is controlled by the controller based on the sensor signal. The EMA automatically sets the EMA2-SGB gear ratio based on the response of the sensors [5].

The electromagnetic force is a function of the magnetic flux density and the supply current, since the length of the coil pack is considered constant in the magnetic field. Using a single coil and multiple coils as shown in Fig. 13.12a, the EMA magnetic flux density can be estimated. The EMA force simulation was performed by simplifying the Gaussian equations, Maxwell's equations, the ampere circuit equation, Faraday's law and the Poynting equation. The EMA has considered a thin coaxial cylindrical conducting shells of radius r and c. The current, I is supplied with number of turns N per unit length.

$$g_\theta = \begin{vmatrix} \mu_0 NIQ/2\pi rl \text{ For } a \le r \le b \\ 0 \qquad \text{Otherwise} \end{vmatrix} \tag{13.13}$$

According to Gauss's law, the electric field takes the form $E = E_r\, e_{\text{r}},$ where,

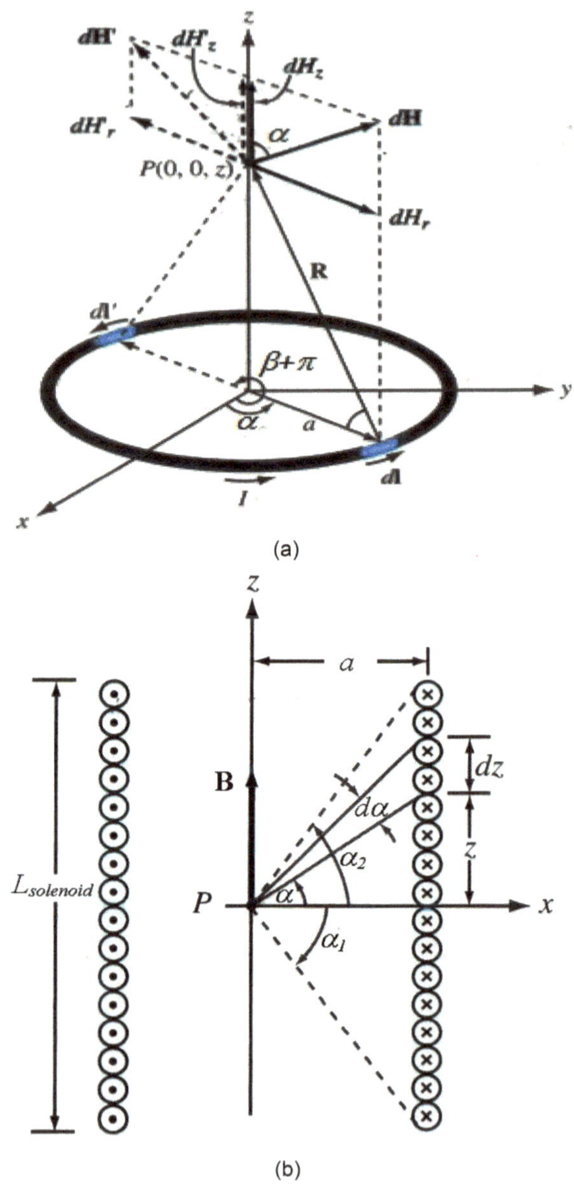

Fig. 13.12 Magnetic flux density **a** single coil, **b** multiple coils [Ulaby, 2005]

$$E_r = \begin{vmatrix} \frac{Q}{2\pi \varepsilon_o r \, l_w} & \text{For } r_i \le r \le r_o \\ 0 & \text{Otherwise} \end{vmatrix} \tag{13.14}$$

Similarly, according to Ampere's circuital law, the initial magnetic field of the EMA due to supply current,

$$B_i = \begin{vmatrix} \mu n I \text{ for } r \le r_o \\ 0 \quad \text{Otherwise} \end{vmatrix} \tag{13.15}$$

The initial momentum density of the electromagnetic field could be,

$$g_\theta = \begin{vmatrix} \dfrac{\mu n I Q}{2\pi l_w r} & \text{if } r_i \le r \le r_0 \\ & \text{Otherwise} \\ 0 & \end{vmatrix} \tag{13.16}$$

Any change in the current flowing in the EMA winding generates an inductive electric field which can be defined as,

$$E_\theta = \begin{vmatrix} -\mu n I r/2 \ \text{ For } r \le r_o \\ -\mu n I r_o^2/2r \ \text{Otherwise} \end{vmatrix} \tag{13.17}$$

F_{emf} is a function of the applied current and the number of coils, as given in Eq. (3.34). Therefore, the electromagnetic force,

$$F_{emf} = \frac{1}{x}\left(\frac{\mu n^2 I^2 V_{EMA}}{2} - I^2\left[\frac{R_c}{l_c} + \frac{R_{rad}}{d(r)}\right]\right)\int_0^t dt \tag{13.18}$$

$B = \mu nI$ ($n = N/L_{sol}$), where $x =$ displacement of solenoid from neutral.

Where, μ is the magnetic permeability (degree of magnetization of a material in response to magnetic field), l_c the length of the core, m, $d(r)$ is the distance of the radiation measuring, and t is the time, s. It is usually determined that the current passing through the coil (loop) creates a magnetic field. Applying Ampere's law to an imaginary coil in the x–y plane of radius r, centred on the z-axis as shown in Fig. 13.12b. The current flowing in the coil can be estimated as,

$$I(r_o) = \begin{vmatrix} 0 & r < r_i \\ \{I * (r^2 - r_i^2)/(r_o^2 - r^2)\} & r_i \le r \le r_o \\ I & r_o > r \end{vmatrix} \tag{13.19}$$

Hence the magnetic flux density can be estimated,

$$B_i(r_o) = \begin{vmatrix} 0 & r < r_i \\ [\mu_o I/2\pi r][(r^2 - r_i^2)/(r_o^2 - r^2)] & r_i \le r \le r_o \\ \mu_o I/2\pi r & r_o > r \end{vmatrix} \tag{13.20}$$

It is generally considered that the magnetic permeability μ_o is 0.006 H/m. It should be noted that when the current (I) 17A is applied to the EMA, the EMA temperature reaches 100 °C, the solenoid efficiency is 0%. Consequently, EMA energy losses are in the form of ohmic heat and electromagnetic radiation.

Electromagnetic Energy Loss

It is assumed that the loss of electromagnetic energy EMA is due to the loss of magnetic hysteresis, losses due to eddy currents and leakage of magnetic flux through the solenoid surface. The first two different types of energy loss are core losses [6].

Leakage magnetic flux through the coil surface results in a significant loss of the electromagnetic field at the screen with the absence of the solenoid surface. The outer loop windings are considered in this study for leakage loss purposes only. Magnetic field loss due to magnetic flux leakage in a solenoid can be calculated using the simplified equation below:

$$H_{\text{in}} = \frac{r_{\text{in}} \iint\limits_{S(r_{\text{in}}, 1_{\text{segment}})} \text{JdS}_{\text{wire}}}{2\pi r^2} \quad (\text{for } r_{\text{in}} \leq r_o) \tag{13.21}$$

$$H_{\text{out}} = \frac{\iint\limits_{S(r_{\text{out}}, 1_{\text{segment}})} \text{JdS}_{\text{wire}}}{2\pi r_{\text{out}}} \quad (\text{for } r_{\text{out}} \leq r_o) \tag{13.22}$$

H_{out} is the main concern for finding flux leaks across the solenoid surface. The H_{in} magnetic field is effective in developing the desired magnetic force to operate the movable pulley, while H_{out} is the cause of the magnetic loss. Solenoid energy loss due to leakage flux can be calculated by the simplified equation below,

$$\begin{aligned} E_{\text{flux leakageloss}} &= N_{\text{out}} \left(\sum_{\text{no.segment}} \frac{\iint\limits_{S(r, 1_{\text{segment}})} {}^{J}/_{2} dS_{\text{wire}}}{2\pi r_o} \right) \\ &= \frac{I(h_o - h_i) L_{\text{sol}}}{4\pi r_w r_o} 1_{\text{single winding}} \end{aligned} \tag{13.23}$$

where N_{out} is the number of the winding on the solenoid surface. J/2 due to the semi-circular copper wire, I is the current supply, h_{o} and h_i are the outer and inner diameter of the solenoid, respectively, L_{sol} is length of the solenoid, $l_{\text{single winding}}$ is the length of single winding r_w is the wire radius and r_{o} is the outer radius where $r_o > r_w$.

13.7 Electromagnetic Actuator Design

The electromagnetic actuator that converts an electric signal into a magnetic field, which produces the linear motion is called an electromagnetic solenoid. The specification of the EMA is shown in Table 13.1 and the developed EMA is shown in Fig. 13.16. When an electric current passes through the coil winding, it behaves as an electromagnet and a plunger is disposed within the coil and drawn to the centre of the coil by the magnetic flux generated inside the coil housing which is fixed to one end of the magnetic disk.

The force and speed of the plunger is determined by the strength of the magnetic flux generated inside the coil. This is independent of the direction of the current in the coil. The force of the solenoid plunger can be significantly increased by increasing the current flow to the solenoid. But too much current can damage the solenoid. Therefore, a thermal control system with a solenoid is required, which is not the main one in this study. The EMA of this study, which relates to the solenoid, is based on the core body, the number of turns of copper wire, the aluminium plunger, and the power supply [7, 8].

The autonomous 2-speed gearbox of an electric vehicle's shifting function has been designed with the electromagnetic actuator (EMA) as shown in Fig. 13.13. This EMA's goal is to perform a direct gear shift using the electromagnetic force that has been produced as a result of the supply current.

Table 13.1 Specification of EMA

Parameters	Values
Coil diameter (dw), mm	1.45
Cross sectional area of wire, mm^2	1.65
Length of solenoid (L_{sol}), mm	75
Solenoid minimum diameter (h_i), mm	10
Solenoid maximum diameter (h_o), mm	50
Packing factor, $f_{\text{pf}} = \frac{N_{\text{actua l}}}{N_{\text{theoretial}}}$	0.525
Number of turns (N), $N = \frac{F_{\text{ema}} * L_{\text{coil}}}{\mu I^2 L_{\text{sol}}}$, turns	163
Length of coil (L_{coil}), $L_{\text{coil}} = \frac{L_{\text{sol}}}{d_w^2}\left[D_0^2 - (D_i + t)^2\right](f_{\text{pf}})$, m	30
Volume of the solenoid coil, $v_{\text{sol}} = \pi\left(h_0^2 - h_{i2}\right) l_{\text{sol}} f_{\text{pf}}$, m^3	2.96×10^{-4}
Mass of solenoid, $m_{\text{sol}} = \rho_{cu} V_{\text{sol}}$, kg $\rho_{\text{cu}} = 8940$ kg/m at room temperature	2.64

Fig. 13.13 Developed EMA for AEM2ST [9]

13.8 Conclusion

The development of a compact and autonomous energy-efficient transmission is a key technological issue for Next-Generation Vehicle (NxGV). An autonomous fuzzy controlled system has been developed associated with the wheel speed sensor and motor sensor to control the power flow to the electromagnetic actuator (EMA) for the smooth operation of the compact and lightweight laboratory-scale AEM2SGB. The test result of AEM2SGB's shift times in the first gear to be 1.5 s with a higher torque of 3000 Nm and 1.8 s in the second gear for 110 km/h with supplying current in the range of 12–16 A. Thus, the upscale AEM2SGB reduces the NxGV's transmission-weight about 40–50% and improves the NxGV performance by 30% and improves battery life by 15%.

Acknowledgements The authors are grateful to the Ministry of Higher Education Malaysia to finance this project under the Prototype Research Grant Scheme (PRGS) 2022. The authors are also grateful to the laboratory technicians who helped to develop this project.

References

1. Hasan, N. (2021). *Development of autonomous electromagnetic two speed gearbox for electric vehicle.* A Master Thesis Dissertation, IIUM Library, International Islamic University Malaysia.
2. Venkatesh, B., Vattikuti, S. P., & Prasad, S. D. (2014). Investigate the combined effect of gear ratio, helix angle, facewidth and module on bending and compressive stress of steel alloy helical gear. *Procedia Materials Science, 6*, 1865–1870.
3. Un-Noor, F., Padmanaban, S., Mihet-Popa, L., Mollah, M. N., & Hossain, E. (2017). A comprehensive study of key electric vehicle (EV) components, technologies, challenges, impacts, and future direction of development. *Energies, 10*(8), 1217.
4. Hayt, W. H., & Buck, J. A. (2006). *Engineering electromagnetics* (7th edn). McGraw-Hill International Edition.

5. Rahman, A., Sany I. I., & Nurul, H. (2022). A fuzzy-logic-controlled two-speed electromagnetic gearbox for electric vehicles. *International Journal of Technology, 13*(2), 167–179.
6. Ataur, R., Afroz, R., & Hassan, A. J. (2022). *Green transportation system.* IIUM Press.
7. Passino, K. M., Yurkovich, S., & Reinfrank, M. (1998). *Fuzzy control* (Vol. 42, pp. 15–21). Addison-wesley.
8. Ivanov, V., Shyrokau, B., Augsburg, K., & Algin, V. (2010). Fuzzy evaluation of tyre–surface interaction parameters. *Journal of Terramechanics, 47*(2), 113–130.
9. Rahman, A., & Hasan, N. (2023) Autonomous electromagnetic two-speed gearbox for electric vehicle. *International Journal of Automotive Technology.* In press.

Chapter 14
Intelligent Control System of Autonomous Transmission

Abstract A lower rpm value for traction (below 200) and a higher rpm value for cruise (above 200) are considered in programming the FLC to control the EMAs and prepare the AEM2SGB to meet the working conditions. The controller sends a signal to the relay to pass the battery current to get the proper force from the EMA to engage the gear automatically. Maximum motor speed of 200 rpm for 1st gear (gr 3.38), the signal from motor speed sensor (MRS) to the controller and the signal from the controller to the relay which is connected with EMA and controller and with ampere booster amplifier. In the 2nd gear EMA the connection is off. While, if the engine speed increases above 200, the second EMA starts working and the first gear EMA connection turns off. Thus, the AEM2SGB works and ensure the EV operation regardless of road profile.

Keywords EV propulsion · Autonomous mode · FLC · Energy boosting

14.1 Introduction

The AEM-2SGB's control strategy has been implemented with a fuzzy controller to make the right decisions based on wheel speed sensor responses. The main areas of operation of the fuzzy controller are for shifting gears, taking into account the dynamic requirements of the vehicle. A fuzzy logic controller (FLC) is the preferred EMA control of the proposed study to control the AEM-2SGB to meet the vehicle load requirements. In this study, the FLC is used to control the power supply to the EMA to base on motor RPM and slope grade to maintain the desired traction RPM of the vehicle under various road conditions. The behavior of the fuzzy controller is examined by making use of the block diagram shown in Fig. 14.1.

A. Rahman et al., *Autonomous Driveline System for Electric Vehicle*, Green Energy and Technology, https://doi.org/10.1007/978-981-96-8418-2_14

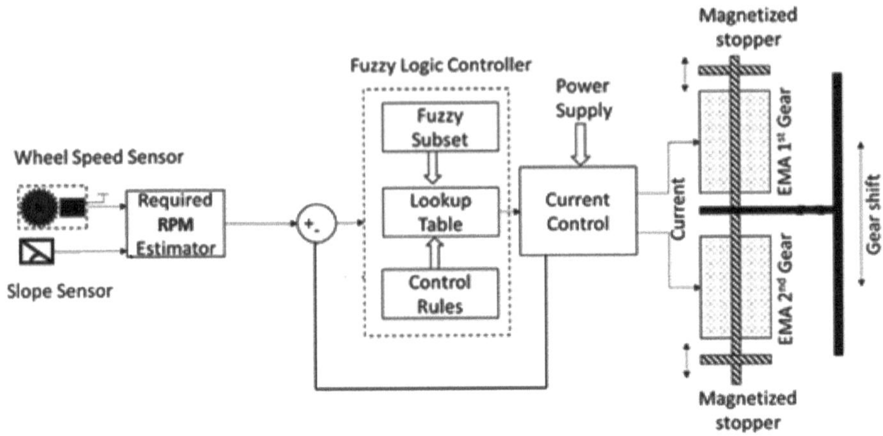

Fig. 14.1 Fuzzy logic controller block diagram for AEM2SGB [1]

14.2 Control Strategy of 2-Speed Gearbox

Fuzzy control system is used for the actuation force development of the EMA by supplying current in the range of 7–16 A based on the wheel speed sensor (WSP) and slope sensor response. The focus of the controlling system is to make the AEM-2SGB an automatic transmission. The fuzzy logic controlled variable is considered as rate at the change of RPM for traction and cruising and change of slope gradient based on the change of RPM while the output variable is considered as power supply (current) to the EMA. The design objective of the fuzzy controller model is to maintain the current supply maximum 17 A to the electromagnetic actuator to actuate the gear shifter from 1st gear to the 2nd gear and vice versa and make the gearbox is an automatic mode. Three fuzzy variables characterize the behavior of the electromagnetic actuator.

Variables of FLC

Three fuzzy variables characterize the behavior of the AEM2SGB system in Fig. 14.2.

Input variables: Motor RPM and slope gradient.

Output variables: Supply current.

For $x_1 =$ Motor RPM (linguistic variable RPM), the universe of discourse is,

$$X = [50, 350]$$

For $x_2 =$ slope gradient from slope sensor (linguistic variable slope), the universe of discourse is,

$$Y = [0, 0.5]$$

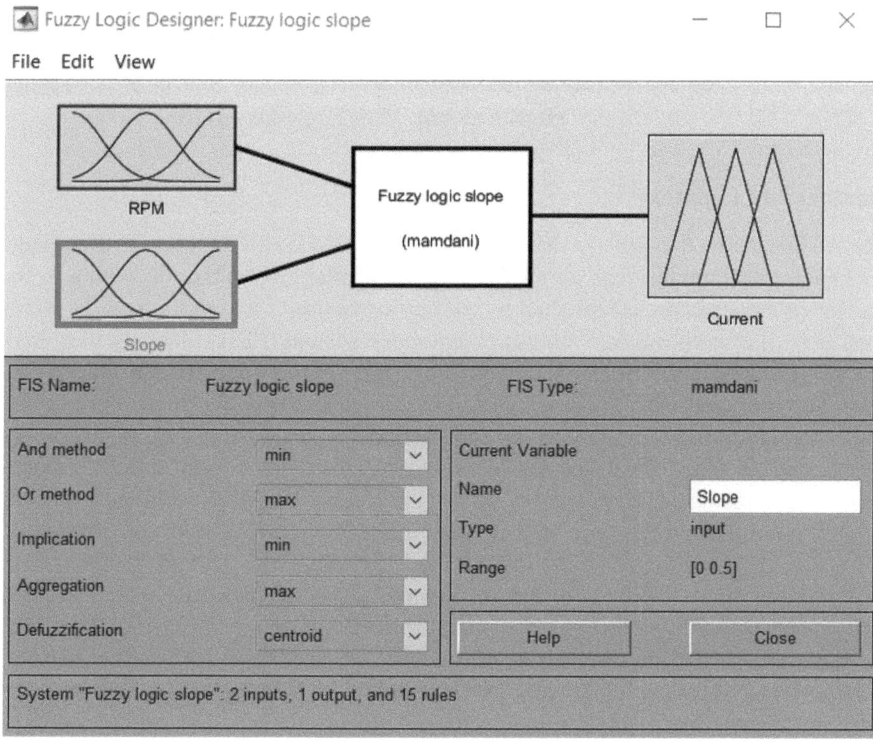

Fig. 14.2 Structure of the adaptive fuzzy logic expert system (FLES)

For z = Current flow from battery (linguistic variable current), the universe of discourse is,

$$Z = [2, 16]$$

Frames for Fuzzy Variables of AEM2ST

The frame of RPM is [RPM, £RPM, X, $\mu_{£RPM}$]. Where, £RPM is known as the set of linguistic value that RPM can take. By using the following fuzzy subsets to introduce the linguistic values,

Traction mode (LOW), Average speed (AVG), Cruising mode (HIGH)

$$£RPM = \{\text{Traction mode, Average speed, Cruising mode}\}$$

The frame of slope gradient is [Slope, £Slope, Y, $\mu_{£Slope}$],

$$£Slope = \{XSmall, Small, Medium, Large, XLarge\}$$

The frame of linguistic variable is [Current, £Current, Z, $\mu_{£Current}$],

$$£Current = \{Very\ Low, Low, Good\ Amount\\ High\ Amount, Very\ High\ Amount\}$$

Membership Function

The aforementioned linguistic variables and values provide the expert with a language for expressing human ideas about the process of making control decisions in the context of the structure established by the choice of inputs and outputs of the fuzzy controller. It also shows how the fuzzy inference system (FIS) is usually triggered. The inference process usually involves two steps: (1) the linguistic variables of all rules are compared with the controller input to determine which rules apply to the current situation, and (2) the inferences (which control actions should be taken) are determined by using the rules that have been defined for current use. Inferences are characterized by a fuzzy set (or sets), which represents the confidence that the input to an object should take on different values. Determination of the conclusion is considered when the rules that are applied to decide what should be the current to the installation. To this end, the recommendations of each rule are considered independently. Later, all recommendations from all rules are combined to determine the flow rate for the installation. At this stage, the degrees of truth (μ) of the rules were determined for each rule using the minimum and then by taking the maximum between the working rules.

The input variables RPM and slope gradient, as well as the output variable current, are restricted to positive values. In Table 14.1 and Fig. 14.3, we show a possible assignment for ranges and triangular membership functions for RPM. The optimization of these assignments is often done through trial and error for achieving optimum performance of FLC.

The following recommendations have been taken into mind in determining the range of fuzzy variables associated with crisp inputs.

Table 14.1 Fuzzy variable ranges for RPM, slope and current

Crisp input range (RPM)	Fuzzy variable (RPM)	Crisp input range (slope %)	Fuzzy variable (slope)	Crisp output range (current)	Fuzzy variable (current)
0–200	Low	(−)10–10	XSmall	0–5	Very low
100–300	Avg	5–20	Small	2–9	Low
200–350	High	13–37	Medium	5.5–12.5	Good amount
		30–45	Large	9–16	High amount
		40–60	XLarge	13–19	VHigh amount

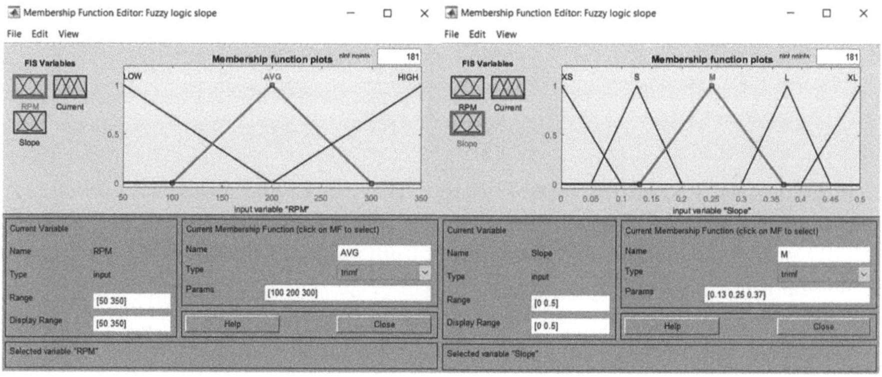

Fig. 14.3 Fuzzy membership function for RPM, slope and current

- Symmetrically distributed fuzzy values of the universe of discourse.
- Use an odd number of fuzzy sets for each variable, so it was assumed that some set is in the middle.
- It is quite typical to use 5 to 7 sets. Adjacent sets overlap (typically from 15 to 25%).

Each rule generates one output function. For more than one rule, the output functions of each rule are aggregated into one output function, and then defuzzified into a single numeric value. Fuzzification of RPM, slope gradient and current is performed using the following functions. These formulas are determined using measurement values

$$\text{Speed}(X) = \begin{bmatrix} X & \text{RPM}_{\min} \leq X \leq \text{RPM}_{\max} \\ 0 & \text{Otherwise} \end{bmatrix} \tag{14.1}$$

$$\text{Slope}(Y) = \begin{bmatrix} Y & \text{Slope}_{\min} \leq X \leq \text{Slope}_{\max} \\ 0 & \text{Otherwise} \end{bmatrix} \tag{14.2}$$

$$\text{Current}(Z) = \begin{bmatrix} Z & \text{Current}_{\min} \leq Z \leq \text{Current}_{\max} \\ 0 & \text{Otherwise} \end{bmatrix} \tag{14.3}$$

In Eqs. (14.1–14.3), X is the first input variable (**RPM**), Y is the second input variable (**Slope**), and Z is the first output variable (**Current**). Prototype triangular fuzzy sets for the fuzzy variables are set up using MATLAB FUZZY Toolbox. These membership functions assist in changing numeric variables into linguistic terms. Within the framework of the present study, the following rules are used to create the input and output membership functions.

There is minimal information for a particular variable and this variable is a responsive indicator, the range of values is divided into numerous identical Triangular membership functions [2],

$$\mu_{\text{triangle}}(x, c_1, c_2, c_3) = \left\{ \begin{matrix} 0; & x \le c_1 \\ \frac{x-c_1}{c_2-c_1}; & c_1 < x \le c_2 \\ \frac{c_3-x}{c_3-c_2} & c_2 < x < c_3 \\ 0; & x \ge c_3 \end{matrix} \right\} \tag{14.4}$$

In Eq. (14.4), x is the any input or output of the input and output set variable, c_1, c_2 and c_3 are the coefficients of membership functions for the described input and output variables. Now, the firing strength of truth values of membership function for individual rules is obtained as.

$$\alpha_{\text{no\_rule}} = \min\{\mu_{\text{linguistic\_variable}}(\text{RPM}), \mu_{\text{linguistic\_variable}}(\text{Current})\} \tag{14.5}$$

where, $\alpha_{\text{no\_rule}}$ is the firing strength of individual rule. Therefore, the firing strength of membership function truth value of output for corresponding rule. There are many rules that in many conditions, for a system whose output is fuzzy, it could be simpler to receive a crisp decision if the output is represented as a single scalar quantity. This conversion of a fuzzy set to single crisp output in order to take action is called defuzzification. This process is necessary because crisp control action is required to actuate the power control for solenoid operation. The final output "Current" through change in RPM and Slope denoted by "$\text{Current}^{\text{crisp}}$," that best represents the conclusions of the fuzzy controller is presented with the implied fuzzy sets. There are many approaches to defuzzification in the literature. Such as the "centre of gravity" (COG) defuzzification method can be used for combining the recommendations represented by the implied fuzzy sets from all the rules [2, 3]. However, an important method for defuzzification named "centre of average" (COA) is considered here, which is known to yield superior result. The output membership values are multiplied by their corresponding singleton values (A_i) and then are divided by the sum of membership values to compute $\text{Current}^{\text{crisp}}$ as follows:

$$\text{Current}^{\text{crisp}} = \frac{\sum A_i \mu_i(\text{Current})}{\sum \mu_i(\text{Current})} \tag{14.6}$$

where, A_i is the subinterval position of ith universe in abscissa under membership function, and μ_i is the firing strength of truth values of rule i for output.

14.3 Creation a Rule Base Model

Now that we have the inputs and outputs in terms of fuzzy variables, we only need to specify what actions to take and under what conditions; those. We need to build a set of rules that describe how the FLC works. These rules are usually in the form of IF–THEN rules, which are developed by a human expert (heuristic). The study of the linguistic variable on the electromagnetic actuator force development can be presented as:

A. The vehicle is in start mode the RPM sensor counting revolution per minute,
   - If the WSP response is LOW but slope position is S and controller action of current flow is GA then the F_{em} develops will be good amount. In this case, the gear engagement (1st gear) from neutral position taken shorter time. If current flow is GA then easy to engage the gear. In slope position M or L current flow needed HA or VHA to engage the gear from neutral to 1st gear.
   - If the WSP response is AVG and slope is M controller action on current flow is GA then the F_{em} develops will be good amount because vehicle already develop acceleration so that the gear engagement will be occur in shorter time. However the XL will be incurred just after the L and the current will be very high amount and the gear engagement will be in very short time from 1st gear to 2nd gear or vice versa.
B. The similar effect can be concluded for the fuzzy by Table 14.2.

Table 14.2 shows the fuzzy performance on the electromagnetic force development for the vehicle in traction and cruising or higher speeds.

Prototype triangular fuzzy sets for the fuzzy variables are set up using MATLAB FUZZY Toolbox as shown in Fig. 14.4.

Figure 14.5 shows the fuzzy membership function. In Fig. 14.5a, when input RPM $= 150$, μ_{LOW} $(150) = 0.33$, μ_{AVG} $(150) = 0.50$, all other membership functions are off (i.e., their values are zero). Therefore, the proposition RPM Sense is LOW is satisfied to a degree of 0.33, the proposition RPM is AVG is satisfied to a degree of 0.50. Figure 14.5b, when input slope grade $= 17\%$, μ_{S} $(17\%) = 0.4$, μ_{M} $(17\%) =$ 0.33; all other membership functions for RPM and slope are off. According to crisp input RPM $= 150$ and slope $= 17\%$, four rules fire. μ_{premise} for the four rules fire firing strength is calculated as,

- $\mu_{RPM*current} = \min(0.33, 0.4) = 0.33$
- $\mu_{RPM*current} = \min(0.33, 0.33) = 0.33$
- $\mu_{RPM*current} = \min(0.50, 0.4) = 0.40$
- $\mu_{RPM*current} = \min(0.50, 0.33) = 0.33$

Table 14.2 Fuzzy performance at different stage of RPM and slope

(Speed sensor) (RPM)	Controller action on current flow to EMA				
	Slope gradient				
	XS	S	M	L	XL
LOW (N-1)	L	GA	HA	VHA	VHA
AVG: (1–2)/(2–1)	VL	L	GA	HA	HA
HIGH (N-2)	VL	VL	L	GA	GA

GA = Highly significant, L/HA = Significant, VL/VHA = Less significant

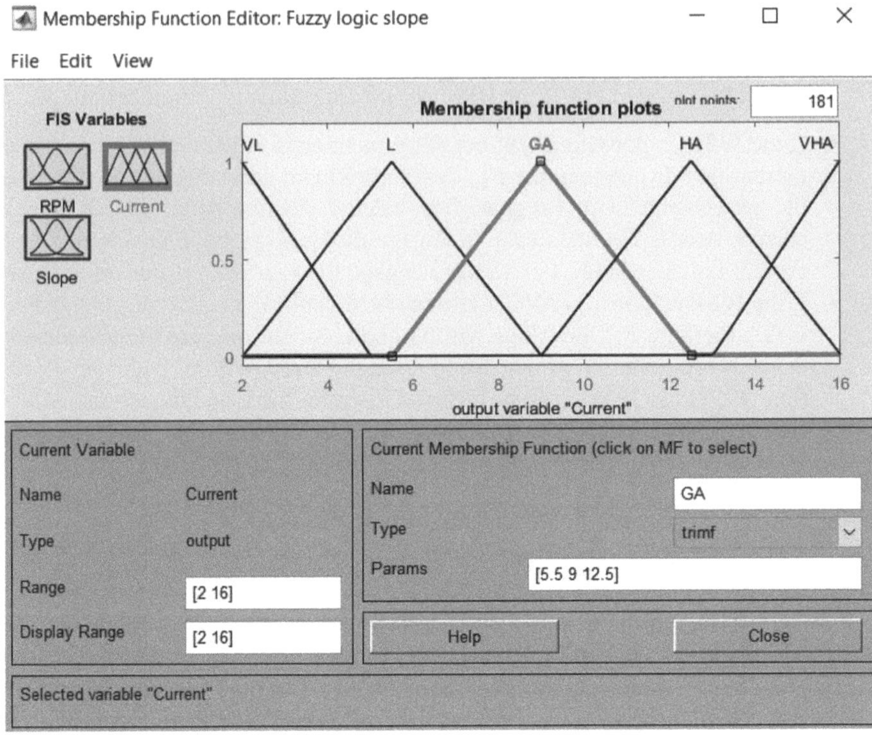

Fig. 14.4 MATLAB fuzzy tool

Figure 14.6 shows the 3-D surface diagram of the fuzzy logic model output due to the variation of inputs variable. In FLC, we don't need to specify all cells in the matrix. No entry means no action is being taken. For example, in the row for HIGH = no HA/VHA, no action is required because current is already at or above the targeted current. The AEM2SGB equipped laboratory scale gearbox performance has been investigated by theoretically and experimentally. Theoretically, it has been investigation by mathematical equations, MATLAB, SOLIDWORK-2019 and Fuzzy approach. Experimentally, it was investigated in Laboratory by developing an experimental simple with DC power supply for motor and ampere booster for developing EMF at the actuator. Detail discussion has been made in Chap. 4.

The AEM-2SGB's automatic control strategy has been implemented with a Fuzzy controller to make the right decisions based on wheel speed sensor responses. The main areas of operation of the Fuzzy controller are for shifting gears, taking into account the dynamic requirements of the vehicle. A fuzzy logic controller (FLC) is the preferred for EMA control of the proposed AEM-2SGB. In this study, the FLC is used to control the power supply to the EMA based on motor RPM and slope grade to maintain the desired acceleration and speed of the vehicle under various road conditions. Figure 14.6 shows the actuation force development of the EMA by

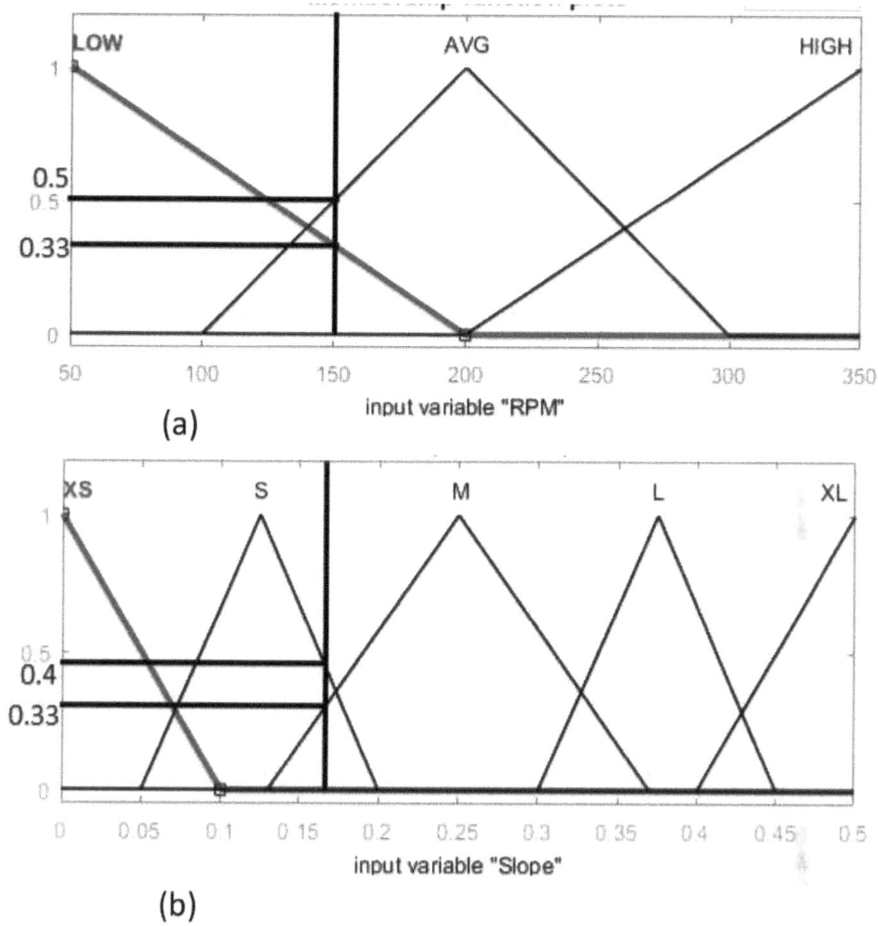

Fig. 14.5 Fuzzy membership function of the MRPM and SS [4]

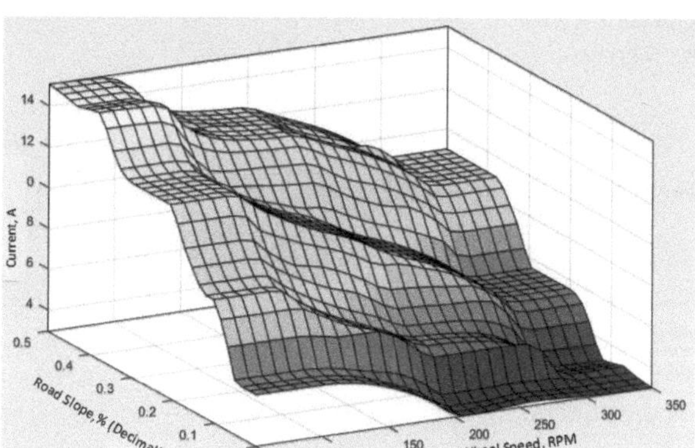

Fig. 14.6 Output variable of the FLC based on vehicle wheel speed (RPM) and slope (decimal) [4]

supplying current in the range of 7–16A based on the motor speed sensor (MSP) and slope sensor (SS) response. The design objective of the fuzzy controller model is to maintain the current supply maximum 17A to the electromagnetic actuator to actuate the gear shifter from 1st gear to the 2nd gear and vice versa.

14.4 Determination of Rule Fire

The singleton fuzzification, combining the fuzzy sets that were created by the fuzzification process to represent the inputs with the premise membership functions for the rules, is particularly simple. It simply reduces to computing the membership values of the input fuzzy sets for the given inputs.

From decision Table 14.3, we observed that the rules that have the premise terms ($\mu_{\text{premise}} > 0$, for all other rules, $\mu_{\text{premise}} = 0$):

- RPM is LOW and slope is Small.
- RPM is LOW and Slope is Medium.
- RPM is AVG and slope is Small.
- RPM is AVG and Slope is Medium.

Determine the applicability of each rule is called “firing”. We say that a rule is fired at time t if its initial membership value at time t is greater than zero. The inference mechanism attempts to determine what the rules are triggered to find out what the rules are relevant to the current situation. The total framework integrates the recommendation of all the rules in order to come to a common conclusion.

According to crisp input RPM = 150 and slope = 17%, four rules fire. μ_{premise} for the four rules (refer to Table 4.3), which amounts to firing strength in each case, can be calculated (from Eq. 3.56) as follows:

- $\mu_{RPM*current} = \min(0.33, 0.4) = 0.33$
- $\mu_{RPM*current} = \min(0.33, 0.33) = 0.33$
- $\mu_{RPM*current} = \min(0.50, 0.4) = 0.40$
- $\mu_{RPM*current} = \min(0.50, 0.33) = 0.33$

Table 14.3 Induced decision table

(Speed sensor) (RPM)	Controller action on current flow to EMA				
	$\mu_{XS} = 0$	$\mu_S = 0.4$	$\mu_M = 0.33$	$\mu_L = 0$	$\mu_{XL} = 0$
$\mu_{LOW} = 0.33$	0	GA	HA	0	0
$M_{AVG} = 0.50$	0	L	GA	0	0
$M_{HIGH} = 0$	0	0	0	0	0

Output Recommendation by Each Rule

It is found that only four cells contain nonzero terms and these cells are considered active. The active cells correspond to the rules [4, 5]:

- RPM is LOW and slope is small: p_1
- Set, current to Good Amount: q_1

i. If p_1 then q_1, μ_{premise} (1) = 0.33. $\mu_{\text{inference}}$ (1) is obtained by 'chopping off' the top of μ_{GA} function of the output variable current, as shown in Fig. 14.7a.

   - RPM is LOW and slope is Medium: p_2
   - Set, Current to High Amount: q_2

ii. If p_2 then q_2, μ_{premise} (2) = 0.33. $\mu_{\text{inference}}$ (2) as shown in Fig. 14.7b.

   - RPM is AVG and Slope is Small: p_3
   - Set current to Low Amount: q_3

iii. If p_3 then q_3, μ_{premise} (3) = 0.40. $\mu_{\text{inference}}$ (3) as shown in Fig. 14.7c.

   - RPM is AVG and slope is Medium: p_4
   - Set current to Good Amount: q_4

iv. If p_4 then q_4, μ_{premise} (4) = 0.33. $\mu_{\text{inference}}$ (4) as shown in Fig. 14.7d.

The recommendations of each rule are considered independently. Later, all recommendations from all rules are combined to determine the flow rate for the installation. At this stage, the degrees of truth (μ) of the rules were determined for each rule using the minimum and then by taking the maximum between the working rules. The input variables RPM and slope gradient, as well as the output variable current, are restricted to positive values. It is concluded that different crisp measurements of RPM and Slope, there are different values of μ_{premise} and $\mu_{\text{inference}}$ function are obtained. Combining all of the recommendations of all the rules, the control actions of fuzzy has been determined. This is done by aggregating (combining operation) of fuzzy sets output by using the Eq. (14.6).

$$\text{Current}^{\text{crisp}} = \frac{\sum A_i \mu_i}{\sum \mu_i}$$
$$\text{with } i = 1, 2, 3, \text{ and etc.}$$

Aggregation of fuzzy set obtained by drawing all derived fuzzy sets on one axis, shown in Fig. 14.7d. This fuzzy set is the desired control action of the controller.

Rule view in fuzzy representing in Fig. 14.8, gives the corresponding output (Current) value for the inputs RPM and slope crisp value of which is 8.75 A. Rule 2, 3,7 and 8 as marked in Fig. 14.8 are fired to represent the system situation for that instant and give the desired output to execute the system.

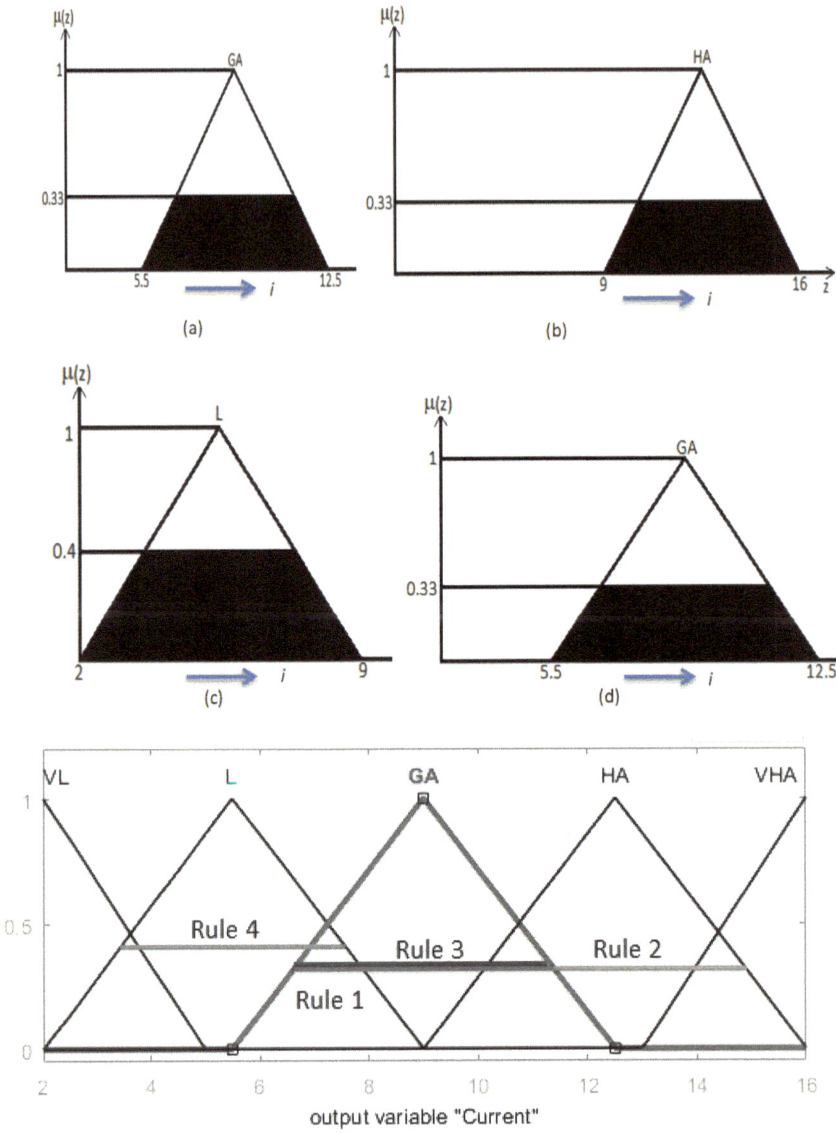

Fig. 14.7 Inference for each rule [4]

This calculation shows the close crisp value for output as shown in Fig. 14.8. However, there is some deviation with result because of using different method of solution which is Centre of area (COA) instead of Centre of gravity (COG) method.

We use motion RPM sensor (MRS) for collecting the data from motor and send it to the controller. Arduino mega 2560 controller has been used for control the gear shifting mechanism. From the FLC rule viewer, we get method and apply to the

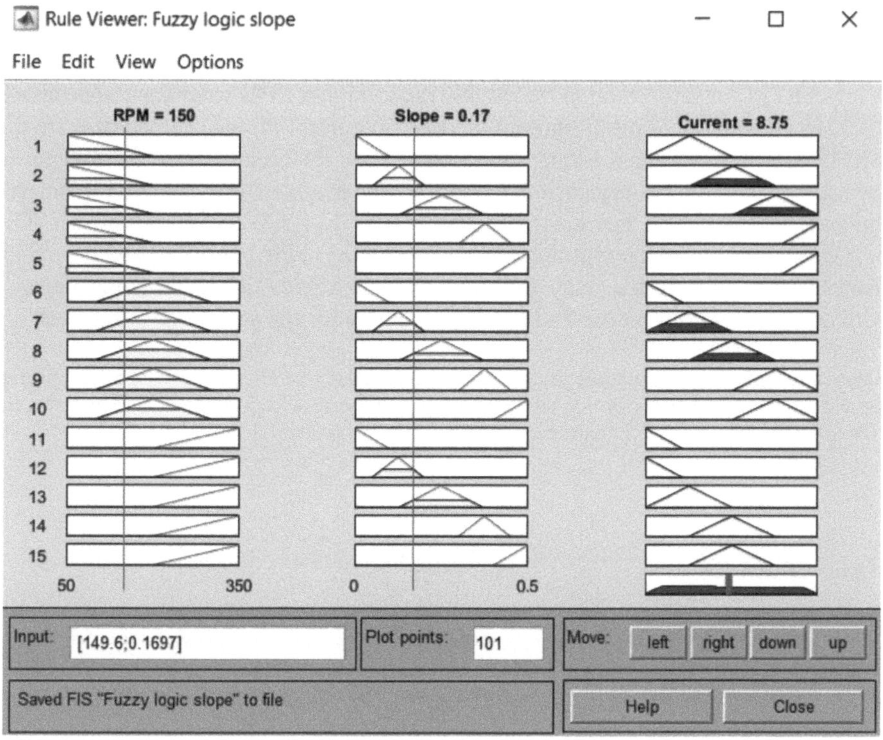

Fig. 14.8 Rule viewer of FIS [1, 5]

arduino controller. For lab scale product, only input variable is RPM from the motor to the controller. Here in controller, we use two types of RPM variable. One is low RPM value for traction (below 200) and High RPM value for cruising (above 200). Maximum motor RPM at 16A is 350 that's why when RPM is below 200 MRS signal to the controller and controller signal to the relay which is attached with EMA and controller and also with ampere booster, signal for 1st gear engagement by passing current on it. In the meantime 2nd EMA connection is off. After that when motor speed is creased to above 200, then 2nd EMA start working and 1st EMA connection is off. The EMA works when either of the motion speed sensor get the reading that satisfy the working condition. Then, the controller will send signal to the relay to pass the current from battery to get the appropriate force by the EMA to engage the gear automatically.

14.5 Conclusion

A lower RPM value for traction (below 200) and a higher RPM value for cruise (above 200) are considered in programming the FLC to control the EMAs and prepare the AEM2SGB to meet the working conditions.. Then the controller sends a signal to the relay to pass the battery current to get the proper force from the EMA to engage the gear automatically. For maximum motor speed of 200 RPM for 1st gear (GR 3.38), the signal from the motor speed sensor (MRS) to the controller and the signal from the controller to the relay which is connected with EMA and controller and with current amplifier, in the 2nd EMA connection for the second gear goes off.

Acknowledgements The authors are grateful to the Ministry of Higher Education Malaysia to finance this project under the Prototype Research Grant Scheme (PRGS) 2022. The authors are also grateful to the laboratory technician helping to develop this project.

References

1. Hasan, N. (2021). *Development of autonomous electromagnetic two speed gearbox for electric vehicle*. A master thesis, IIUM Library.
2. Passino, K. M., Yurkovich, S., & Reinfrank, M. (1998). *Fuzzy control* (Vol. 42, pp. 15–21). Addison-wesley.
3. Ivanov, V., Shyrokau, B., Augsburg, K., & Algin, V. (2010). Fuzzy evaluation of tyre–surface interaction parameters. *Journal of Terramechanics, 47*(2), 113–130.
4. Rahman, A., Sany, I. I., & Nurul, H. (2022). A fuzzy-logic-controlled two-speed electromagnetic gearbox for electric vehicles. *International Journal of Technology, 13*(2), 167–179.
5. Rahman, A., & Hasan, N. (2023). Autonomous electromagnetic two-speed gearbox for electric vehicle. *International Journal of Automotive Technology* (In press).

Chapter 15
Design Optimization of AEM2ST

Abstract Two-speed transmission develops higher torque in first gear and faster speeds in second gear than the single-speed transmission. The first gear ratio provides more torque, allowing the car to maintain traction in starting mode and slope climbing, while the second gear ratio allows the vehicle to reach a maximum speed of 120 km/h with an engine speed of 5,500 rpm. The maximum torque that can be achieved with the first gear (gear ratio 3.38) is approximately 2,380 Nm at a speed of 60 km/h, while the second gear forward (gear ratio 1.39) is of approximately 120 km/h. The gear changing operation has been carried out with electromagnetic actuators controlled by a fuzzy logic controller for efficient response depending on the road profile.

Keywords Gear shifting · Fuzzy logic controller · Electromagnetic actuator · Vehicle's propulsion

15.1 Introduction

The efficiency can be as high as 90% in a very compact size. The first and second gears are helical gears that operate at high speeds on a large scale and provide less noise and wear and longer lifespan. Reverse gear is a spur gear used to slide the gears while the idle gear slides on the other two gears. Gear-shifts will be handled by the synchronization system to switch between the first and second gears. Synchromesh is now used in most manual transmission systems because of its ability to absorb shocks, provide smoother gear changes and better acceleration. Switching operations are performed by electromagnetic actuators, which are based on controllers and sensors.

Gears z1 to z6 helical gear used as forward gears and z7 to z9 spur gear used as a reverse gear. Gear teeth with a helix angle distribute the load and stress to other teeth due to the simultaneous engagement of several gear teeth.

A. Rahman et al., *Autonomous Driveline System for Electric Vehicle*, Green Energy and Technology, https://doi.org/10.1007/978-981-96-8418-2_15

Gear z3 and z5 have a feature in which it is attached to the engagement ring for a synchronizer. Gear z7–Gear z9 will be a spur gear allowing the idler gear to slide in reverse gear as shown in Fig. 15.1.

For this design, the input shaft does not use the output transmission line (gear ratio 1:1), so the input shaft and output shaft will be partially processed. The prototype will also separate the input and output shafts to avoid confusion and clearly show how it works. With relatively simple arrangements, such as shaft transmissions with fewer gears, transmission noise can be reduced for a longer lifespan because in the gearbox we use lubrication oil.

The function of the synchronizer is to match the speed of rotation of the ring with the output shaft or gear cone. It consists of several main components, like synchronizer hub, slip, locking ring, clutch gear and engagement ring shown in Fig. 15.2.

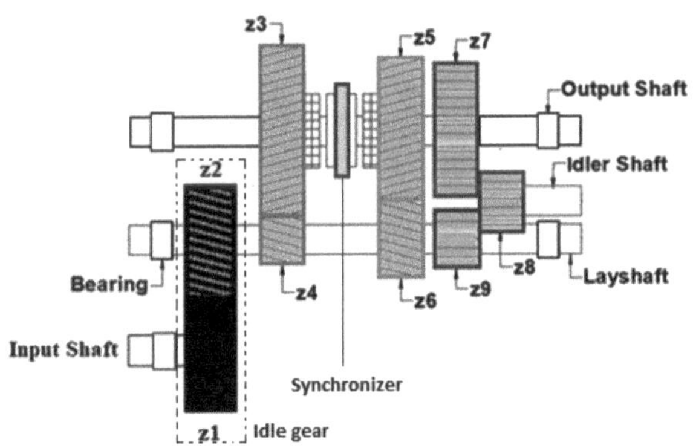

Fig. 15.1 Gear transmission sketch [1]

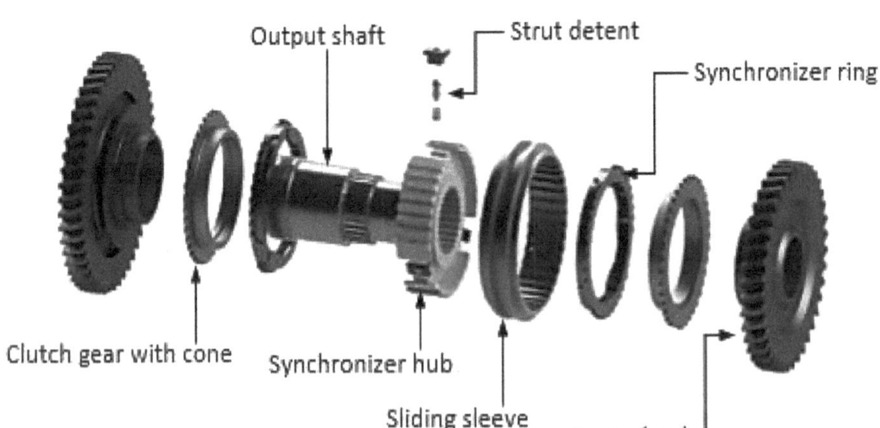

Fig. 15.2 Single cone-type synchronizer [1]

Table 15.1 Gear ratio for the vehicle in traction in hilly track

Vehicle speed (m/s)	Slope (°)	Tractive force, F_R (N)	Torque requirement, T_w (N m)	Lowest gear ratio
5	0	244.76	56.29	0.13
	5	1952.33	449.04	1.07
	10	3645.12	838.38	1.99
	15	5310.23	1221.35	2.90
	20	6935.00	1595.05	3.79
	25	8507.04	1956.62	4.65
	30	10,014.39	2303.31	5.47
10	0	273.42	62.89	0.15
	5	1981.00	455.63	1.08
	10	3673.79	844.97	2.01
	15	5338.90	1227.95	2.92
	20	6963.66	1601.64	3.80
	25	8535.70	1963.21	4.66
	30	10,043.05	2309.90	5.49
15	0	321.20	73.87	0.18
	5	2028.77	466.62	1.11
	10	3721.56	855.96	2.03
	15	5386.67	1238.94	2.94
	20	7011.44	1612.63	3.83
	25	8583.48	1974.20	4.69
	30	10,090.83	2320.89	5.51

15.2 Classical Calculation

This section presents the calculation of the vehicle traction performance as shown in Table 15.1, in terms of gear ratio, traction and cruising force, wheel torque and vehicle speed. The torque of the motor will be mentioned to the power graph of HPEVS AC Electric Motor Drive Systems (AC-50/AC-51) as shown in Fig. 15.3.

15.3 Gear Estimation

Case 1: Gear Ratio for Traction

Vehicle dynamics performance for different gear ratios are shown in Fig. 15.4. When car in traction mode, due to air resistance traction force is different in various traction speed. When the velocity of the car increasing, aerodynamic force also increasing.

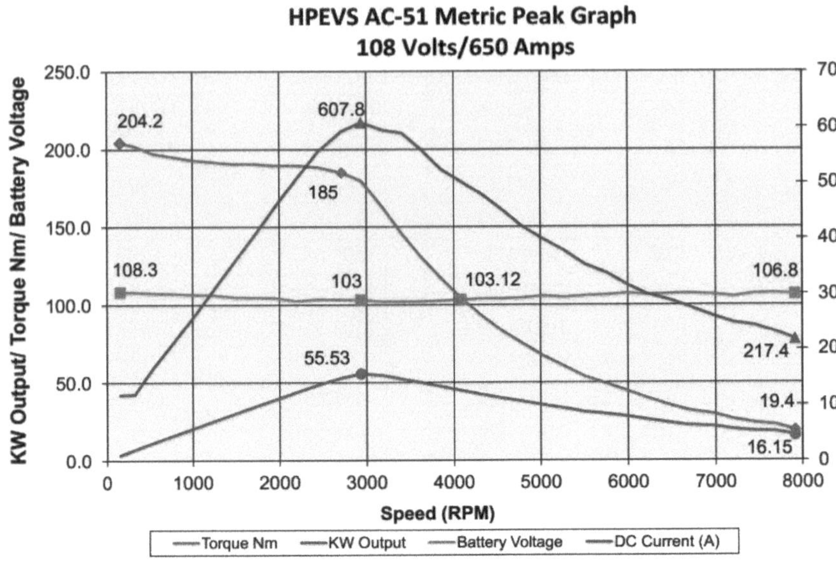

Fig. 15.3 AC51 at 108 V/650 Amps

Then when car is going to cruising mode traction force getting decreasing due to constant acceleration of a car. The lowest gear ratio based on the initial mode of road conditions, it will usually be for the average slope of about $\theta = 20°$ with a value of the gear ratio between about 3.24 and for starting the car we need higher traction force with higher gear ratio 3.9 for propel the vehicle.

15.4 Case 2: Gear Ratio: Cruising

If a car maximum speed is 120 km/h, assuming $a =$ constant for cruising, here, $\theta = 0$, and wheel speed of 1061 rpm. By referring to Fig. 15.3, the motor torque produced at 5500 rpm is around 53.88 Nm, considering the wheel torque 238.31 Nm, transmission efficiency 85%, differential speed ratio 4, and the computed value of gear ratio is 1.3.

Table 15.2 shows that vehicle is required higher traction force with higher gear-ratio 3.9. However, a very high gear ratio is not very suitable for an electric vehicle as it can cause too much acceleration when the vehicle is driven from standstill. The gear ratio considered a choice of gear ratio for thrust, which is 3.5.

Figure 15.5 shows that the motor torque 200 Nm with gear ratio 3.5, the wheel torque could be developed about 7900 Nm, which is suitable for the vehicle in acceleration. For other gear ratio in different torque not acceptable for lighter gearbox. 3.5 gear ratio with motor torque 200 Nm cover the starting mode and traction mode. Although acceleration is quite low and the time required to reach a higher speed

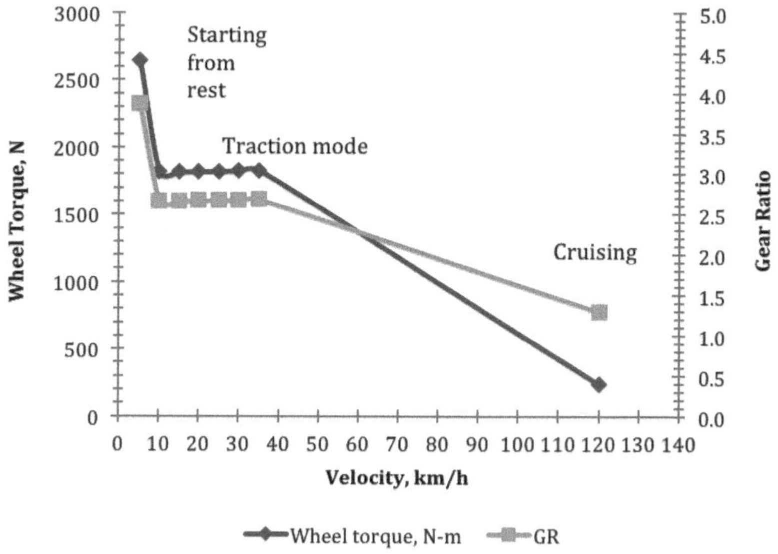

Fig. 15.4 Wheel torque and gear ratio at different speed [2]

Table 15.2 Gear-ratio for different motor torque

Motor torque	Gear ratio	Wheel torque, Nm	Traction force, N
200.00 (traction torque)	1.00	680.00	2266.67
	1.50	1020.00	3400.00
	2.00	1360.00	4533.33
	2.50	1700.00	5666.67
	3.00	2040.00	6800.00
	3.50	2380.00	7933.33
	4.00	2720.00	9066.67
160.00 (traction torque)	1.00	544.00	1813.33
	1.50	816.00	2720.00
	2.00	1088.00	3626.67
	2.50	1360.00	4533.33
	3.00	1632.00	5440.00
	3.50	1904.00	6346.67
	4.00	2176.00	7253.33
50.00 (cruising speed)	1.00	170.00	566.67
	1.50	255.00	850.00
	2.00	340.00	1133.33
	2.50	425.00	1416.67
	3.00	510.00	1700.00
	3.50	595.00	1983.33

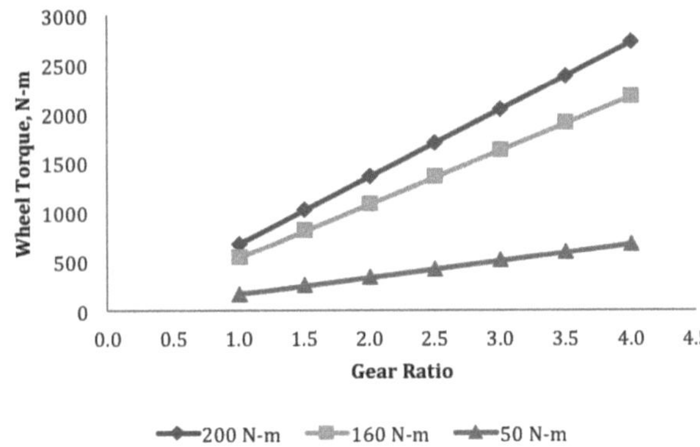

Fig. 15.5 Wheel torque versus GR in different motor torque [1]

is increased, this is more beneficial because it reduces battery power consumption, provides better driving comfort and increases driving safety. For design advantage, a smaller gearbox can be created, resulting in a smaller and lighter gearbox.

Proof: Traction Gear Ratio 3.5

It is known from the power flow of the driveline, that the power at the drive axle is developed from the power input of the motor. It can be modeled as,

$$\eta * T_m \omega_m = T_w \omega_w \tag{15.1}$$

where, T_m and T_w are the motor torque and wheel torque respectively. While, w is the transmission efficiency ω_m and ω_d is the angular speed of the motor and drive shaft, respectively. The angular speed ω_m and ω_d are estimated using the equation: $\omega_p = \frac{\omega_m}{GR}$ *and* $\omega_d = \frac{\omega_p}{4}$

By considering the ω_m equals to 500 rpm and ω_d and ω_p are computed as 14.96 and 3.74 rad per sec, respectively.

However, using the gear ratio 1.3, ω_p as 443 rad/sec and ω_d as 110 rad/sec, the vehicle cruising speed is estimated as 120 km/h.

Gear ratio that will be used as the second transmission is about 1.3. The maximum target speed does not exceed 120 km/h, therefore a gear ratio of 1.3 is permissible, as this provides a high speed and significant wheel torque.

The Car traction force is estimated as,

$$F_t = ma + f_r W \cos\theta + W \sin\theta + \left(\rho_A * C_D * A_f * \left(\frac{V}{3.6}\right)^2\right)/2 \tag{15.2}$$

Using, $f_r = 0.014$, $\theta = 0°$, $\rho_A = 1.225$ kg/m^3, $CD = 0.36$, $A_f = 2.87$ m^2 and $m =$ 1500 kg, $rw = 0.4$ m, the traction force on flat road at constant speed with 110 km/h, is computed as 836 N. The corresponding torque is computed as 335 Nm.

15.5 Vehicle Mobility

Case 1: Vehicle Cruising

Vehicle mobility on flat road with constant speed of 110 km/h, the motor is considered as 80 Nm, the corresponding gear ratio can be computed about 1.4.

Case 2: Vehicle in Traction

Vehicle start from the rest, the traction force is estimated as 1059 N the motor torque is considered as 190 Nm, the gear ratio is required can be computed about 2.0.

Case 3: Vehicle on Hilly Track Slope of 20°

Given that vehicle speed, $v = 5$ km/h, $\sin\theta = 0°$, and wheel speed of 44.2 rpm. By referring to Fig. 15.3, the motor torque produced at 500 rpm when stationary to move is around maximum torque 200 Nm for AC-51 motor (starting mode), the gear ratio required 3.9.

For a rear-wheel drive vehicle, this is an advantage in terms of increased traction force. While, a front wheel drive vehicle, the traction force will decrease during acceleration for the speed of 0 to 35 km/h in 7 s in 15° slope then we found the acceleration in different time,

$$a = \frac{dv}{dt}$$

For the vehicle speed range 0–30 km/h in 7 s, the acceleration of the vehicle is estimated as 1.38 m/s^2. While, the traction force of the vehicle is estimated as 6101 N.

For vehicle speed, $v = 10$ km/h, $\sin\theta = 15°$, and wheel speed of 88.41 rpm. By referring to Fig. 15.3, the motor torque produced at 500 rpm when stationary to move is around maximum torque 200 Nm for AC-51 motor.

Using wheel torque of 1815.85 Nm, differential speed constant 4 and transmission efficiency 85%, the traction gear ratio is estimated as 2.67 for the vehicle acceleration speed of 10 km/h in 2 s. For vehicle speed, $V = 35$ km/h, $\sin\theta = 15°$, and wheel speed of 309.46 rpm. The motor torque produced at 1000 rpm when stationary to move is around maximum torque 196 Nm for AC-51 motor and motor develops torque at the driving wheel of 1830 Nm.

Considering the vehicle operational time 7 s at speed 35 km/h, Traction gear ratio when slope angle 15° needs 2.74 and the traction gear ratio required 3.23 requires to accelerate the vehicle over the slope 20°. Table 15.3 shows an example of gear ratio calculation for transmission when the vehicle will be in drive mode on hilly track at different speeds. The vehicle will have good traction without any interruption if the maximum gear ratio is optimized to 3.3.

Force required for the vehicle to be in cruising 794 N at gear ratio 1.3 and wheel speed required 1061 rpm and/or 120 km/h.

Table 15.3 Traction mode gear ratio in relation of vehicle speed and slope angle

Velocity, km/h	Slope angle	Traction force, F_t (N)	Wheel torque, T_w (N m)	GR
5	0.00	8829.00	2648.70	3.9
10	5.00	3532.52	1059.75	1.56
	10.00	4803.24	1440.97	2.12
	15.00	6053.20	1815.96	2.67
	20.00	7272.88	2181.86	3.21
15	5.00	3537.88	1061.36	1.56
	10.00	4808.61	1442.58	2.12
	15.00	6058.56	1817.57	2.67
	20.00	7278.24	2183.47	3.21
20	5.00	3545.39	1063.62	1.56
	10.00	4816.11	1444.83	2.12
	15.00	6066.07	1819.82	2.68
	20.00	7285.75	2185.72	3.21
25	5.00	3555.04	1066.51	1.57
	10.00	4825.77	1447.73	2.13
	15.00	6075.73	1822.72	2.68
	20.00	7295.40	2188.62	3.22
30	5.00	3566.84	1070.05	1.57
	10.00	4837.57	1451.27	2.13
	15.00	6087.52	1826.26	2.69
	20.00	7307.20	2192.16	3.22
35	5.00	3580.78	1074.23	1.58
	10.00	4851.51	1455.45	2.14
	15.00	6101.47	1830.44	2.69
	20.00	7321.14	2196.34	3.23

15.6 Transmission Torque Simulation

We can determine the output torque and output speed by changing the torque and motor speed. However, the results obtained are not entirely accurate due to the relationship of the motor torque curve with the unreliable speed. As shown in Fig. 15.3, the torque speed curve of the HPEVS AC drive system (AC-51) is presented as a line graph, and each information can only be obtained with real which is inaccurate value but close enough for choose. The results obtained will be estimates and the data roughly included will reflect the desired results.

The results obtained are based on 85% transmission efficiency and presented in Table 15.4

Torque of the wheel can be estimated as,

Table 15.4 Wheel torque and speed based on motor drive systems (AC-51)

Motor torque (N m)	Wheel torque (N m)			Wheel speed (RPM)			Vehicle speed km/h		
	1st gear	2nd gear	Reverse gear	1st gear	2nd gear	Reverse gear	1st gear	2nd gear	Reverse gear
200	2380	884	1904	35.71	96.15	44.64	4.04	10.87	5.049
193	2296.7	853.06	1837.36	71.43	192.31	89.29	8.08	21.75	10.098
192	2284.8	848.64	1827.84	107.14	288.46	133.93	12.12	32.62	15.147
190	2261	839.8	1808.8	142.86	384.62	178.57	16.16	43.5	20.196
188	2237.2	830.96	1789.76	178.57	480.77	223.21	20.2	54.37	25.245
177	2106.3	782.34	1685.04	214.29	576.92	267.86	24.24	65.25	30.294
138	1642.2	609.96	1313.76	250.00	673.08	312.5	28.27	76.12	35.343
108	1285.2	477.36	1028.16	285.71	769.23	357.14	32.31	87	40.392
83	987.7	366.86	790.16	321.43	865.38	401.79	36.35	97.87	45.441
69	821.1	304.98	656.88	357.14	961.54	446.43	40.39	108.75	50.49
54	666.4	247.52	533.12	392.86	1057.69	491.07	44.43	119.62	55.539

$$T_w = T_m n_g n_d \eta_t \quad (15.3)$$

where, T_w is the wheel torque, Nm, T_m motor torque Nm, n_g gear ratio, n_d differential speed constant, and η_t transmission efficiency.

The different wheel torque and rpm in different velocity based on our 1st gear ratio 3.5 and 2nd gear ratio 1.3 and reverse gear ratio 2.8. If motor torque is considered as 190 Nm, the wheel torque is estimated for the first gear and second gear as 2261 Nm and 848.6 Nm, respectively.

Vehicle Dynamic Torque

The first gear ratio is appropriate for accelerating an electric vehicle from a stop or for driving on a slope that requires for the development of more wheel torque. The maximum torque of 200 to 190 Nm can only be achieved in the lower range of motor speeds (500–1,500 rpm). This means that the acceleration and slope phases are only applicable up to the specified rpm. The first gear ratio provides more torque, allowing the car to maintain traction in the starting mode, while the second gear ratio allows the vehicle to reach a top speed of 120 km/h at a motor speed of 5,500 rpm.

The first gear is found very suitable for propelling an electric vehicle from a static mode or for driving on a slope. The maximum motor torque of 200 to 190 Nm is recorded in the speed range of 500 to 1500 rpm. Figure 15.6 shows the acceleration performance of the EV using the first gear. While, Fig. 15.7 shows the cruising speed motor rpm maximum 120 km/h where motor RPM is 5500 rpm because maximum velocity is 120 km/h. Meanwhile, different motors give different results. To accelerate the EV to a higher speed, second gear is a better choice than first gear, is not only for the higher traction power development but for saving transmission energy as well. It is noticed that the motor unable to develop the good traction power when vehicle is in start from the rest and climbing on slope using a single gear ratio. However, it is excellent at higher to propel the EV faster.

For 3500 rpm, with 138 Nm wheel torque power and current need same for different velocity because of different gear ratio with constant voltage (108 V). The same method is used to find power, current, time and velocity at different wheel speed and torque.

Figure 15.8a shows that when car is in traction mode power and current need higher than when car is in cruising mode. In traction, power and current is increasing

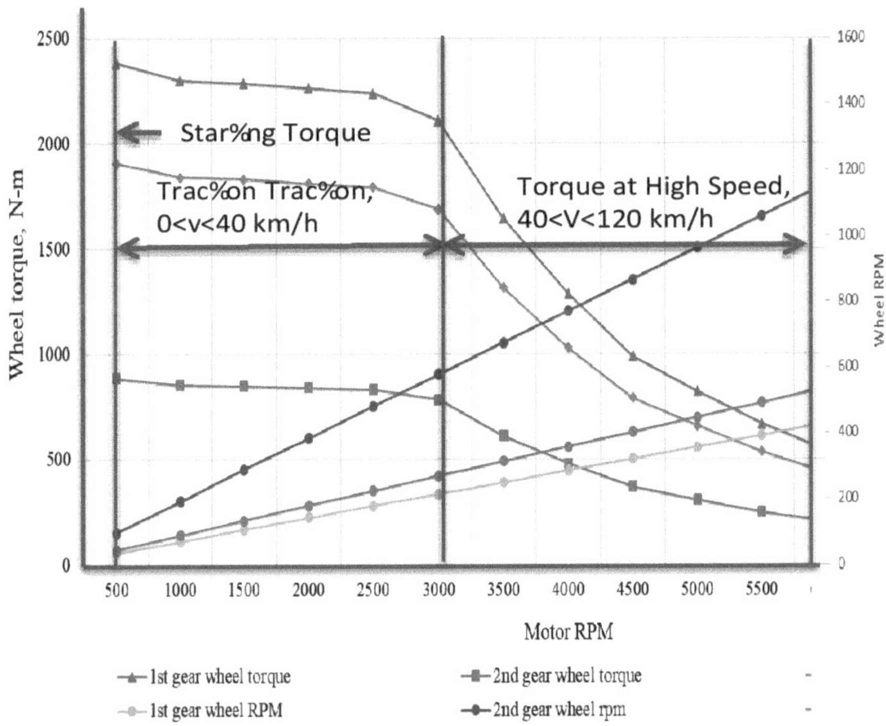

Fig. 15.6 Simulated performance of the EMA2SGB [3]

Fig. 15.7 Wheel torque and RPM [3]

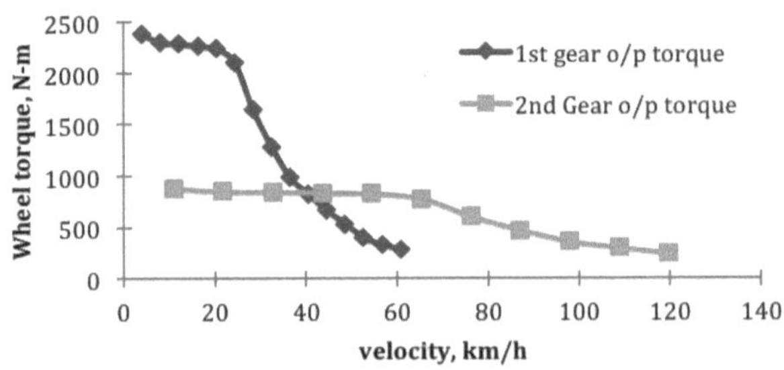

with motor rpm with maximum velocity about 60 km/h. When car is shifting to the 2nd gear with same power and current based on vehicle maximum speed of 120 km/h.

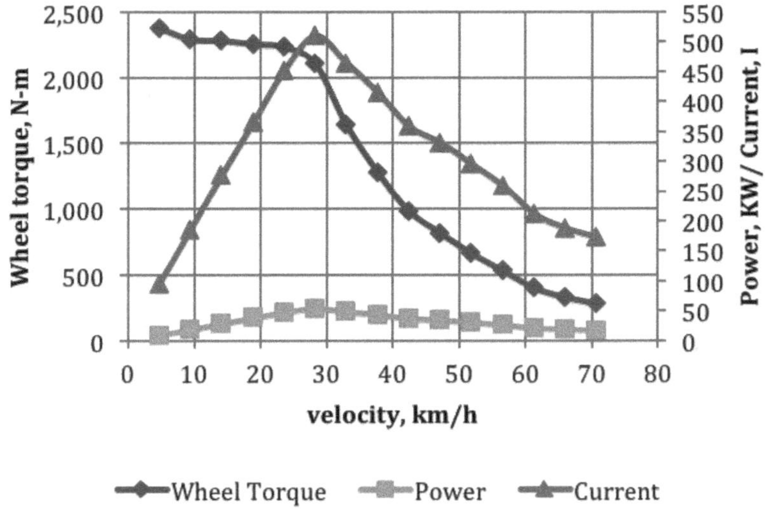

(a)

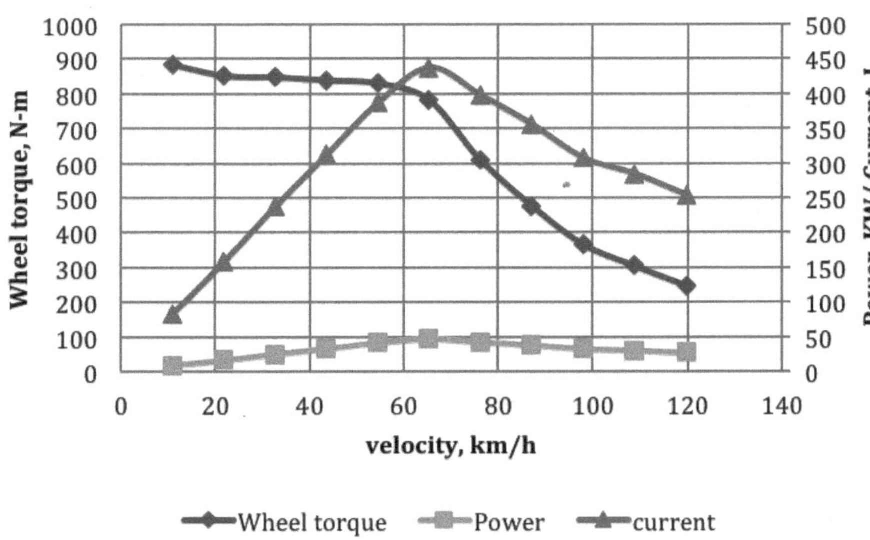

(b)

Fig. 15.8 Torque, power and current vs velocity **a** for 1st gear, **b** for 2nd gear [3]

Capacity of battery 33 KWh is equipped with our test EV, we use this for calculation how far can car go with this current based on HPEVS AC Electric-Motor Drive Systems (AC-51). For maximum speed of a car we consider 120 km/h for motor 5500 RPM and motor torque 53.88 N m. For estimating the battery charge:

$$Q = \frac{E}{V} \tag{15.4}$$

where, Q is the total charge Ah, E energy (33 kWh), i is the battery terminal voltage, t is the time for the battery to deliver the energy 305.6 Ah to the motor until the end of the store energy.

Figure 15.8b shows that when gear shifting to 1st to 2nd gear, then power and current decrease with increasing of velocity. If we see 20° slope calculation power and current need more because of more traction force needed to overcome the friction. According to HPEVS AC Electric-Motor Drive Systems (AC-51), Motor efficiency measured by,

$$\eta_m = \frac{P_o}{P_i} * 100\% \tag{15.5}$$

For the maximum speed of a car is 120 km/h, the motor efficiency is computed as 82.1% and transmission loss about 17.9%.

15.7 Comparison of Transmission Performanc

Comparison of torque characteristics of one and two speeds transmission has been made in this chapter, based on simulation result. However, the experimental results are given in the comparison of single-speed and two-speed transmission in Chap. 16.

Each transmission has considered to be powered by the HPEVS (AC-51) AC induction motor with a maximum rotational speed of 7500 rpm and maximum torque of 200 Nm. A comparison of two-speed and one-speed transmissions is shown in Fig. 15.9. These two types of transmission use the same motor of maximum torque 200 Nm and speed of 7500 rpm. It is capable of accelerating the EV to maximum speed of 120 km/h. For a single speed transmission system, the single gear ratio is obtained from the average of the gear ratios for the two-speed transmission. It shows that the transmission only develops a vehicle speed of up to 95 km/h in a single gear ratio. For the speed of 120 km/h using a single gear ratio, it loses balance of torque and acceleration and also control. On the other hand, two-speed transmission satisfies the requirements of high torque in first gear and high speed in second gear and provides more efficient gear change by a motor. Thus, the two speed transmission provide the smooth mobility of the EV with minimum transmission losses and make EV energy efficient.

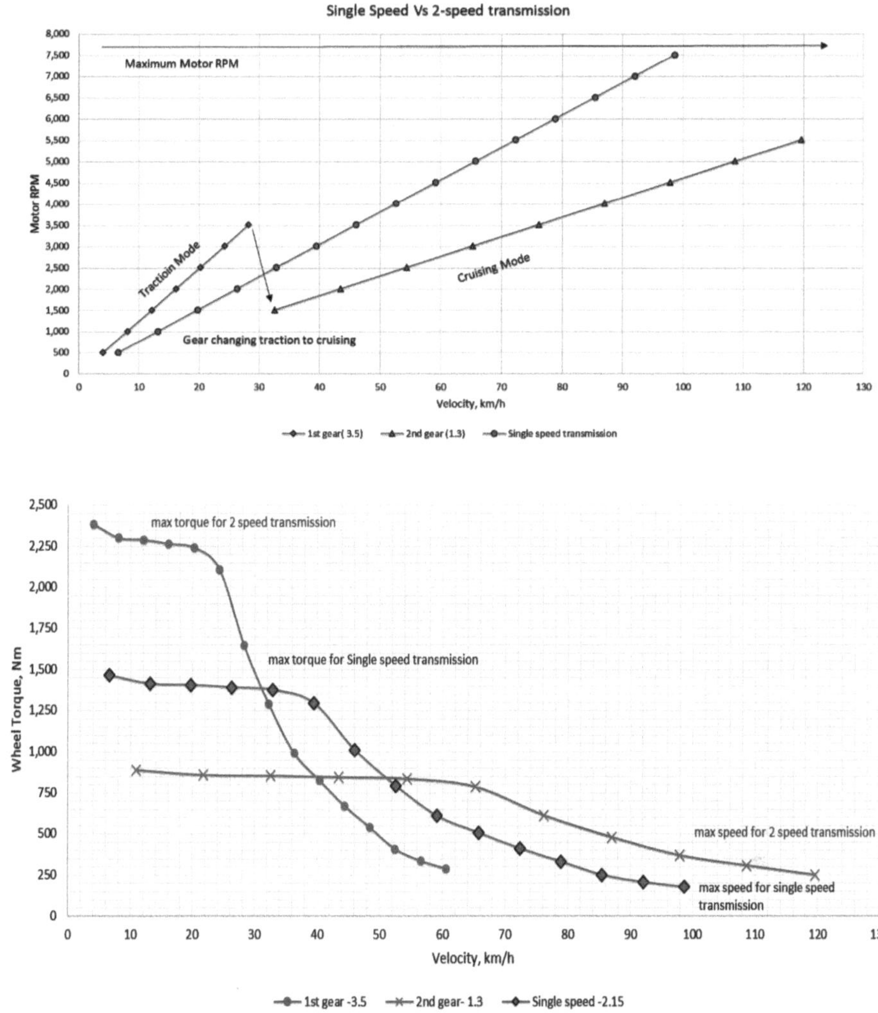

Fig. 15.9 Torque comparison single speed and 2-speed transmission [1, 3]

Figure 15.10 shows the comparison output torque and vehicle speed of the old prototype (Prototype 1), which have the first and second gear ratio of 1.842 and 0.865 while the new prototype (Prototype 2) has first and second gear ratio of 3.5 and 1.3. By comparing forward gears for both Prototype 1 and Prototype 2 as shown in Fig. 15.10, it can be seen that the torque produced by both forward gears by Prototype 2, are very low compared to the new prototype, with almost 65% reduction of torque. The maximum torque that can be achieved by the Prototype 1 is around 2380 Nm while the maximum torque produced by the Prototype 2 approximately 775 Nm [5].

The gear ratio of an EV is usually very high. Therefore, it can be deduced that the gear design of the Prototype 1, forward gear ratio could provide torque in traction

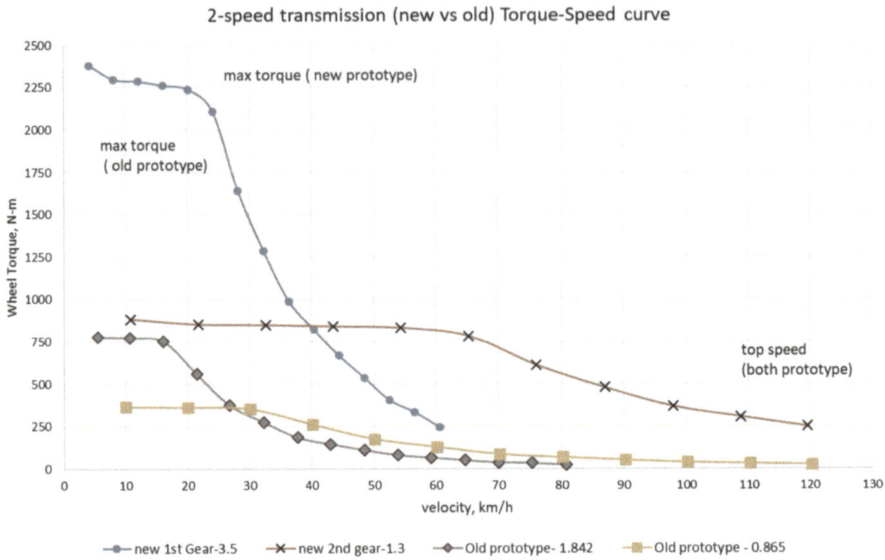

Fig. 15.10 Torque-speed comparison between Prototype 1 and Prototype 2 [1, 4]

almost twice the torque from the Prototype 2, which can provide a better acceleration and slope climbing capability. The first forward gear can propel the EV up to 60 km/h for Prototype 1 and 70 km/h for Prototype 2 and top speed that can be achieved by both second forward gear is around 120 km/h [3, 6]. It also be concluded that the Prototype 1, provides a better acceleration of an EV to reach a higher speed and better slope climbing capability.

15.8 Conclusion

The gear ratio of an EV is essential to be in high speed or cruising mode. However, it is not efficiently develop the traction torque for the vehicle to climb if the EV uses a single speed gear ratio. Therefore, lower gear ratio (3.0 to 3.5) is used full for the EV to meet the traction torque. This study has prototype an AEM2SGB with gear ratios of 1.39–3.38 to meet the vehicle propulsion torque and speed requirement.

Acknowledgements The authors are grateful to the Ministry of Higher Education Malaysia to finance this project under the Prototype Research Grant Scheme (PRGS) 2022. The authors are also grateful to the laboratory technician helping to develop this project.

References

1. Hasan, N. (2021). *Development of autonomous electromagnetic two speed gearbox for electric vehicle*. A master thesis, IIUM Library.
2. Rahman, A., & Hasan, N. (2023) Autonomous electromagnetic two-speed gearbox for electric vehicle. *International Journal of Automotive Technology* (In press).
3. Rahman, A., Sany, I. I., & Nurul, H. (2022). A fuzzy-logic-controlled two-speed electromagnetic gearbox for electric vehicles. *International Journal of Technology*, *13*(2), 167–179.
4. Venkatesh, B., Vattikuti, S. P., & Prasad, S. D. (2014). Investigate the combined effect of gear ratio, helix angle, facewidth and module on bending and compressive stress of steel alloy helical gear. *Procedia materials science*, *6*, 1865–1870.
5. Hayt, W. H., & Buck, J. A. (2006). *Engineering electromagnetics* (7th edn). McGraw-Hill International Edition.
6. Un-Noor, F., Padmanaban, S., Mihet-Popa, L., Mollah, M. N., & Hossain, E. (2017). A comprehensive study of key electric vehicle (EV) components, technologies, challenges, impacts, and future direction of development. *Energies*, *10*(8), 1217.

Chapter 16
Autonomous Transmission and Vehicle Mobility

Abstract Through fast actuation response and a discernible decrease in transmission loss due to precise gear position maintenance without interruption or whining noise, AEM2SGB improves vehicle acceleration gear shift time. Its smaller, lighter transmission, rapid gear changes, and superior EV mobility regardless of road profile give it advantages over manual and automatic transmissions. Presenting the electromagnetically driven 2-SGB's performance a prototype designed to guarantee the precise gear ratio linked to speedy acceleration is the aim of Chapter 16. A sophisticated fuzzy logic controller (FLC) uses the solenoid's electromagnetic force to move the plunger back and forth in order to forecast how the EV will behave overall in relation to the road profile. The EM2SGB controls the vehicle dynamics by minimizing energy consumption and ensures the propulsion of the electric vehicle in operating status.

Keywords EM2SGB · EV propulsion · Energy potential · Gear shifting

16.1 Introduction

An actual size EMA was built to function with the 2-SGB using the results of the analysis, and a smaller EMA was designed and manufactured to validate the suggested concept. An iron core is used in the schematic model of a laboratory-scale EMA, as seen in Fig. 16.1, to boost the magnetic energy stored and provide a stronger EMA force. A non-magnetic (aluminum) plunger is connected to an iron stopper or mechanism to serve as the actuator's force object.

An electromagnetic actuator's (solenoid's) size, coil diameter, winding count, maximum current carrying capacity, and supply voltage are all critical design elements. For the AEM2SGB system to maintain traction and cruising power development for the EV, the EMA dimensions are crucial for modifying the gear ratio.

A. Rahman et al., *Autonomous Driveline System for Electric Vehicle*, Green Energy and Technology, https://doi.org/10.1007/978-981-96-8418-2_16

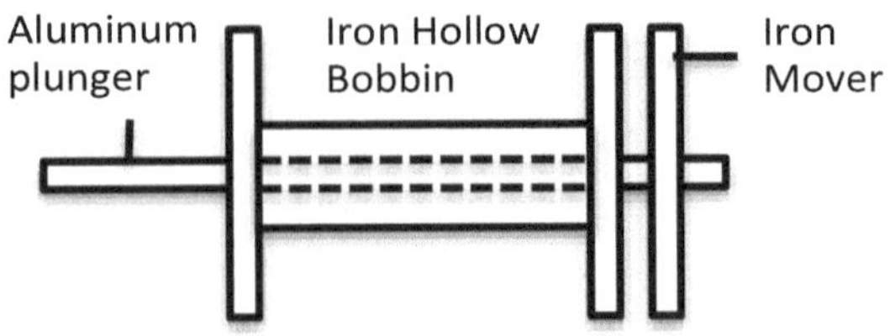

Fig. 16.1 Electromagnetic actuator [1]

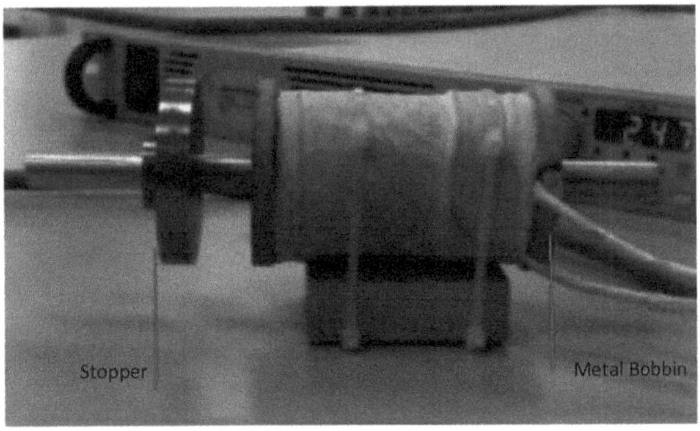

Fig. 16.2 Lab scale EMA for the proof of concept

16.2 Development of EMA

In this investigation, the gear ratio falls between 3.5 and 1.3. To create the AEM2SGB, a laboratory scale was selected. Increasing the coil's diameter and number of turns typically has an impact on a solenoid's electromagnetic field. To create the desired electromagnetic force, or gear shifting force, the EMA's coil diameter and number of turns are considered. To achieve the appropriate electromagnetic force within the control current range, the EMA winding's coil diameter is estimated to be 1.45 mm, and its total number of turns is 175 turns.

For the proof of concept design, the tiny solenoid actuator seen in Fig. 16.2 is produced. Iron is used to make the coil that the copper wire is coiled around in order to boost the collected magnetic flux density and provide a stronger electromagnetic force. According to the formula $F = BlI$, the resultant force is directly proportional to the magnetic flux density. Details of the EMA windings are displayed in Table 16.1.

16.3 AEM2ST Performance: Theoretically

EMA magnetic force performance had been conducted theoretically by using Microsoft Excel to show the results of power supply (Current). RPM and slope gradient in graphical analysis. The lab scale EMA parameters are considered: mass, m = 2.64 kg, length = 75 mm, copper wire diameter = 1.45 mm, wire length 30 m,

Table 16.1. Specification of the lab scale EMA.

Description	Unit	Values
Coil diameter (d_w)	mm	1.45
Cross sectional area of wire	mm^2	1.65
Length of solenoid (L_{sol})	mm	75
Solenoid bobbin inside diameter (h_i)	mm	10
Solenoid bobbin outside diameter (h_o)	mm	50
Packing Factor $= \frac{105(actualwinding)}{200(theoreticalwinding)}$	–	0.525
Number of turns, $N = \frac{L_{sol}[h_o-(h_i+2t)]}{2d_w^2} xf_{pf}$	Turns	163
Length of the coil, $L_w = \frac{L_{sol}\left[h_o^2-(h_i+2t)^2\right]}{2d_w^2} xf_{pf}$	m	30
Volume of the solenoid coil, $Vol^m = \pi\left(h_o^2 - h_i^2\right)L_{sol}xf_{pf}$	m^3	2.96×10^{-4}
Mass of solenoid, $m_{sol} = \rho_{cu}Vol^m$, $\rho_{cu} = 8940^{kg}/_{m^3}$	kg	2.64

and total coils 163 turns. The purpose of the fuzzy logic controller (FLC) approach is to forecast the AEM2SGB system's dynamic behavior and performance over time. Energy usage is reduced by using EMA actuator control systems with FLC, which use the power supply to create the desired electromagnetic force. A regulated variable and a controlled variable make up the EMA control system.

A simulation of the AEM-2SGB's characteristics for shift times in the first and second gears is presented in Tables 16.2–16.3. To explain the linguistic value of electromagnetic force, Table 16.2 displays the fuzzy subsets of Xsmall (XS), Small (S), Medium (M), Large (L), Xlarge (XL), and Extreme Large (EL). When transitioning from the second to the first gear and from the first to the second gear, the linguistic values for the sensor's voltage, the universe of discourse, V, are equal to 036 V, and the linguistic values for the current, the universe of discourse, I, are equal to 0–16 A. This finding suggests that shifting from the second to the first gear takes less EMA time than shifting from the first to the second gear. The rationale is that less axial force is needed to engage the synchronizer with the gear that has the first gear's lower spin. It is more obvious that the higher rotation of the gear—which takes longer because of the resistance of the slot sitting—is what causes the shifting from the first to the second gear.

The cone's inertia is 0.061 for shifting in the 0–1st gear and 0.018 for shifting in the 1–2nd gear; its friction coefficient is 0.1; and its mean diameter is 65 mm. The cone angle for engagement is determined to be 7.3 for better results for the number of cone-1 (single cone synchronizer). The gear cone and synchronizer ring can have angular speed variations ranging from 50 to 200 RPM. As seen in Fig. 16.3, the shifting force has been computed for the various gear shifting times.

Table 16.2. Performance of the fuzzy logic controller for 1st gear shifting

Input range		Fuzzy variable	Shifting time from second gear to first gear (Traction) (ms)
Sensor signal (V)	Current flow (A)		
0–1	0–.5	XS	300
0.5–1.5	5.0–7	S	300
1.0–2.5	6.5–8.5	M	200
2 .5–3.5	8.5–10.5	L	151
3.0–4.5	10–12	XL	92
4.0–5.0	12–16	EL	39

Table 16.3. Performance of the fuzzy logic controller for 2nd gear shifting.

Input range		Fuzzy variable	Shifting time from first gear to second gear (Higher speed) (ms)
Sensor signal (V)	Current flow (A)		
0–1	0–.5	XS	300
0.5–1.5	5.0–7	S	300
1.0–2.5	6.5–8.5	M	150
2.5–3.5	8.5–10.5	L	131
3.0–4.5	10–12	XL	120
4.0–5.0	12–16	EL	66

Figure 16.4 illustrates how the synchronizer's incremental shifting time lowers the axial force required to engage the gear. In order to modify the speed, the first gear shifting requires higher axial force against the traction force, whereas the second gear shifting requires less force. For the EMA's power supply, which is managed by the controller based on the RPM motion sensor's reaction for first gear shifting and for cruising or greater speed, a 24 V DC lead acid battery is taken into consideration.

The Fem $\geq$ Fax has been estimated utilizing the electromagnetic force in the simulation research of AEM2SGB. As the vehicle's traction torque changes, so does the electromagnetic force. In order to overcome the clamping force and achieve the appropriate gear ratio, the electromagnetic force increases as the torque does. Relative permeability for iron (0.2 impurity), μ_r is 5000 (as this material is chosen for the plunger because of its higher magnetic permeability). Thus, permeability for such a metal is $\mu = \mu_o \mu_r = 1.26 \times 10^{-6} \times 5000 = 6.3 \times 10^{-3}$ in N/A. This relationship (Eq. 3.22) helps us know how much EMF force can be supplied to the solenoid to achieve a given value of the current. Therefore, based on Eq. 3.22, the number of turns of the coil is N = 163 turns, the length of the solenoid is $L_{sol} = 0.075$m and the radius of the inner solenoid interior is r = 0.02m. A sample estimation of electromagnetic force:

Electromagnetic Magnetic flux:

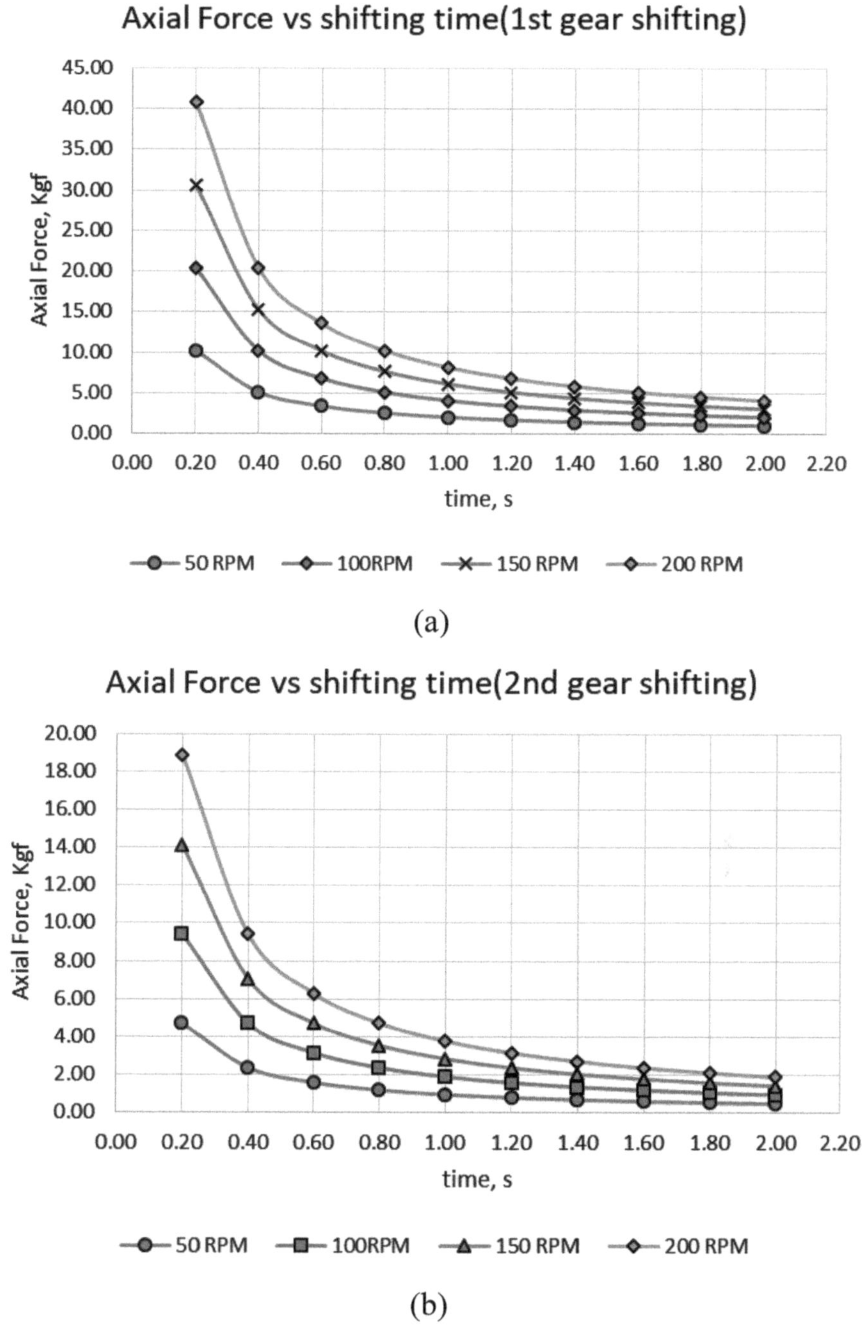

Fig. 16.3 Axial force versus shifting time **a** for 1st gear **b** 2nd gear shifting [2, 3]

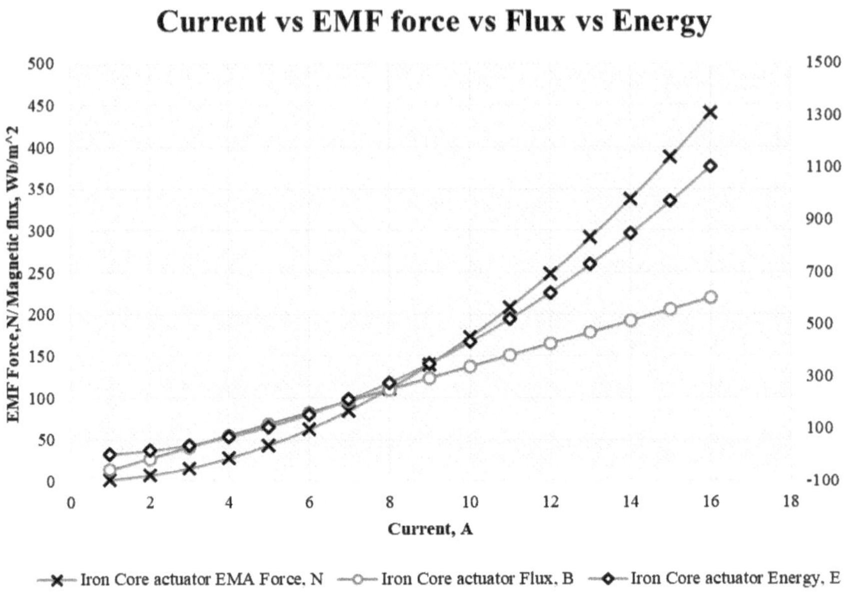

Fig. 16.4 Electromagnetic force and energy at different current supply [1, 4]

$$B = \frac{6.3 \times 10^{-3} \times 163 \times 16}{0.075} = 219.04\ \text{Wb/m}^2$$

Electromagnetic force:

$$F_{em} = 163\frac{6.3 \times 10^{-3} \times 16^2}{0.075}2\pi \times .02 = 440.32\ \text{N}$$

Magnetic energy stored in the solenoid coil:

$$E_{eng.\,L_i} = \frac{6.3 \times 10^{-3} \times 163^2 \times 16^2 \times 2.96 \times 10^{-4}}{0.075} = 1103.51\ \text{J}$$

The electromagnetic force, flux and electromagnetic energy can be calculated as supply current in the range from 1A-16A.

Figure 16.5 illustrates the minimum and maximum electromagnetic force needed for different supply currents and winding counts. Additionally, it symbolizes the intended electromagnetic force to guarantee EMA actuation for the current investigation. The magnetic field that is produced around the solenoid really stores this energy. A portion of the energy is transformed into thermal energy rather than mechanical energy to assist the plunger in moving [1, 5]. This indicates that the plunger's smooth operation is facilitated by the residual magnetic energy. It should be mentioned that eddy current heat loss is the source of the heat generated during the power delivery. Eddy currents cannot be prevented, however they can be lessened by lamination and a cooling system or heat sink.

A solenoid-actuated 2-SGB has been developed for test bench purposes based on design specifications and theoretical analysis. It is composed of an engine, an alternator that serves as a power generator to supply power to the EMA in order to develop the desired electromagnetic force that is equal to or greater than shifting force, and two sets of solenoid devices that adjust the gear ratio and maintain distance from the magnetic disk. Since the given AEM2SGB for this study is designed for lab size, an electrical motor has been used to create the driving RPM for the gearbox. Figure 16.6 illustrates the design of the laboratory-scale EMA. The main aim of EMA developments in laboratory scale is to carry out a parametric study of the electromagnetic actuator (EMA) to optimize the design of full-scale AEM2SGB for the car [6].

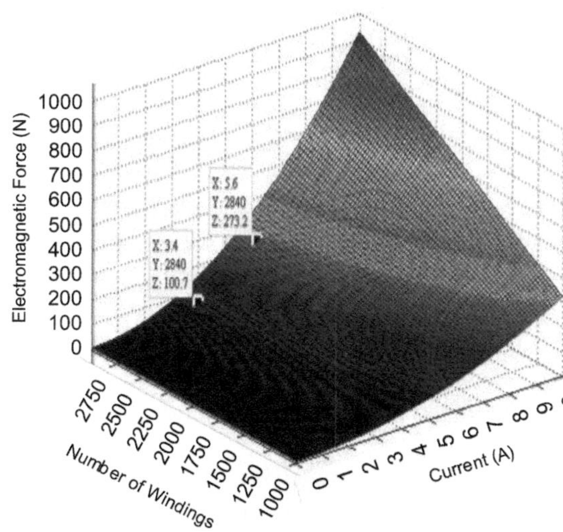

Fig. 16.5 Electromagnetic force versus number of windings and current [1, 4]

Fig. 16.6 Developed EMA [4]

16.4 AEMA2SGB Performance: Experimentally

As seen in Fig. 16.7, while the EMA2SGB is being tested in the self-operating mode, the controller switches power between EMAs according to the motor speed input. To attain a gear ratio of 1.3, the controller shifts the power system to the EMA of the second gear when the motor speed hits 200 rpm or above. The current flow rises to 16 A when the motor speed reaches 500 rpm. In order to reach a gear ratio of 1.3, the gear changer must engage the second gear due to the larger electromagnetic force produced by the EMA's enhanced current supply. The controller shifts the power system to the EMA for the first gear when the motor speed drops to 200 rpm, with a gear ratio of 3.5.

When a current flow of 7 A is observed, the first gear engages due to the EMA's decreased electromagnetic force. Depending on the amount of current flowing, the gear shifter needs a greater axial force to engage the synchronizer, allowing the second gear to move at faster rates. When moving in first gear or at slower speeds, the gear shifting force is less. The experiment was run both with and without an FLC using a 24 V DC power system.

Static Mode

A full-scale EMA to 2-SGB passenger automobile system with an automatic controller and a manual smart controller was parametrically investigated using a laboratory-scale EMA simulation research, as illustrated in Fig. 16.8a. At the laboratory level of EMA, experiments were conducted both with and without a controller. The following EMA parameters were taken into consideration: the gap is approximately 7 mm, the effective area is 0.000188 m^2, the number of turns (N) is 163, the magnetic permeability of the iron core is approximately 0.0063 N/A2, and the electromagnetic force is greater than the clamping force. The created EMA controls in two modes, one being manual and the other being automatic, as shown in Table 4. 4.

In manual mode, a switch is used in place of a relay that makes a direct connection to the ampere booster. In order to transmit the current to the EMA to engage the gear, the first gear switches on when the RPM sensor indicates the value for traction mode,

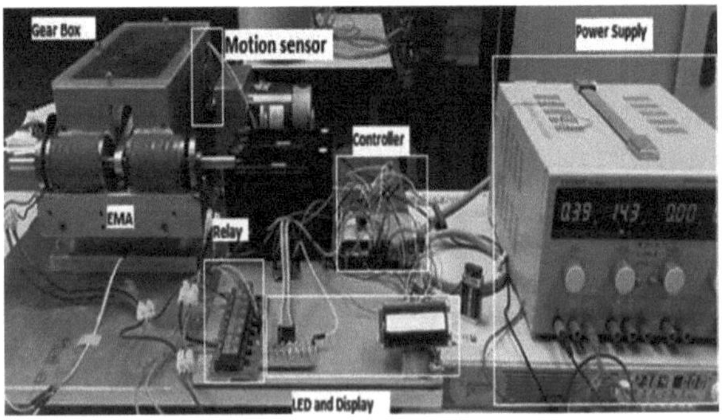

Fig. 16.7 Experimental set-up for AEM2SGB performance investigation

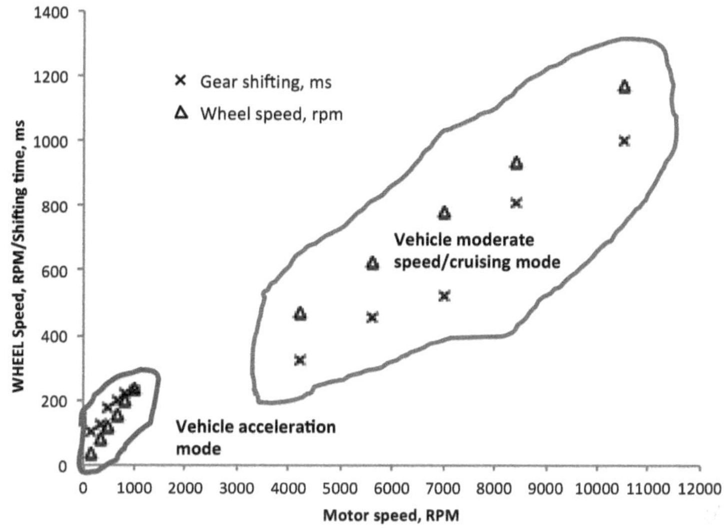

(a) Static mode

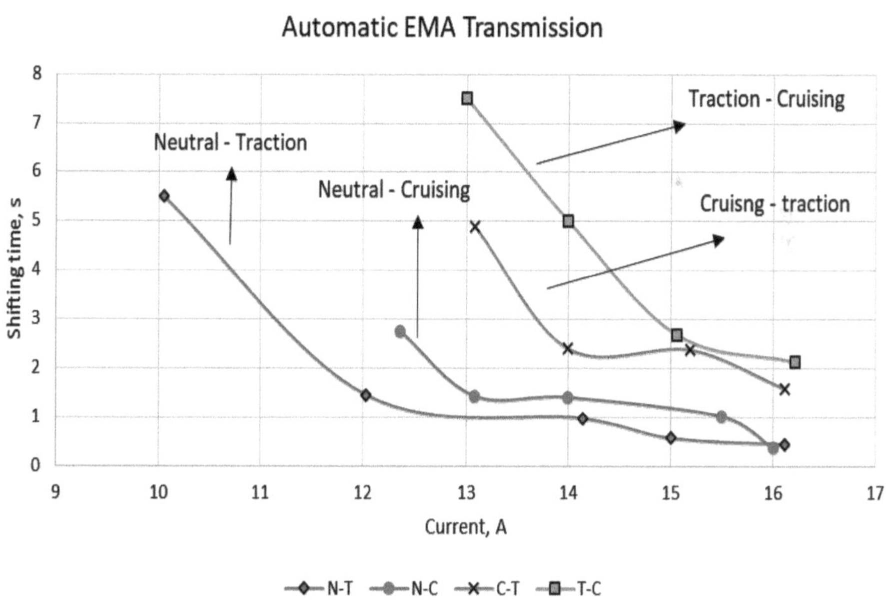

(b) Dynamic mod

Fig. 16.8 AEM2SGB dynamic testing with varying loads [3] and the second gear switches on when the RPM sensor indicates cruising mode. An 8 channel relay (30 V DC) is used in automatic mode to help manage the current and send it to the EMA in various vehicle modes.

Dynamics Mode

To ascertain the effectiveness of the fuzzy autonomous system, dynamic testing was performed with mechanical loads on the drive shaft of the AEMA2SGB, as illustrated in Fig. 16.8b. The autonomous fuzzy system employs the motor RPM sensor to shift gears from second to first, indicating that the vehicle is in traction, when the applied mechanical force is larger and the motor speed is lower. The vehicle's slope resistance was taken into account while calculating the mechanical load, which is defined as, $F = W(f_r + \sin\theta)$. The load was varies based on grade resistance $\theta = \sin^{-1}(G)$, G is the grade in percentage.

The EMA laboratory scale's performance when working with a load—which is thought to be the force that propels the gear shifter from the first to the second gear or vice versa—as displayed in Fig. 16.9. The performance of the EMA has been examined through the development of a two-speed gearbox on a laboratory scale. It takes more than 5 N of axial force to move the gear shifter from the first gear to the second, and less than 3 N to move it from the second gear to the first. An electromagnetic force of 3.5 N for a 6 A supply current and 5.2 N for a 10 A supply current is created by the EMA.

This suggests that the EMA's electromagnetic force growth is proportionate to the flow of current. Because of the longer gear-shifting time, the EMA takes longer to accelerate and produces less electromagnetic force as a result of the decreased current flow. The car thus slows down and begins to lose traction. The diagram's AB zone, commonly referred to as the dead zone, represents the slow acceleration brought on by timing shifts from first to second gear. Traction losses are frequent in this area since the EV's motor output does not reach the driving wheels. There is a slight change in acceleration due to the vehicle's brief activity.

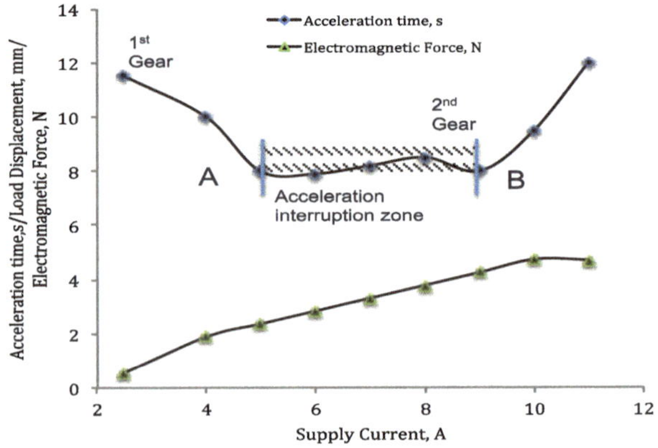

Fig. 16.9 Performance of the EMA to maintain acceleration time [4]

The vehicle's ascending slope, speed, and the efficiency of the fuzzy autonomous system are all intimately correlated with the shifting time. The synchronizer's simpler engagement in the first gear on a 30% slope results in a minimum gear shifting time of 110 ms, while the synchronizer's challenges at a motor speed of 11,000 rpm, or 90 km/h, result in a maximum gear shifting time of 1200 ms. Table 16.4 shows the suggested EMA2SGB's performance both with and without a fuzzy controller. There is a 28 mm shifter travel to engage the first or second gear.

The wheel speed sensor and motor speed sensor are connected to the controller in order to perform the controller function. Due to system limitations, only the motor speed sensor is utilized. For vehicle traction, where the gear spins more slowly, a lower motor speed is taken into consideration; for a vehicle moving quicker, where the gear spins more quickly, a higher motor speed is taken into consideration.

As seen in Fig. 16.9, the results indicate that 2-SAEMT (2-SAEGB) has a power savings advantage over 5-SMT and SDCT. 5-SMT and SDCT quantitative data, including shifting time, transmission weight, and transmission power losses, were extracted from published studies [7–9]. For instance, the EV will save 0.93*356 or 331 Wh in transmission loss and 1.1*0.375 or 412 Wh in weight savings for a 90 km/h EV if the 5-SMT is swapped out for a 2-SAEMT. Consequently, 743 Wh of gross battery power savings are realized (Table. 16.5).

In contrast, the EV will save 1,245 Wh and 1,500 Wh of battery power if a 2-SAEMT is used in place of the SDCT and SAT. As a result, 743 Wh of power is

Table 16.4. Performance of the laboratory-scale EMA2SGB with an EMA and 163 turns

EMA2SSGB—without fuzzy controller			EMA2SSGB—fuzzy controller		
Neutral to first gear (Motor speed, 0–200 rpm)					
Voltage, V	Current, A	Time, ms	Voltage, V	Current, A	Time, ms
24	10	754	24	10	916
24	11.1	627	24	11.1	241
24	12	167	24	12	161
24	14	111	24	14	95
24	16	68	24	16.11	65.5
First gear to second gear (motor speed, 200–500 rpm)					
24	12	965	24	13.08	930
24	14	500	24	14	364
24	15.35	527	24	15.19	260
24	15.98	369	24	16.11	137
Second gear to first gear (motor speed, 500–75 rpm)					
24	13	973	24	13	650
24	14	404.6	24	14	259
24	15	375.3	24	15.1	167
24	16	297.2	24	15.98	110

Table 16.5. Advantages of 2-SAEMT for the EVs

Gear shfting mode	Transm. type	Vehicle dynamic parameter				2-SAEMT ES[c]	
		[a]WT	[b]ST	Acceler.	[c]TL	RST	W_S
		(kg)	(ms)	(m/s^2)	(Wh)	%	%
1st gear to 2nd gear	5-SMT	48	150	214.14	356	92.29	37.50
	2-SMT	33	800	276.03	190	85.54	9.09
	2-SEMT	36	850	283.41	202	86.39	16.67
	2-SAEMT	30	116	1735.56	28	0.00	0.00
	4-SAEMT	91	150	405.96	356	92.29	67.03
	7-SDCT	89	200	2977.82	48	42.17	66.29
2nd gear to 1st gear	5-SMT	48	700	458.86	166	83.48	37.50
	2-SMT	33	770	286.79	183	84.98	9.09
	2-SEMT	36	800	301.13	190	85.54	16.67
	2-SAEMT	30	110	1825.02	26	0.00	0.00
	4-SAT	91	140	434.96	332	91.74	67.03
	5-SMT	48	150	214.14	356	92.29	37.50
	2-SMT	33	800	276.03	190	85.54	9.09
	4-SAT	91	140	434.96	332	91.74	67.03
	7-SDCT	89	320	1861.14	54	46	66.29

Notification:

- 5-SMT—five speed manual transmission, AT—automatic transmission,
- AEMT—autonomous electromagnetic transmission, DCT—dual clutch transmission EMT—electromagnetic manual transmission,
- W_T—Transmission weight, ST—Shifting time, TL—Transmission losses, RST—Reduction shifting time, W_S—Weight saving,
- a-actual values, b-experimental values, c-calculated
- ES[d]—Energy saving of EV mass 1300 kg

saved overall. A similar conclusion may be drawn if the 2-SAEMT is used in place of the SDCT; the EV will save 1,245 Wh in total power. This will result in a 5% longer battery life, a 10% reduction in battery charging power, and an additional 6–8 km of EV travel.

The overall efficiency of the vehicle is significantly impacted by the drivetrain's performance. As the drivetrain becomes more efficient, the car uses less fuel. Figure 16.10 compares the EMA-2SGB to the 2-dcMGB [10] in terms of top speed, battery power savings, traction losses, and acceleration time. The lengthier gear shifting time, which is partially caused by the conversion of rotational motion into linear motion, has made it more difficult for the 2S-dcMGB, Porsche Tycan 10s, and Tesla 12s to accelerate. The car thus starts to lose traction and slow down.

Although it is simple to think that the electric motors in Tesla and Porsche need more battery power to satisfy uphil's requirement for rapid traction power, this isn't

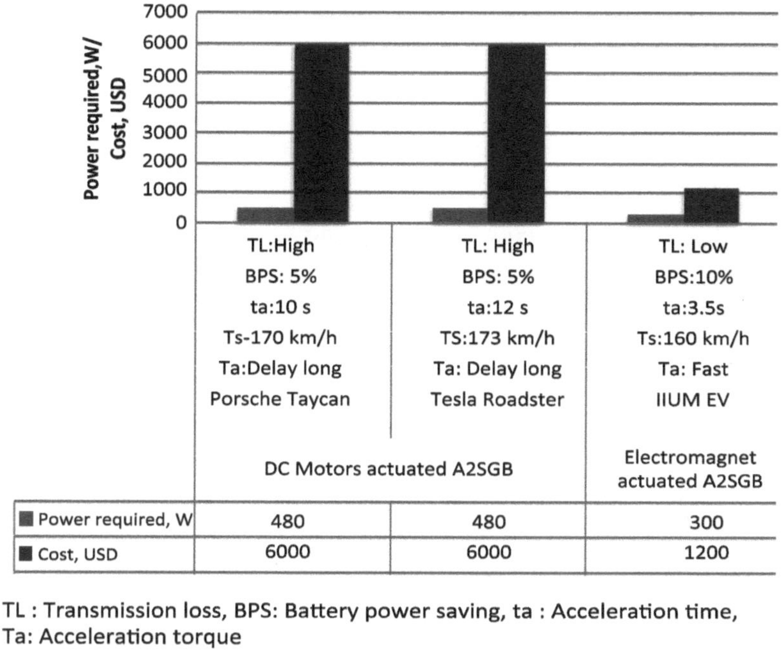

Fig. 16.10 Comparison of 2-SAEGB over the 2-dcMGB [10]

necessarily the case. However, both EVs save 5% on battery power by using a 2s-dcMGB instead of a single motor dc motor gearbox (1s-dcMGB). In contrast, the EMA-2SGB has little traction power losses and accelerates 2.85 times faster than the Porsche 2-dcMGB and 3.45 times faster than the Tesla Roadstar. Because of its quicker gear change, the EMA2SGB has thereby saved about 10% of battery power. Additionally, the 2S-dcMGB uses 37.5% more power to provide the gear-shifting force than the EMA-2SGB. In comparison to a 2S-dcMGB, the EMA-2SGB has a reduced price and development cost based on laboratory scale development cost.

16.5 Conclusion

Because of the effectiveness of the fuzzy controlling mechanism that uses sensors and magnets, the potential of AEM2-SGB is evident when shifting from the first to the second gear and from the second to the first gear. Because of its 37%–66% weight reduction, 40%–90% reduction in transmission loss, and 5% increase in battery life, the AEM-2SGB is a potential gearbox for EVs. The autonomous fuzzy intelligence system of the AEM-2SGB enables it to change gears for traction in 110 ms from the second to the first and in 116 ms from the first to the second.

As a result, transmission loss is decreased by 42% to 93% and the problem of vehicle traction control interruption is resolved. Regarding energy savings, an EV will save 743 Wh, 1,245 Wh, and 1,500 Wh of battery power if the 5-SMT, 7SDCT, and 4SAT are swapped out for 2-SAEMT, respectively, in comparison to other possible gears. Consequently, a 10% decrease in battery charging frequency will result in an extra 6–8 km of EV mileage and a 15% increase in battery life.

Acknowledgement The authors express their gratitude to Eastern Michigan University for granting permission to continue this project and to the Ministry of Higher Education Malaysia for funding it under the Prototype Research Grant Scheme (PRGS) 2022. The laboratory technician who assisted in the development of this project is also appreciated by the authors.

References

1. Sorniotti, A., Pilone, G. L., Viotto, F., & Bertolotto, S. (2011). A novel seamless 2-speed transmission system for electric vehicles: Principles and simulation results. *SAE International Journal of Engines, 4*(2), 2671–2685.
2. Sorniotti, A., Subrahmanyan, S., Turner, A., Cavallino, C., Viotto, F., & Bertolotto, S. (2011). Selection of the optimal gearbox layout for an electric vehicle. *SAE International Journal of Engines, 4*(1), 1267–1280.
3. Rahman, A. & Hasan, N. (2023). Autonomous electromagnetic two-speed gearbox for electric vehicle. *International Journal of Automotive Technology*. (in press).
4. Rahman, A., Sany I.I., & Nurul, H. (2022). A fuzzy-logic-controlled two-speed electromagnetic gearbox for electric vehicles. *International Journal of Technology 13*(2), 167–179.
5. Polychronis, S., Nikolaos, C., Tsourveloudis, L. D., & Emmanuel S. K. (2019). Experimental research of transmissions on electric vehicles' energy consumption. *Energy, 12*(388), 2–15.
6. Jaffar, A. H., & Rahman, A. (2019). Study on intelligent control system of electromagnetic continuously variable transmission system. *International Journal of Recent Engineering and Technology, 7*(6S), 47–53.
7. National Research Council. (2015) *Cost, effectiveness, and deployment of fuel economy technologies for light-duty vehicles*. National Academies Press, Washington, D.C.
8. Ho-Chang, J., & Deok, J.K. (2015). Analysis of energy consumption performance for electric vehicle considering transmission shifting pattern. EVS28, KINTEX, Korea, May 3-6, 2015.
9. Faid, S. (2015). A highly efficient two-speed transmission for electric vehicles. *EVS28*, 3–6.
10. Porsche Taycan (2019). Retrieved July 2, 2020, from https://www.caranddriver.com/news/a28903274/porsche-taycan-transmission/.

Part V
Intelligent Electromagnetic Continuously Variable Transmission

Chapter 17
Continuously Variable Transmission (CVT) Revolution

Abstract Engine (or motor) power is smoothly and effectively transferred to the car's driving wheel by a continuously variable gearbox (CVT). A number of CVT models have already been created to reduce transmission losses without sacrificing acceleration time. We developed a novel electromagnetic actuator for CVT because most of them have limitations in their actuation mechanism. By closely examining the kinematics of the CVT clamping forces and electromagnetic forces of the EMA, an analytical model of the electromagnetically actuated CVT is created. The desired CVT ratio will be ascertained by means of the speed and throttle position sensors. The fuzzy logic controller has been utilized to control the actuation because of its ease of use, resilience, and lack of need for an exact mathematical model.

Keywords Electromagnetic CVT · Fuzzy logic controller · CVT ratio · Robustness and energy efficient

17.1 Introduction

In any road situation, the transmission is a device that distributes power produced by an automobile's engine or electric motor to the wheels in order to move the vehicle forward or backward at a specific pace. When using a manual transmission, a gearbox with a gear shifting system allows various engine-to-wheel ratios. When shifting gear ratios, the engine crankshaft must be disengaged from the gearbox via a friction clutch. Different gear ratios are automatically provided by an automatic gearbox using a torque converter.

A modern automated gearbox known as a continuous variable transmission (CVT) provides a varied gear ratio between the engine and the drive shaft by means of two cone-shaped pulleys connected by a metal belt. By adjusting the pulley diameter in accordance with the vehicle speed and road conditions, different gear ratios are accomplished. As a result, vehicles with CVT systems used fuel more effectively [1]. As more automakers use CVTs to improve fuel efficiency, such as in the new Saga (Proton, 2017), the new City (Honda, 2017), the Teana (Nissan, 2017), and the

A. Rahman et al., *Autonomous Driveline System for Electric Vehicle*, Green Energy and Technology, https://doi.org/10.1007/978-981-96-8418-2_17

Audi A5 2.0 TDI (Wikipedia, 2017a), the CVT will continue to replace traditional gearing systems in the years to come.

A research was carried out to examine the performance of continuously variable transmissions, automatic transmissions, and manual transmissions [2]. The time it takes to accelerate to 100 km/h is 10.20 s for manual transmissions, 10.76 s for automatic transmissions, and 7.85 s for CVTs, according to simulation results on cars weighing 1500 kg and having an engine capacity of 3000 cc. Furthermore, environmentally friendly automotive technology manufacturer The Torotrak Group (2007) asserted that its CVT transmission reduced harmful emissions and fuel consumption by roughly 19% when compared to a car with a conventional gearbox.

To produce enough clamping force for the required CVT ratio and torque, the current pulley-based CVT used the electro-hydro-mechanical (EHM) actuation technology [3]. Using hydraulic pressure, the EHM moves and keeps the pulley sheave in the proper position. One way to convey motion and/or force is through hydraulics, often known as fluid power [4]. Electronically, solenoid valves regulate the flow of oil in both direction and quantity. This method uses more gasoline since the hydraulic pump continuously supplies the system with pressurized oil while the engine is running.

High power consumption, power loss, and temperature rise for extended operation are some of the main problems that make CVTs that use the EHM actuation mechanism less efficient, according to [5, 6]. As a result, the secondary pulley sheave actuation loses input power when the hydraulic fluid's viscosity drops below the ideal level due to an excessively high temperature. As a result, the pulley sheave's holding force is unstable and impairs the car's reaction and performance, particularly when hill climbing.

According to a study by the authors [4, 7] on hydraulic control for the CVT, the system is quite good at creating enough pressure, but it has trouble keeping the pulley movable sheave at the proper position because of hydraulic pressure saturation. Because the hydraulic control system requires high pressures and flow rates, fluid pumping losses are one of the main causes of torque loss in current CVTs. Additionally, the study suggested attaching a preloaded linear elastic spring to the movable secondary sheave to ensure a low clamping force in the event that the hydraulic system fails. On the other hand, the secondary pulley's movable sheave's camping force feels spongy when climbing the slope. Therefore, the car cannot accelerate with the traction force alone.

A few unfavorable features of CVTs, such as the unpleasant "rubber band" sound and the sluggish reaction of the movable sheaves for quick acceleration time under hard acceleration, have also made them annoying to many fans and reviewers [8]. The Toyota Wish has three primary issues, according to customer complaints on the CVT: rattling noise, reduced torque while climbing hills, and delayed acceleration when moving from a stop. The car has been towed to the workshop due to sensor failures, which are another source of complaints. Consumer forums and blogs for cars may have complaints [9–11].

Therefore, this study on the design and development of a novel CVT actuation system is highlighted by the issues with traditional hydraulically actuated CVTs. This study developed the novel “Electromagnetic Actuation (EMA)” mechanism for the CVT system, which increases the traction force at the tire-road interfaces and stiffens the clamping force on the secondary pulley in state-of-condition. As a result, the car accelerates quickly and reduces transmission losses, which leads to a 15% reduction in the vehicle’s overall propulsion power. This is linked to a slow acceleration speed and a delay in reaching a higher gear ratio.

The new EMA-CVT also eliminates maintenance issues, enhances vehicle transmission losses, reduces transmission weight by around 15%, and prevents the development of any rattling sounds. For instance, the hydraulic operated CVT requires hydraulic oil replacement every 50,000 km driven. Therefore, the cutting-edge EMA-CVT can be viewed as an eco-friendly technology for the automobile sector.

17.2 EMA-CVT Development

Compared to traditional gearboxes, the continuously variable transmission (CVT) is more effective at transferring power from the engine to the wheels because it may offer an endless variety of gear ratios depending on the speed of the vehicle and the road conditions. The hydraulic losses, however, make CVTs that are actuated by a hydraulic system inefficient. The hydraulic pump’s 30% overdesign to provide adequate clamping force is one of the causes of hydraulic losses [12]. In order to maintain the CVT’s sheave in the appropriate position, the engine must continuously give power to the hydraulic pump, which will impact fuel consumption.

The main causes of the low efficiency are the huge oil pump displacement and the very high clamping forces used to prevent substantial belt slip, both of which cause the oil pump to lose a significant amount of torque [13]. Additionally, when the temperature of the hydraulic fluid is too high, its viscosity drops below the ideal level, which causes input power losses. The car’s reaction and performance are impacted by the pulley sheave’s unstable holding force. According to customer complaints on the CVT specifically for the Toyota Wish, the car has three primary issues: rattling noise, decreased torque while climbing hills, and poor acceleration when moving from a stop.

Additionally, complaints have been made regarding sensor failures that need the automobile to be towed to the repair. Therefore, the goal of this research is to create a full-scale EMA prototype for the CVT that has a central manual controlling system (CMCS) and sensor-based control system. The size of the Toyota Wish automobile will be the precise basis for the prototype’s design. In order to guarantee that the EMA operates with roughly 80% power to activate the CVT system, CMCS will be implemented in the event that sensors fail. With funding from the Research Management Centre, International Islamic University Malaysia (RMC, IIUM), the authors [9, 14] created a ¼ scale EMA-CVT actuation. According to the experimental results, the car can accelerate in 3.5 s at startup, which is 60% faster than the hydraulic CVT.

Nevertheless, the prototype’s CVT actuation responsiveness was deemed to be excellent for the car’s cruising speed but subpar for climbing hills. To enhance acceleration time, clamping force, and fuel consumption, the existing CVT actuation system must be optimized, or a new, more efficient EMA mechanism must be adopted.

Scope on Focusing on EMA-CVT Development

The study's research topics have been categorized as follows in order to maximize the EMA-CVT's design and development, making the vehicle dynamic and energy-efficient.

- To provide a parametric analysis and optimization model for EMA for the pulley actuation of the passenger car CVT.
- To create a movable sheave for the passenger car's CVT by means of an electromagnetic actuator (EMA).
- To maintain the required gear ratio and CVT torque, an intelligent method to regulate the EMA's electrical power flow must be developed.
- Using the constructed test-rig, the i-EMA-CVT will be tested, evaluated, and validated to actuate the secondary's movable sheave to push the vehicle under various load and road conditions.

The goal of this project is to design an intelligent electromagnetic actuator for a full-scale car's CVT system by optimizing the actuator's settings. This might be a novel approach for a CVT actuation system to reduce acceleration time and increase overall power consumption. In order to maintain the power required by the CVT transmission under different road conditions, this work also focuses on integrating the fuzzy logic control (FLC) system with the electromagnetic actuation system. Power control, acceleration time, and transmission performance have all been analyzed and assessed. A summary of how the study's goal and scope align is provided in Figure 17.1.

A laboratory test-rig has been used to evaluate the EMA-CVT model that was constructed. A thorough assessment looks at how well EMA-CVT performs as well as how consumers engage with the car as an automated gearbox. Additionally, EMA-CVT on-road situations will undergo recurrent dynamic testing, including intelligent system functioning testing, which will inform the testing and evaluation methodology. It is, however, outside the purview of this investigation. The driving cycle rule [UN/EEC/MP.2 /2019] will be the basis for the EMA-CVT's safety. Figure 17.2 shows the steps taken to conduct research on each of the study scopes.

The EMA-CVT has been benchmarked by comparing the experimental results with the reference published data. Showing the EMA-CVT's potential as a cutting-edge, energy-efficient automatic transmission system for automotive technology is the goal of benchmarking.

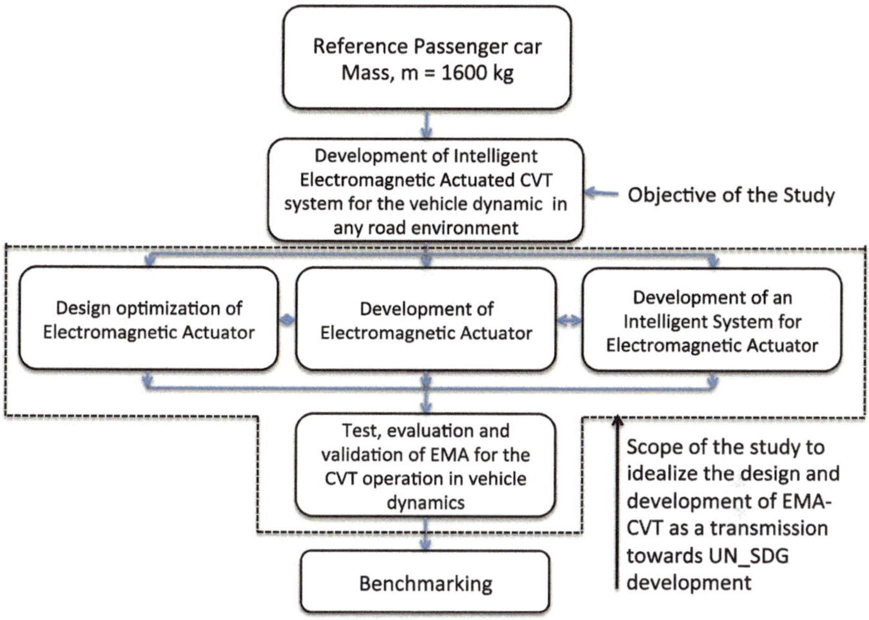

Fig. 17.1 Scope of the study in line with the objective [1]

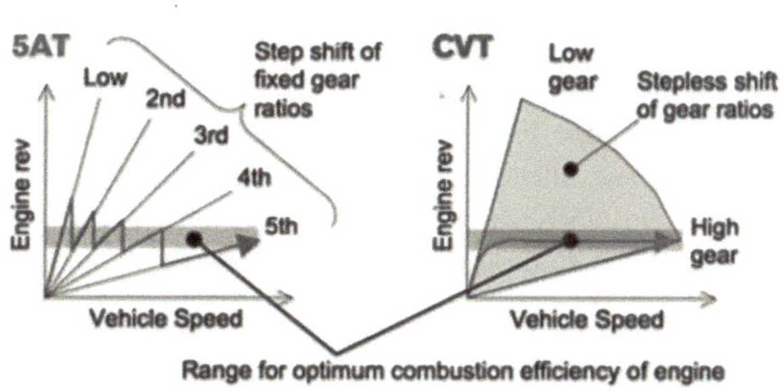

Fig. 17.2 Comparison of conventional gear box and the CVT [15, 16]

17.3 PRE-View of CVT

The ultimate objectives for engineers designing car transmissions are smooth speed changes, a larger range of speed ratios, and a low-maintenance, straightforward system. Conventional automatic transmissions are becoming more advanced in terms of performance and driveability. However, the increasingly common CVT units do not use the conventional gears; instead, they use a "shiftless" belt-drive system that can adjust the drive ratio to best suit the torque curve of the engine and the driving conditions, particularly for light vehicles. A CVT transfers power from an engine to the driveshaft without a clear gear ratio by alternating steplessly between its lowest ratio and its highest ratio, which is an overdrive.

The engine should be able to maintain its most efficient rpm. When the car accelerates from a halt, the engine speed rapidly increases from idle to the target rpm. In order to create acceleration, the gearbox ratio continuously varies while the engine maintains the same speed [17].

A comparison between CVT and traditional manual or automatic transmission systems is shown in Fig. 17.2. Engine speed fluctuates during gear changes in both manual and automated systems because power is transferred from the engine to the drive wheels via a predetermined series of gear combinations. Conversely, CVT allows for a continuous range of gear ratios, from the highest to the lowest. For changeable driving situations, a CVT enables the engine to run at or close to its optimal specific fuel consumption rate through smooth and stepless gear changes [16]. The engine may always be run at its most fuel-efficient operating point thanks to the CVT idea, which enables the engine speed to be controlled independently of the vehicle speed.

CVTs can therefore significantly reduce the production of dangerous exhaust pollutants and improve fuel economy. The CVT's seamless speed shift, wider-range speed ratio, and straightforward operation allow it to produce more torque, power, and rapid acceleration than traditional transmissions. Studies and research have been done on CVT use and how it relates to sport utility vehicle (SUV) fuel economy. Fuel efficiency has increased by up to 20% [16].

The automobile industry need a comfortable, dependable, and efficient transmission for cars, even though traditional automatic transmissions are still useful. Because of its transmission capacity, comfort, dependability, longevity, and efficiency, CVT is favored above other automatic transmissions [18]. Additionally, it eliminates the requirement for a torque converter or clutch to disconnect the car from the engine. During torque shifting, the powertrain's power flow does not need to be stopped, allowing for smooth acceleration and a quick, stepless reaction to the driver's demands without jerking. Furthermore, no matter how fast the vehicle is moving, CVT maintains it in its ideal power range.

17.4 Types of CVT System

The early 1900s saw the first documented use of CVT. Daimler and Benz, the pioneers of the German automobile industry, installed a rubber V-belt CVT in their first gasoline-powered vehicle in 1886. As time draws near, the researchers continue their investigation into the evolution of contemporary CVT. CVT was created primarily for two reasons, according to Emery Hendriks, general manager for research and development at Van Doorne: First off, the most recent CVTs are matching, if not surpassing, the sophisticated and expensive gears of today in terms of fuel economy and driving performance.

Second, high-efficiency steady-state engines that are optimized for CVTs and operate in a narrow revolution band are being considered by vehicle engineers due to stricter government limits on fuel consumption and exhaust pollutants. Then, in

Holland, the Van Doorne brothers successfully introduced a belt CVT into the vehicle industry in 1955. The running diameter of the pulley was changed to change the ratio of the transmission, and a strengthened rubber belt is utilized for torque transfer. Belt type CVT is seeing rapid growth in the market. A push belt CVT has been installed in over 5 million cars since Van Doorne's Transmission (VDT) began producing belts in 1985.

A modified version of this transmission, the steel belt transmission, is currently the most widely used CVT. A launching device (torque converter or clutch), variator system, gear train, control system, and final drive reduction are features that are common to all CVT types, according to the authors [19, 20].

Pulley-Based CVT

The most popular CVT for continuously variable transmissions is the pulley-based CVT, commonly referred to as the Van Doorne System, as seen in Fig. 17.3. A belt rides between two sheaves that face one another in a variable-diameter pulley transmission. Since this is based on a variable system, a change in the ratio results in a change in the transmission. Each pair of pulleys has one movable part that moves in unison. The pulley's radius is determined by the separation between the moving sheaves and the fixed pulley [20]. The maximum pulley diameter is the single restriction on the unlimited number of gear ratios that can be produced by varying the pulley's radii.

The process of shifting from low to overdrive ratio by varying the primary and secondary pulleys' radii is depicted in Fig. 17.4. Accordingly, a CVT can theoretically operate through an endless number of "gear ratios" at any given engine or vehicle speed. The belt's rotating sheave cone and immovable pulley cone create friction, which transfers power [21]. The belt experiences a tremendous amount of stress where it makes contact with the cone due to the friction used to transfer power. When rotating continuously, the belt wears down. Therefore, this technology works well for light-engine vehicles and can only be implemented in vehicles with 1.2-L or less engines.

The V-belt that is utilized is composed of either chain, metal belt, or fabric reinforced rubber. Any belt should be strong enough and stiff enough to transfer loads

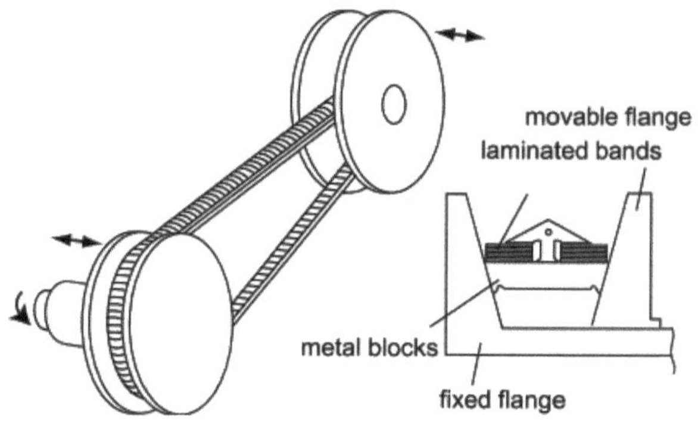

Fig. 17.3 Pulley based CVT with metal V-belt [20, 21]

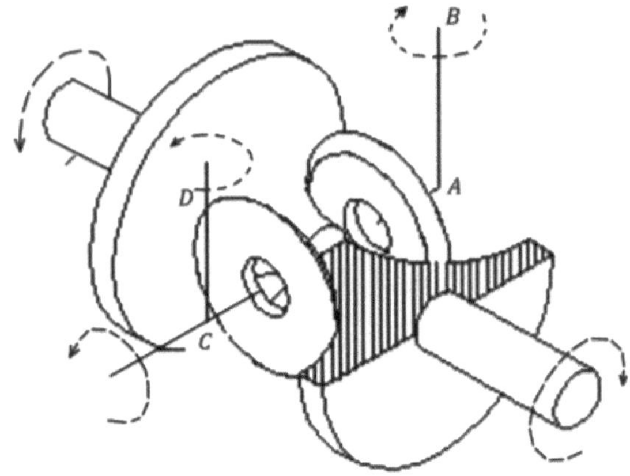

Fig. 17.4 Toroidal traction mechanism [22, 23]

from the driving pulley to the driven pulley. Additionally, these belts need to be made flexible enough to maintain the smallest possible minimum rolling radius. Tractional effort is transferred via a frictional force between the angled belt face and the pulley sheaves. A particular controller mechanism, such as a mechanical linkage, hydraulic, electromechanical, etc., controls the movement of the pulley sheaves in order to employ the pulley-based CVT system for cars.

The pulley-based CVT system requires a starting device, extra gearing, and a hydraulic system to be completed. For low power designs, the majority of manufacturers use electronic power clutches and wet plate clutches. However, the centrifugal effect is used in simple CVTs, which are found in small vehicles like scooters and snowmobiles, to move the pulley sheave.

Toroidal Traction Drive CVT

Toroidal CVT can also be described as a roller-based CVT uses power rollers and discs to transfer engine power, as seen in Fig. 17.4. The discs serve as an input drive and an output drive, respectively. Wheels connect the two discs and transfer engine power to the drive shaft. Wheels spin on two axes, slanting up and down around the vertical axis and spinning on the horizontal axis. An impact of ratio variation is produced by this motion. With this configuration, a small number of rotations on the input disc result in a large number of rotations on the output disc, which lowers speed but increases torque and vice versa.

The main challenge in creating this CVT, meanwhile, is the traction rolling element material's limited capacity to tolerate high contact pressures and temperatures, which have a substantial impact on efficiency [23]. Inaccurate torque transmission results from contact angle variations caused by bulk deformation of the contact component.

Variable Diameter Elastomer Belt CVT

The CVT type depicted in Fig. 17.5 employs a flexible, flat belt that is fixed to movable supports. These supports can alter the gear ratio and radius. The supports,

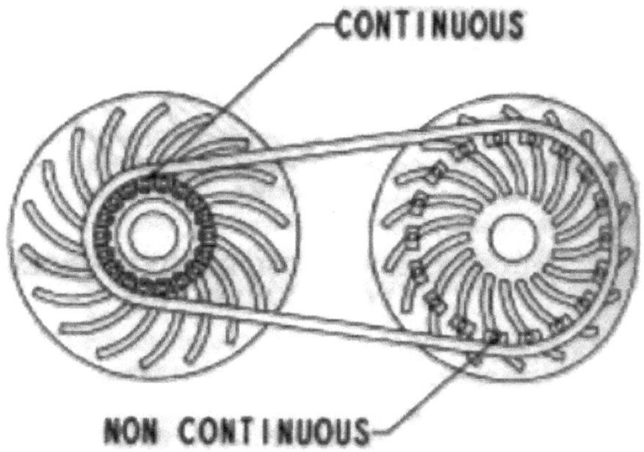

Fig. 17.5 Variable diameter elastomer belt CVT [24, 25]

however, split apart to create a discontinuous gear path at high gear ratios. This may result in issues with slide and creep [24]. Research and development has been focused on push belt CVTs because of this intrinsic defect.

17.5 Review on Actuation System

The two changeable conical pulleys that make up a CVT variator have a set center distance and are joined by a chain or metallic belt in the middle [12]. A variety of gear ratios are provided via a set of actuators that change the conical pulley's position. Through its smooth speed shift, larger range of speed ratios, and straightforward mechanism, the CVT system's actuator mechanism has been improved through extensive study to meet its primary purpose of providing faster acceleration, more torque, and more power than previous transmissions. Hydraulic actuated CVT has trouble maintaining the pulley's sheave at its precise location, but it can supply sufficient pressure. Pneumatic actuated-CVT face problem with the compressibility property of the air create fluctuation in the position of the sheave of pulley [24].

Hydraulic or Electro-Hydro-Mechanical Actuation System

The most popular CVT at the moment is a pulley-based CVT, which uses an electro-hydro-mechanical (EHM) actuation system to adjust the CVT ratio and apply sufficient clamping force to the V-belt. Since the EHM actuation system relies on hydraulic pressure, a suitable hydraulic pump is needed to provide constant pressure while the system is operating, particularly to keep the belt tension constant and avoid slippage, as seen in Fig. 17.6. This is typically accomplished by continuously drawing power from the automobile's engine, which is an inefficient use of energy and eventually results in the car needing more fuel, as seen in Fig. 17.7.

The hydraulic fluid medium transfers the force and pressure from the master cylinder to the CVT clamping mechanism. Valves regulate the flow of the primary

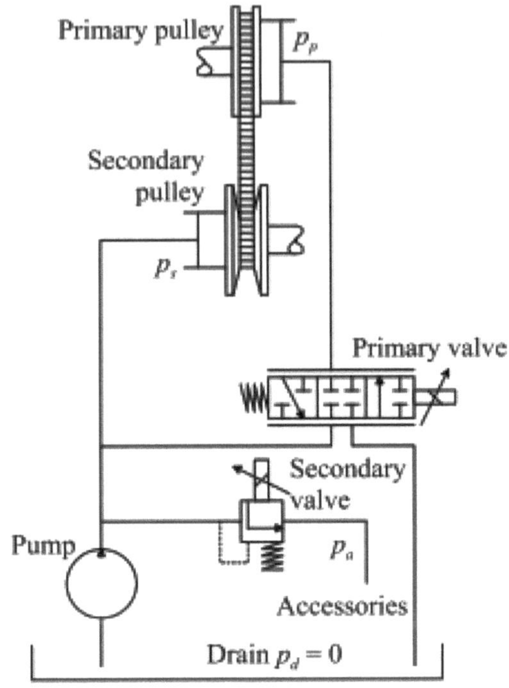

Fig. 17.6 Variator with hydraulic actuation system [7, 25]

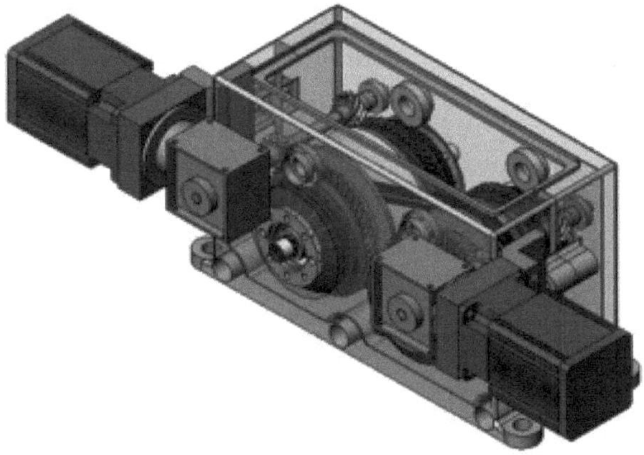

Fig. 17.7 Pulley-belt CVT with EM actuation system [3, 4]

and secondary liquid pressures. To prevent slippage between the intermediate element and the pulley sheaves, a line pressure is supplied by the secondary pressure, also known as the secondary clamping force [12].

Electromagnetic Actuation System

An electromagnetic actuator (EMA) is a coil that is wrapped around a movable metallic core and has a length significantly longer than its diameter. When an electric current is conducted through it, the coil creates a uniform magnetic field in a volume of space. By allowing the core to be moved in and out of the center, the coil's form allows it to change its inductance and turn into an electromagnet. IIUM has created an EMA-CVT on a ¼ scale to enhance the pulley's response time, traction torque control, and fuel efficiency.

A fuzzy logic controller also be incorporated into the EMA CVT to guarantee a seamless gear shift in a variety of driving situations. The EMA-CVT innovation from 2009 to 2019 is depicted in Fig. 17.8. According to testing findings, the EMA generates electromagnetic forces of 301 N at a supply current of 3.37 amp, resulting in an acceleration time of 2.5 to 3.5 s and an electromagnetically actuated CVT system that is extremely energy efficient [26]. The issue, however, arose from the plastic bobbin softening when the supply current exceeded 10 A.

17.6 Basic Terminology of Electromagnetic System

Table 17.1, which show typical magnetic values and their units, provides a summary of some basic words commonly used in discussions of electromagnetism for easy reference.

Magnetomotive Force

This is the electric circuit's equivalent of electromotive force, often known as voltage or EMF. It uses ampere-turns (AT) as its unit. NI Ampere-turns MMF produced by a coil carrying I amps with n turns. Identified by the fancy letter, it can be computed as,

$$F_m = N \times I \tag{17.1}$$

Magnetic Flux

Magnetic flux is a measure of the number of force lines and it similar to current in an electric circuit. Its unit is the Weber (Wb), and it is designated by the Greek letter Phi (φ)

$$\varphi = B * A_{\varphi} \tag{17.2}$$

where, A_{Φ} is the area of the magnetic field and B is the magnetic flux density.

<table>
<tr><td>2009-2010
EMA-CVT Concept (Lab scale 1)</td><td>2011-2012
EMA-CVT bobbin Testing with higher ampere supply (Lab scale 2)</td></tr>
<tr><td>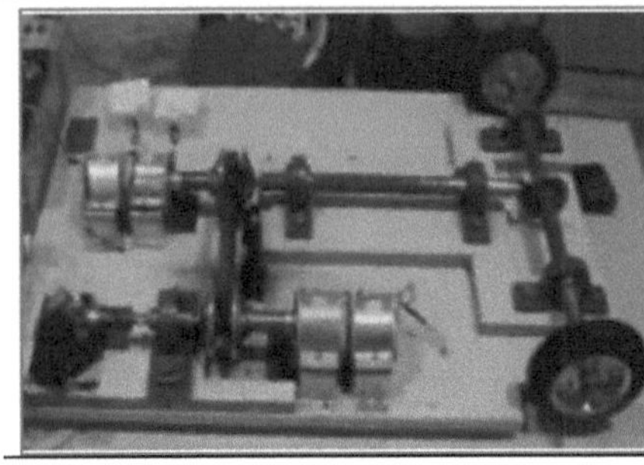</td><td></td></tr>
<tr><td>• Feasibility study of EMA-CVT for the vehicle alternative transmission
• EMA-CVT system outperforms the HyA-CVT system due to its quick actuation</td><td>• Very fast gear changing in 100-110 ms at high current supply
• Plastic bobbin/housing has bended due to heat gain</td></tr>
</table>

<table>
<tr><td>2016-2017
Metal bobbin EMA for CVT
(Lab scale 3)</td><td>2017—2019
Upscale EMAT -Proof of Concept for the CVT of Passenger Car</td></tr>
<tr><td>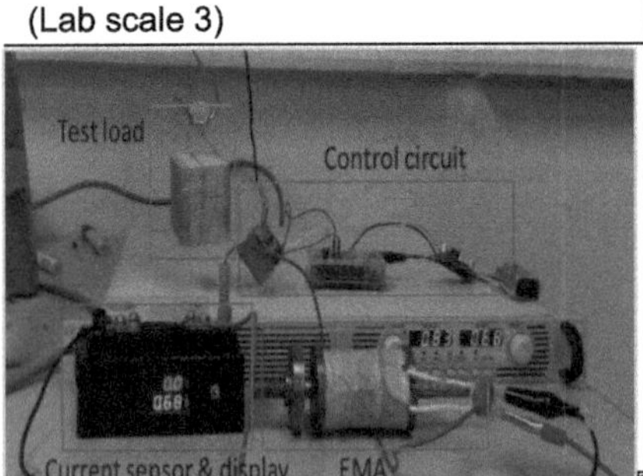
</td><td></td></tr>
<tr><td>• Feasibility study of the metal bobbin EMA for CVT actuation for the vehicle alternative transmission
• EMA can sustain with high temperature and prevent the creep speed of gear changing</td><td>• Very fast gear changing in 80-110 ms at current supply of 12-15A
• Lighter and provides fast acceleration.
• Provides Uniform Acceleration. Acceleration time 3-5s which is 61%, 45% and 30% less than AT, MT and Conventional CVT.</td></tr>
</table>

Fig. 17.8 Electromagnetic CVT innovation [7, 9, 26, 27]

Table 17.1 Magnetic quantities and unit

Quantity	Symbol	Unit	Abbreviation
Magnetic flux	φ	Weber	Wb
Inductance	L	Henry	H
Magnetic flux density	B	Tesla (Wb/m2)	T
Magnetic field strength	H	Ampere per meter	A/m
Magnetomotive force (mmf)	$\Im$ or F_m	Ampere turns	A-t
Energy	W	Joule	J
Force	F	Newton	N

Magnetic Flux Density and Magnetic Flux Strength

Whenever a magnetic flux ϕ exists in a body or component, it is due to the presence of magnetic field intensity or strength H, given by

$$H = F_m/l \tag{17.3}$$

where, F_m is the magnetomotive force acting on the component, and l is the length of the component. Its unit is ampere per meter (A/m). The flux density is the flux divided by the cross sectional area of the magnetic conductor that the flux is transported in. Its unit is the Weber-per-metre-squared (Wb/m^2) or *Tesla* (T), and it is designated by the letter B. There is a definite relationship between the flux density and magnetic field intensity of any material. In free space, B and H are related by the constant μ_0, known as the permeability of free space:

$$B = \mu_0 H \tag{17.4}$$

Magnetic Flux Linkage

Magnetic flux linkage is the total magnetic flux linking a given circuit or conducting structure. If the structure consists of a single conductor with multiple loops, for example a solenoid, the λ is equal to the flux linking all loops of the structure. It is given as

$$\lambda = N\varphi \tag{17.5}$$

Permeability

Permeability describes how much a material resists magnetic flux flow, and so is equivalent to the resistivity of electrical conductors. It has units of Tesla-metres-per-Ampere or Henry-per-meter, and is designated by the Greek letter Mu (μ). The permeability of free space, μ_0, has a fixed value of $4\pi \times 10^{-7}$ [henry/meter].

Relative Permeability

The flux density in a material also depends upon the magnetic field intensity to which it is subjected. Its value is given by:

$$B = \mu_0 \mu_r H \tag{17.6}$$

A material is classified according to the nature of its relative permeability μ_r, which is actually related to the internal atomic structure of the material. Permeability of a particular material can be defined as $\mu = \mu_r \mu_0$ [Henry/meter]. Most non-magnetic materials are classified as either *Paramagnetic* (μ_r slightly greater than 1.0) or *Diamagnetic* (μ_r slightly less than 1.0). Practically, μ_r considers 1.0 for these materials [28]. Magnetic (known as *Ferromagnetic*) materials include certain forms of iron and its alloys in combination with cobalt, nickel, aluminium, and tungsten. They are easy to magnetize since they have a high value of relative permeability [29]. This is the permeability of a conductor relative to that of free space. For example, soft iron has a relative permeability of around 6000, which means its *actual* permeability is $6000 \times 4\pi \times 10^{-7} = 0.0075$ Tm/A.

Electromagnetic Force

Conservation of energy principles stated that the energy neither be created nor be destroyed, rather it transforms from one to the other form. In electromagnet circuit as shown in Fig. 17.9, most of the energy is stored in the airgap separating between the two ferromagnetic bodies [28, 29]. The energy that stored in the airgap, also could be termed as the coupling field, then be converted to mechanical energy. Considering the conservative portion of the system in Fig. 17.8, they could be written as [28].

$$\text{Input electircal energy} = \text{Mechanical work done} + \text{Increase in stored energy} \tag{17.7}$$

Symbolically the above equation could be written as

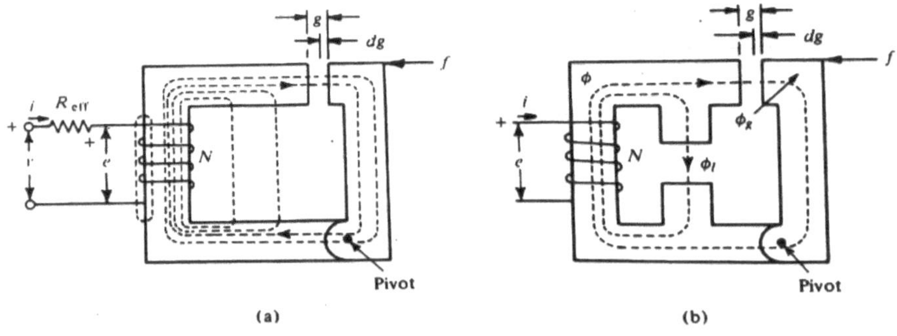

Fig. 17.9 Electromagnet **a** approximate flux path **b** simplified flux path [29, 30]

$$v_i\,dt = F\,dg + d\omega_\varphi \tag{17.8}$$

where, F is the force of electrical origin, dg is the differential length of the gap g, and w_φ is the increase in stored magnetic energy. From Faraday's law, voltage v can be expressed in terms of flux linkage λ as

$$v = \frac{d\lambda}{dt} \tag{17.9}$$

So that Eq. (17.8) becomes:

$$F\,dg = -d\,W_\varphi + i\,d\lambda \tag{17.10}$$

In an electromechanical system either (i, g) or (λ, g) may be considered independent variables. If we consider (i, g) as independent, the flux linkage λ is given by $\lambda = \lambda(i, g)$, which can be expressed in terms of small changes as

$$d\lambda = \frac{\lambda}{i}di + \frac{\lambda}{g}dg \tag{17.11}$$

For the stored energy $W_\varphi = W_\varphi(i, g)$, so that

$$d\,W_\varphi = \frac{W_\varphi}{i}di + \frac{W_\varphi}{g}dg \tag{17.12}$$

When substitute Eqs. (17.10) and (17.11) in (17.9), yield

$$F\,dg = \left(\frac{W_\varphi}{x} + i\frac{\lambda}{g}\right)dg + \left(\frac{W_\varphi}{i} + i\frac{\lambda}{i}\right)di \tag{17.13}$$

Because the incremental changes di and dx are arbitrary, F must be independent of these changes. Thus for F to be independent of di, its coefficient must be zero. Consequently, Eq. (17.12) becomes:

$$F\,dg = -\frac{\partial\,W_\varphi}{\partial g}(i, g) + i\frac{\partial\lambda}{\partial g}(i, g) \tag{17.14}$$

Which is the force equation and hold true if i is independent variable. If on the other hand, λ is taken as the independent variable, that is, if $i = i(\lambda, g)$ and $W_\varphi = W_\varphi(\lambda, g)$, then

$$d\,W_\varphi = \frac{\partial\,W_\varphi}{\partial\lambda}dg + \frac{\partial\,W_\varphi}{\partial g}dg \tag{17.15}$$

which, when substituted into Eq. (17.10), gives

$$F\,dg = -\frac{\partial\, W_{\varphi}}{\partial g}dg - \frac{\partial\, W_{\varphi}}{\partial \lambda}d\lambda + i\,d\lambda$$

Because $\partial\, W_{\varphi} \big/ \partial\lambda = i$, finally the above equation becomes:

$$F\,dg = -\frac{\partial\, W_{\varphi}}{\partial g}(\lambda, g) \tag{17.16}$$

The quantity of $i\,d\lambda$ in Eq. (17.11) is called the electromagnetic differential energy. Its relationship between the stored energy in the gap and the mechanical energy generated by the electromagnet could be presented [28] as:

$$i\,d\lambda = dw_{\varphi} + \partial W_{Mech.} \tag{17.17}$$

The stored energy in the gap could be specified into two types namely the airgap field and the leakage field:

$$d\,W_{\varphi} = d\,W_{\varphi g} + d\,W_{\varphi l} \tag{17.18}$$

Also

$$\begin{aligned} \lambda &= N\phi = N\phi_g + N\phi_1 \\ d\lambda &= Nd\phi = Nd\phi_g + Nd\phi_1 \end{aligned} \tag{17.19}$$

So that, Eq. (17.17) could be written as

$$Ni\;\; d\varphi_g + Ni\;d\,\varphi_l = d\,W_{\varphi\,g} + d\,W_{\varphi\,l} + d\,W_{mech} \tag{17.20}$$

If current is adjusted so that the airgap flux remains constant while the airgap is shortened by the difference dg, then $Ni\;\; d\varphi_g \;=\; 0$ and all the electromagnetic differential energy input is stored in the leakage field and we have $Ni\;\; d\varphi_l = d\,W_{\varphi l}$, so

$$d\,W_{\varphi\,g} + d\,W_{mech} = 0$$

$$d\,W_{\varphi\,g} = -d\,W_{mech} \tag{17.21}$$

The energy stored in the field of the airgap as simplified [29], when the fringing is neglected, to be:

$$W_{\varphi\,g} = Vol_g\,\frac{B_g^2}{2\,\mu_o} = A_g\,g\,\frac{B_g^2}{2\,\mu_o} \tag{17.22}$$

Vol_g is the airgap volume, A_g is the airgap face area, and B_g is the airgap magnetic flux density. When φ_g is constant, B_g must also be constant and

$$dW_{\varphi g} = \frac{B_g^2}{2\,\mu_o} d\,Vol_g = \frac{A_g\,B_g^2}{2\,\mu_o} dg \tag{17.23}$$

The force equation can be written as,

$$F = \frac{A_g\,B_g^2}{2\,\mu_o} \tag{17.24}$$

By substituting the magnetic field strength $H = F_m / l_e$ into magnetic flux density equation $B = \mu_0\,H$, we get from [30]:

$$B = \frac{F_m\,\mu_0}{g} = \frac{N\,i\,\mu_0}{g} \tag{17.25}$$

Substituting into Eq. (17.24, yield the electromagnetic force equation

$$F = \frac{(N\,i)^2\,\mu_o\,A_g}{2\,g^2} \tag{17.26}$$

According to the equation above, the electromagnetic force is dependent on the electromagnet's effective area, the number of coil turns, the provided current, and the distance between the electromagnet and the object being pulled.

17.7 Ema Power Flow Control Strategy

Power supplies can be divided into two categories, which will be covered in more detail below: switching mode power supply and linear regulated power supply. Different control strategies were needed for both power supplies in order to manage the voltage level and current flow.

Linear Regulated Power Supply

A DC power supply, sometimes referred to as a linearly regulated power supply (LRPS), receives an AC input and produces a steady DC output voltage. The block diagram of a standard regulated DC power supply is displayed in Fig. 17.10. It includes voltage control, rectification, filtration, and a step-down transformer. The ac voltage will be reduced by the step-down transformer to the desired output voltage. The transformer step-down ratio will determine the output voltage.

The full wave bridge rectifier, which has four diodes, will then receive the step-down voltage. The arrangement of the diodes ensures that the AC voltage output is unidirectional and that the filter input displays the dc voltage. The rectifier produces

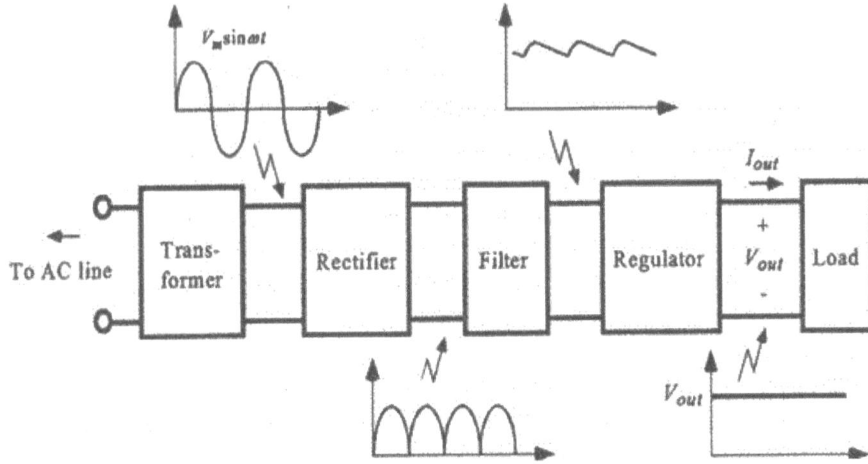

Fig. 17.10 Typical components of linear regulated power supply [3, 31]

a pulsing dc voltage with a high ripple content as its output voltage. As a result, a capacitor is employed as a filter, producing a nearly constant dc value with reduced ripple. The voltage regulator is the final block. It is employed to keep the output voltage constant despite variations at the input supply. Transistor series regulators, as well as fixed and variable IC regulators like the LM7805 and LM317, are examples of the regulator.

LRPS offer benefits and drawbacks depending on how the circuit is used. Since the LRPS can only lower an input voltage to produce a lower output voltage, the circuit cannot be used to step up the voltage. Additionally, the linear regulator's efficiency is only between 35 and 65 percent due to the significant power loss caused by the regulator. For instance, when a 5.0 V regulator receives 12 V and 100 mA of current (or 1200 mW of power) at its input, it needs to waste roughly 700 mW of power in order to give 500 mW to the load. Thus, the efficiency is just roughly 42%. Small applications over 10 watts are inefficient and not cost-effective due to the linear regulator's loss from heat dissipation [32].

Switching Mode Power Supply

The Switching Mode Power Supply (SMPS), which has an efficiency of 65 to 95 percent, is far more efficient in terms of power conversion than LRPS. Nevertheless, SMPS's drawback is that its design is far more intricate. Additionally, in many applications, the output voltage needs to be free of noise from the switching process. Many applications demand the use of both LRPS and SMPS, despite the fact that they differ greatly from one another. To eliminate noise from the final output, for instance, a linear regulator may be employed after an SMPS has controlled the beginning voltage. Forward-mode (also known as a buck converter) and boost-mode (also known as a boost converter) are the two fundamental forms of SMPS [30]. Additionally, both kinds could be combined and referred to as buck-boost converters.

Buck Converter SMPS

The buck converter's basic configuration is depicted in Fig. 17.11. Its output side filter is made up of a capacitor C and an inductor L. The pulse width modulation (PWM) signal supplied by a controller governs the switching process. The two distinct phases of operation for the buck converter are when the power switch (SW) is turning on and when it is shutting off. The input voltage is fed straight into the L-C filter components' input when the switch is turned on.

Because of the inductor's residual flux, the inductor current starts to increase linearly from an initial current. The voltage on the inductor's input side attempts to flow to earth when the switch is off, but is blocked by diode D when it becomes forward biased. After then, the stored energy keeps going to the output via the inductor and diode [33].

Boost Converter SMPS

The basic diagram of a boost converter, which consists of the same parts as a buck converter but in a different configuration, is shown in Fig. 17.12. The inductor's current increases from zero when the switch is turned on since the input voltage source is connected directly to it. The energy in the inductor's core is then entirely transferred to the output capacitor when the power switch is turned off [33].

Fig. 17.11 Buck converter basic diagram [32, 33]

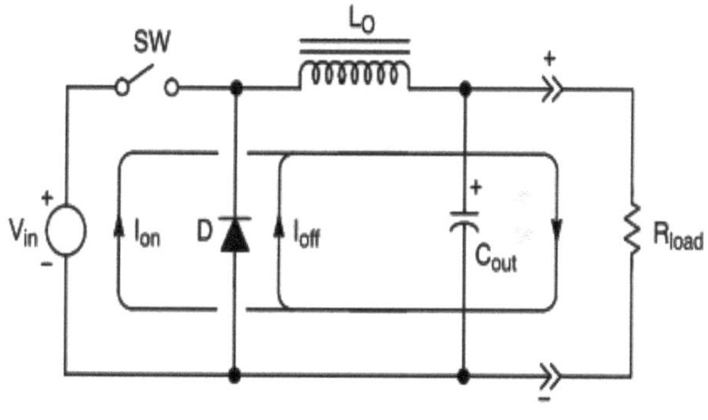

Fig. 17.12 Boost converter basic diagram [32, 33]

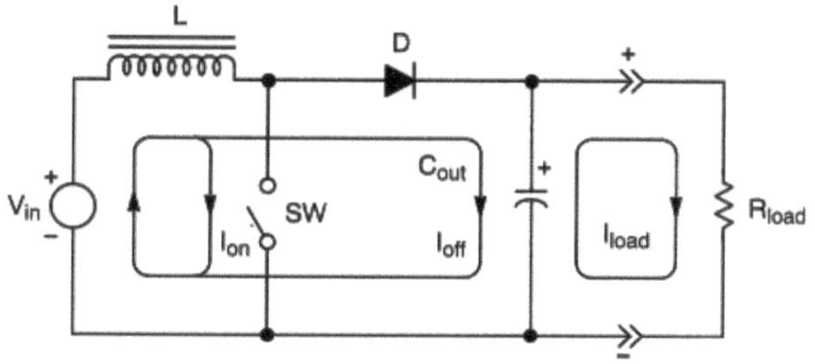

17.8 Control System for Vehicle

A control system is a tool used to change a system's behavior so that it always operates in a desired manner. For instance, one may like to keep the temperature and pressure in a reactor vessel in a chemical process plant at certain levels, or a driver may want to keep their vehicle moving at a specific speed on a highway despite hills or unfavorable winds. All of these scenarios demand for a control system to automatically accomplish the intended outcome without the need for human involvement [32].

The discrepancy, or error, between the plant's actual and intended output is how an automatic control system operates. The controller implements this decision-making process, which is known as a control law. A close-loop control system for an automobile's auto speed control is demonstrated in Fig. 17.13. Sensors provide the controller with information about the real speed, and an actuator modifies the gas pedal's position to carry out the control decisions.

Figure 17.14 shows the step response is a popular analysis method for figuring out a system's typical performance. The peak time (T_P) and rise time (Tr) are used to measure it. 0–100% increase time index is utilized for underdamped systems that overshoot. The peak time is not specified if the system is overdamped, and the 10–90% rise time Tr1 is typically utilized. The settling time Te and the percentage overshoot indicate how closely the actual response resembles the step input. Semiconductor [31] defines the percentage overshoot as follows:

$$P.O = \frac{M_{pt} - fv}{fv} * 100\% \tag{17.27}$$

where M_{pt}, is the peak value of the time response, and fv is the final value of the response. Normally, fv is the magnitude of the input, but many systems have a final value significantly different from the desired input magnitude. The settling time, T_S is defined as the time required for the system to settle within a certain percentage of the specified error band.

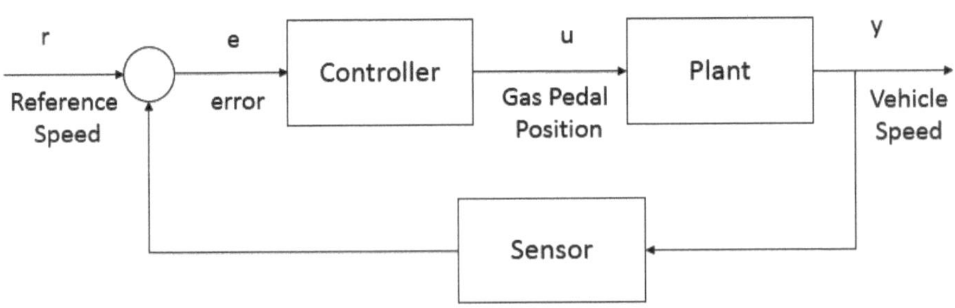

Fig. 17.13 Close-loop control system configuration with car auto cruise control

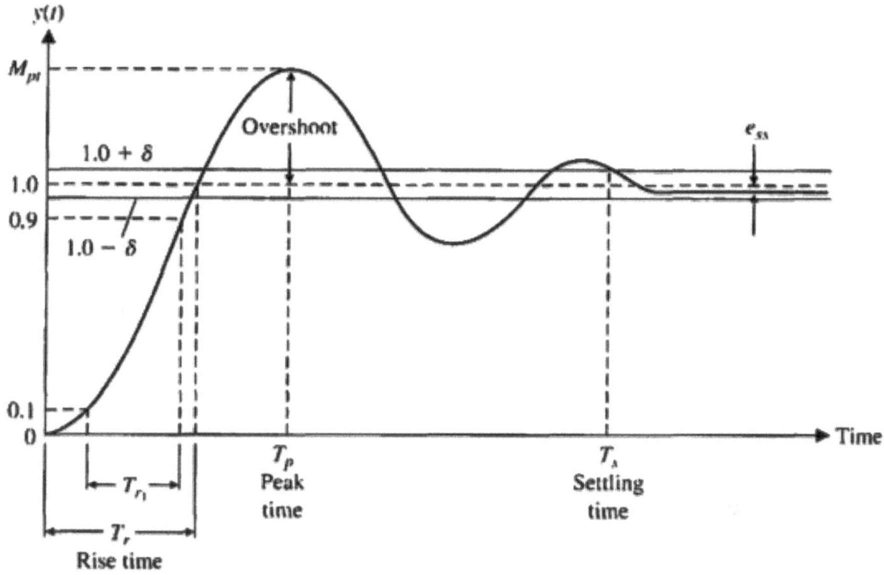

Fig. 17.14 Step response of a control system [3, 31]

17.9 Control System for Cvt

For comparatively straightforward applications, such as shift timing and quality, the first electronic control for automatic gearboxes was launched in the 1980s. A throttle position sensor was utilized to determine the vehicle's engine load, and a speed sensor was employed in place of a governor (speed limiter) in this control system.

Fuzzy Logic Controller

The number of applications of fuzzy logic has grown dramatically because of its robustness, simplicity, and lack of a precise mathematical model. Applications for fuzzy logic are numerous and range from consumer goods like cameras, camcorders, washing machines, and microwave ovens to industrial process control, medical instrumentation, portfolio selection, and decision-support systems. Fuzzy logic is a very effective technique for swiftly and effectively handling imprecision and nonlinearity systems since it makes decisions using common sense. Fuzzy logic is not a panacea, though [34, 35].

The foundation of a fuzzy logic controller (FLC) is fuzzy logic, which is far more similar to human thought processes. The foundation of a fuzzy logic controller (FLC) is fuzzy logic, which is far more akin to natural language and human thought than the conventional logical system. This is far more in line with natural language and human thought than the conventional logical approach. As a result, the FLC offers an algorithm that transforms the expert knowledge-based linguistic control approach into an automatic control strategy.

In this sense, the FLC methodology is especially helpful when processes are too complicated for traditional control techniques to analyze or when the information sources are imprecise or unclear. One of the pioneers of fuzzy logic, M. Sugeno, attempted to provide a method that offers a methodical design process of an FLC and guarantees its stability inside fuzzy control theory as explained [32] in order to get around the difficulty and complexity of creating fuzzy control rules.

Even though fuzzy logic is a relatively new and systematic science it was first presented by Lotfi A. Zadeh over 50 years ago—the ideas behind it have been around since the beginning of human history. The first step in a control process is the mapping of input to output by humans. A practical method for mapping an input space to an output space is discovered to be fuzzy logic. The goal of contemporary fuzzy logic control systems is to simulate human decision-making. For example, using data on the vehicle's speed and motor effort, a fuzzy logic system may determine the appropriate gear ratio to provide the right amount of power. The father of fuzzy logic, Lotfi A. Zadeh, famously said: "Fuzzy logic is faster and cheaper, but in almost every case you can build the same product without fuzzy logic" [34, 35].

The fuzzy logic control technique seems to be the best control methodology for complicated and highly non-linear systems whose exact mathematical model is hard or impossible to obtain but which human specialists can describe. Fuzzy rules are linguistic variables that represent the dynamic behavior of the controlled system. Human expert knowledge is used to formulate the set of fuzzy control rules. A FL controller works on the same idea as a human operator, which is to modify the input signal while focusing solely on the system output [36]. Table 17.2 shows the basic control summery based on the publish research output.

17.10 Conclusion

This chapter summarizes a number of studies on CVT system enhancement. Due to its smaller and lighter transmission unit, CVT has an advantage over manual and automatic transmissions. The main challenges at the moment are economical power and fuel economy, rapid acceleration, and emission management. In essence, the issues are closely related to precise actuation failure. A novel EMA actuation method developed for this study would be able to partially address those problems. Additionally, CVT systems are currently quite popular because to their smooth and jerk-free transmissions, which improve driveability.

In contrast to hydraulic actuation systems, the EMA idea seems to offer a notable reduction in transmission loss by maintaining precise position without fluctuation and a significant improvement in the traveling time of car acceleration through fast actuation response. Using an intelligent actuation system can result in a significant reduction in power usage.

Table 17.2 Summary of relevant works on CVT control system

References	Type of control	Remarks
Cholis [37]	Proportional, integral and derivative (PID)	The controller was used to control the electro-mechanical single acting pulley rubber V-belt continuously variable transmission for a scooter
(Supriyo et al., 2015)	Fuzzy-PID controller	Fuzzy logic controller was used to improve the fix gain PID controller in regulating the DC servomotor position of electromechanical type CVT
Ye et al. [38]	Fuzzy logic controller (FLC)	FLC has advantage over a nonlinear system, robust against parameter variations, and does not require an accurate mathematical model. With linguistic variables, a FLC algorithm is easier to understand
Pulles et al. [39]	Proportional and integral (PI)	PI is a part of PID controller. In this paper, it was used to minimize the CVT belt slip
Zhang et al. [40]	Fuzzy-PID based on a particle swarm optimization (PSO)	The controller was used to maintain the desired speed of single motor actuation system for CVT (SM-CVT)
Cholis et al. [41]	Proportional and derivative (PD)	Ziegler-Nichols tuning method was used to find the gain of the controller. PD controller improve the overshoot percentage and steady state error

Acknowledgements The project was funded by the Prototype Research Grant Scheme (PRGS) 2016 by the Malaysian Ministry of Higher Education, for which the authors are thankful. The authors also acknowledge the funding provided by the International Islamic University Malaysia for the development of prototype of EMA-CVT and the Eastern Michigan University, Ypsilanti, Michigan, USA, for enabling the finalization of EMA-CVT for commercialization.

References

1. Hassan, A. J. (2020). *Development of intelligent electromagnetic continuously variable transmission.* [A PhD Thesis Dissertation]. IIUM Library, International Islamic University Malaysia.
2. Denton, T. (2011). *Automobile mechanical and electrical system.* Elsevier.
3. Dorf, R. C., & Bishop, R. H. (2011). *Modern control systems* (12th ed.). Pearson.
4. Tawi, K. B., Supriyo, B., Ariyono, S., Husain, N. A., Hamid, A. R. A., Azlan, M. A., & Kob, M. S. C. (2015). Design of electro-mechanical dual-acting pulley continuously variable transmission. *Journal of Mechanical Engineering and Sciences (JMES), 8*(June), 1332–1342.
5. Ang, K. K., Quek, C., & Wahab, A. (2002). MCMAC-CVT: A novel on-line associative memory based CVT transmission control system. *Neural Networks, 15*(2), 219–236. https://doi.org/10.1016/S0893-6080(01)00143-5
6. Tawi, K. B., Mazali, I. I., Husain, N. A., Kob, M. S. C., Supriyo, B., & Ariyono, S. (2012). Development of electro-mechanical actuation system for continuously variable transmissions in automotive applications. Industrial Research Workshop and National Seminar 2012.

7. Pesgens, M., Vroemen, B., Stouten, B., Veldpaus, F., & Steinbuch, M. (2006). Control of a hydraulically actuated continuously variable transmission. *Vehicle System Dynamics, 44*(5), 387–406.
8. Pesgens, M., Vroemen, B., Veldpaus, F., & Steinbuch, M. (2003). Control of a continuously variable transmission in an experimental vehicle. *Ifac*, (2000).
9. Rahman, A., Sharif, S. B., Mohiuddin, A. K. M., Rashid, M., & Hossain, A. (2014). Energy efficient electromagnetic actuated CVT system. *Journal of Mechanical Science and Technology, 28*(4), 1153–1160.
10. Review Kereta Malaysia. (2012). Masalah/problem Proton Preve. Retrieved April 14, 2017, from http://keretamy.blogspot.my/2012/05/proton-preve-common-problem.html
11. Zareen. (2015). *Proton preve CVT exec: Review pemilik sebenar*. Retrieved April 14, 2017, from http://zaren88.blogspot.my/2015/02/proton-preve-cvt-exec-review-pemilik.html
12. Zarul, Yunos, S., & Khan, A. Q. A. (2014). Khas untuk pemilik Proton Prevé dan Suprima S: Jom bagi ulasan mengenai kereta anda di sini! Retrieved April 14, 2017, from http://www.funtasticko.net/khas-untuk-pemilik-proton-preve-dan-suprima-s-jom-bagi-ulasan-mengenai-kereta-anda-di-sini/
13. Rothenbühler, Y. (2009). New slip synthesis and theoretical approach of CVT slip control. *4337*, 159.
14. Nissan Global. (2019). XTRONIC CVT. Retrieved October 1, 2019, from https://www.nissan-global.com/EN/TECHNOLOGY/OVERVIEW/cvt.html
15. Rahman, A., Sharif, S. B., Hossain, A., Mohiuddin, A. K. M., & Zahirul A. (2012). Kinematics and nonlinear control of an electromagnetic actuated CVT system for passenger vehicle. *Journal of Mechanical Science and Technology*, *26*(7), 2189–2196. https://doi.org/10.1007/s12206-012-0542-0
16. Bavay, J.C., & Verdun, J. (1991). Alliages fer-silicium. *Technique de l'ingenieur, 1.*
17. Birch, T., & Rockwood, C. (2006). *Automatic transmissions and transaxles* (3rd ed.). Prentice Hall.
18. Burke, M., Briffet, G., Fuller, J., Heumann, H., & Newall, J. (2003). Powertrain efficiency optimization of the torotrak infinitely variable transmission (IVT). *SAE International*. https://doi.org/10.4271/2003-01-0971
19. Bullinger, M., Pfeiffer, F., & Ulbrich, H. (2005). Elastic modelling of bodies and contacts in continuous variable transmissions. *Multibody System Dynamics, 13*(2), 175–194.
20. Penning, B., Drogen, M. V., Brandsma, A., Ginkel, E. V., & Lemmens, M. (2006). Van Doorne CVT fluid test: A test method on belt-pulley level to select fluids FOR push belt CVT application. In *Continuous variable transmission (CVT)* (1st ed., pp. 217–231). SAE International.
21. Kong, L., & Parker, R. G. (2008). Mechanism and Machine Theory Steady mechanics of layered, multi-band belt drives used in continuously variable transmissions. *43*, 171–185. https://doi.org/10.1016/j.mechmachtheory.2007.02.003
22. Nishizawa, H., Yamaguchi, H., & Suzuki, H. (2005). Friction characteristics analysis for clamping force setup in metal V-belt type CVTs. *R&D Review of Toyota CRDL, 40*(3), 14–20.
23. Delkhosh, M., Foumani, M. S., Boroushaki, M., Ekhtiari, M., & Dehghani, M. (2011). Geometrical optimization of half toroidal continuously variable transmission using particle swarm optimization. *Scientia Iranica, 18*(5), 1126–1132.
24. Tanaka, H., & Machida, H. (1996). Half toroidal traction-drive continuously variable power transmission. *Proceedings of the Institution of Mechanical Engineers, Part J: Journal of Engineering Tribology, 12.*
25. Shafie, A. A., & Ali, M. H. (2009). Development of an efficient CVT using electromechanical System. *World Academy of Science, Engineering and Technology.*
26. Rahman, A., Jaffar, A. H., & Sany, I. I. (2022). IoT-based electromagnetic actuator for CVT: Basic design and prototyping. *International Journal of Powertrains, 11*(1), 38–61.
27. Rahman, A., Sazzad, B. S., & Hossain, A. (2012). Kinematics and non-linear control of electromagnetic actuated CVT system. *Journal of Mechanical, Science and Technology., 26*(7), 2189–2196.

28. Sharif, S., Rahman, M. A., Mohiuddin, A. K. M., Rahman, M.A., Sharif, S., Mohiuddin, A. K. M., Rashid, M. M., & Hossain, A. (2013). Study on the development of a fuzzy logic control electromagnetic actuated CVT system. *International Journal of Engineering Systems Modelling and Simulation*, *5*(4), 217–225. ISSN 1755-9766
29. Boldea, I., & Nasar, S. A. (1997). *Linear electric actuators and generators* (1st ed.). Cambridge University Press.
30. Leander, W., Matsch, L., & Morgan, J. D. (1987). *Electromagnetic and electromechanical machines* (3rd ed.). John Wiley & Sons.
31. Semiconductor, O. (2014). Switch—mode power supply reference manual. *Technology*, *56*(8), 968–9A1. http://www.onsemi.com/pub_link/Collateral/SMPSRM-D.PDF
32. Antsaklis, P., & Gao, Z. (2005). Control systems. In *The electronics engineers' handbook* (5th ed.). Mc Graw Hill.
33. Clarke, R. (2017). Magnetism: Quantities, Units and Relationship. Surrey University. https://doi.org/10.1016/j.scient.2011.08.005. http://info.ee.surrey.ac.uk/Workshop/advice/coils/terms.html#example
34. MathWorks. (2019). *Subsystem basics*. Retrieved October 12, 2019, from https://www.mathworks.com/help/simulink/ug/what-is-a-subsystem.html
35. The Mathworks. (2012). *MATLAB: Fuzzy logic toolbox users' guide*. Natick, MA. Retrieved from http://www.mathworks.com/products/fuzzy-logic/demos.html
36. Dorf, C., & Robert, H. B. (2011). *Modern control systems Richard* (12th Ed.). Pearson Prentice Hall.
37. Cholis, N. (2013). Electro-mechanical single acting pulley rubber V-belt continuously variable transmission (EMSAP RVB-CVT) for scooter. Universiti Malaysia Pahang.
38. Ye, M., Liu, Y. G., & Cheng, Y. (2016). Modeling and ratio control of an electromechanical continuously variable transmission. *International Journal of Automotive Technology*, *17*(2), 225– 235. https://doi.org/10.1007/s12239
39. Pulles, R. J., Bonsen, B., Steinbuch, M., & Veenhuizen, P. A. (2005). Slip controller design and implementation in a continuously variable transmission. In *Proceedings of the 2005, American Control Conference, 2005*, (pp. 1625–1630). https://doi.org/10.1109/ACC.2005.1470200
40. Zhang, L., Shen, H., Holly, L., Rojas, S., & Wang, S. (2017). Theoretical analysis and test of CVT using a single motor actuation system, *32*, 62–70.
41. Cholis, N., Ariyono, S., & Priyandoko, G. (2015). Design of single acting pulley actuator (SAPA) continuously variable transmission (CVT). *Energy Procedia*, *68*, 389–397. https://doi.org/10.1016/j.egypro.2015.03.270
42. Kim, J. H., & Oh, S. J. (2000). A fuzzy PID controller for nonlinear and uncertain systems. *Soft Computing*, *4*(2), 123–129.
43. Altas, I. H., & Sharaf, A. M. (2007). A generalized direct approach for designing fuzzy logic controllers in Matlab. *Simulink Gui Environment*, *1*(4).

Chapter 18
Electromagnetic CVT Development

Abstract A thorough understanding of the CVT's movable sheave actuation based on vehicle dynamics under various load and driving conditions has been used to develop intelligent EMA-CVT. The real CVT ratio can be found using the Crankshaft Position Sensor (engine RPM), the CVT Pulley Position Sensor, and the CVT Secondary Pulley Revolution Sensor. By using DCSS to manage the power supply to the EMA, the FLC then adjusts the CVT appropriately based on the error (the difference between the goal and actual ratio). This work developed a fuzzy intelligent system to forecast the dynamic behavior of the EMA system in order to determine the current control for the EMA actuation when the CVT is operating. Using digital control switching and a fuzzy logic controller, the power supply has been managed to set the targeted CVT gear ratio. The CVT has a built-in failsafe mode to guard against damage in the event that some of the sensors fail. In order to prevent the central manual switching system from damaging the CVT and to fix the gear ratio equal to the first gear, the system switched to manual mode during failsafe mode. The time required to accelerate the car is found to be between 1.8 and 3.1 s. The EMA generates electromagnetic force between 108 and 301 Nm for supply currents between 6 and 9.33 amps.

Keywords EMA system dynamic behaviour · Digital control switching system · Central switching system · Energy effectiveness

18.1 Introduction

The intelligent EMA-CVT was designed and developed using a thorough understanding of the movable sheave actuation of the CVT based on vehicle dynamics under various load and driving conditions. The CVT actuation of the PROTON PREVE Model served as the basis for the actual size of the electromagnetic actuator. In order to study EMA parametrically using MathLAB simulation modeling, this chapter presents the modified mathematical models that have been developed based on the simplification of magnetic induction, electromagnetic energy and momentum,

A. Rahman et al., *Autonomous Driveline System for Electric Vehicle*, Green Energy and Technology, https://doi.org/10.1007/978-981-96-8418-2_18

electromagnetic radiation, and relativity and electromagnetism [1–3]. Understanding the vehicle dynamics, electromagnetic force generation, and CVT's operating principle simplifies the mathematical model. The creation and testing of laboratory-scale EMA has led to the design optimization of EMA.

A temperature sensor placed around the EMA while providing it with current (in amperes) has been used to test electromagnetic interference (EMI). To conduct the testing and evaluation of the EMA on CVT operation, the chapter also introduces a TEST-BED that was created as a result of this study. A thorough understanding of the CVT gear ratio, electromagnetic power regulation, and vehicle dynamics under all driving conditions has been used in the test-bed's creation. Because the PROTON PREVE vehicle was not available, the EMA was tested using the Toyota Wish real CVT. For all road conditions, the traction force generated at tire-road contacts was replicated using a load cell sensor with a commensurate weight. This chapter concludes by introducing a fuzzy logic controlling system that maintains the CVT gear sheave movement at the proper position for different traction torque development for the vehicle propulsion on the different road conditions by regulating an appropriate power level of the electromagnetic actuator.

18.2 Design Consideration

The electromagnetic actuator's required clamping force on the CVT's movable sheave has been established by taking into account the traction torque needed at the driving wheel under various load and road conditions. For a vehicle to propel itself past the force of gravity and motion resistance (rolling motion resistance, aerodynamic resistance, and slope resistance), traction torque generated at the driving wheel is crucial. Few presumptions are made on the design of the EMA. The assumptions are:

- It was assumed that the vehicle would be operating on an asphalt surface road with a rolling resistance of 0.013 and a coefficient of road adhesion of 0.85 in order to simulate its performance [1].
- According to Wong's (2008) technique, the Toyota Wish's frontal area is 2.0 m^2, while Wikipedia (2019a)'s data on the Toyota Prius, its replacement model, shows that the drag coefficient is 0.29.
- The Toyota Wish CVT states that the spring force that activates the CVT's primary pulley is 300 N/mm.
- The road infrastructure in Malaysia considers a maximum road gradient of 20% and a cruising speed of 110 km/h.

18.3 Mathematical Model

The EMA-CVT was developed taking into account the dimensions and size of a full-scale Toyota Wish automobile. The CVT pulley clamping force is estimated while accounting for the vehicle's traction force. The electromagnetic force generated by the EMA is equal to or greater than the clamping force due to its design. By integrating the fuzzy logic controller with the EMA-CVT, the system is able to intelligently regulate the power flow that it requires.

18.3.1 Vehicle Traction Force

It is necessary for a vehicle to produce a sufficient tractive force to overcome the forces of gravity, tire rolling resistance, and aerodynamic drag. A vehicle's acceleration is determined by the forces acting upon it and is determined by Newton's second law, which may be expressed as,

$$\begin{aligned} a &= \frac{F(t) - \Sigma R}{m} \\ \mathit{with}\ \ F(t) &= \frac{1}{r_w}\left(T_e N_g N_D \eta_t\right) \\ \Sigma R &= R_r + R_a + R_g \end{aligned} \tag{18.1}$$

where, m is the total mass of the vehicle in kg, *F(t)* is the instantaneous traction force of the vehicle in *N*, and R_r, R_a, and R_g are the motion resistance of the vehicle due to rolling, aerodynamics and grade. The road load resistive forces are normally the rolling resistance between tires and road interfaces can be defined as $R_r(t) = f_r W$ R_{roll} R_{roll} R_{roll}, air resistance, $R_a = 0.5\rho_a C_d A_f v^2$ and drag resistance or uphil grading resistance, $R_g = WG$, where G≡Sinθ and θ is the slop of the road. However, the traction force for the vehicle in starting mode can be modeled as,

$$F(t@\mathit{start\,from\ rest}) = \mu W \tag{18.2}$$

while the traction force at the cruising mode can be defined as,

$$F(t@\text{cruising}) = 0.5\rho_a C_d A_f v^2 \tag{18.3}$$

where, ρ_a is the air density in kg/m^3, C_d is the coefficient of aerodynamic resistance, A_f is the frontal area of the vehicle in m^2 and v is the travelling speed of the vehicle relative to the wind in m/s. Rolling and grade resistance are the primary causes of vehicle traction on the highest grade. Since the speed was lower and more stable, the aerodynamic drag could be disregarded. The aerodynamic drag causes the vehicle's traction force while it is in cruising mode, which is when the road is flat and the speed is extremely high 110 km/h according to Malaysian traffic regulations. *Rg* is therefore thought to be negligible, and the vehicle propels at a constant state speed while also taking into account the gravitational force. In comparison with aerodynamics, the effects of rolling motion resistance are negligible.

The traction torque required for a vehicle to propel on road at a desired speed can be estimated as,

$$T_w(t) = (ma + \Sigma R)r_w \tag{18.4}$$

Gear ration of the transmission system required to develop the traction torque $T_w(t)$ at the driving wheel of the vehicle to propel can be estimated as,

$$N_g = \frac{T_w(t)}{T_e(t)N_d\eta_t} \tag{18.5}$$

where N_g is the gear-ratio, N_d is the final drive ratio, η_t is the transmission efficiency. The propulsion power required to propel the vehicle at maximum speed can be estimated as,

$$P_p(t) = T_w(t)\omega(t) \tag{18.6}$$

where (*t*) is the wheel's angular speed. The vehicle is in beginning mode, and the (*t*) is very low, a high amount of traction is needed. Consequently, a greater gear ratio is required in this form of transmission in order to increase the vehicle's wheel torque. Furthermore, in order to drive the vehicle at moderate and greater speeds, the intermittent gear ratio is required.

The purpose of the study is to provide the highest torque and torque for a vehicle going at a moderate pace by creating the EMA-CVT, which can produce the required traction torque at the wheel while negotiating the road profile by altering the gear ratio from maximum, intermittent, and minimum. In order to solve the problems with the current CVT, manual transmission, and automatic transmission, electromagnetic actuators (EMA) are being introduced in this study.

The engine can run closer to its optimal efficiency line because to the CVT system's ability to give an endless number of gear ratios between the top and lowest limits [2]. More torque will be generated at the pulley system's output shaft at higher gear ratios (GR). However, when the engine runs at the same speed with greater GR, the car speed will be slower than at lower GR. The overall gear ratio of this CVT model in terms of diameter, speed and torque can be modelled as [3]:

$$GR = \frac{N_P}{N_S} \; or \; \frac{T_S}{T_P} \; or \; \frac{D_S}{D_P} \tag{18.7}$$

where, N_s is the secondary pulley speed, N_P is the primary pulley speed, T_P is the torque in primary pulley, T_s is the torque in the secondary pulley, T_w is the torque of the final drive, T_e T_e is the torque at engine, D_s is secondary pulley diameter, and

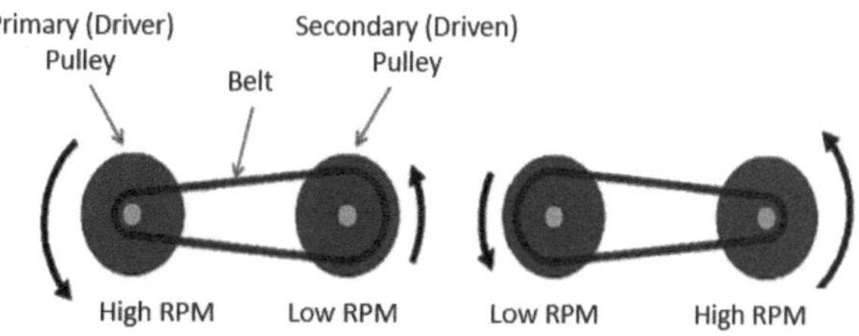

Fig. 18.1 CVT pulley diameter for under-drive and over-drive (O/D) situation [4]

D_p is the primary pulley diameter. Figure 18.1 shows example of pulley's position at different gear ratios.

A CVT system has an endless number of speed ratios because of the pulley's infinite diameter variation, but no precise speed ratio or gear ratio can be found. However, Fig. 18.1 shows that there are three fundamental functioning conditions for CVT gear ratio configuration:

- Input pulley is smaller than the output pulley, the underdrive (GR > 1) occurs. The available torque from the driving pulley is increased by this combination of gear ratios.
- Input pulley is greater than the output pulley, there is an overdrive (GR < 1). This causes the driving gear to rotate more quickly.
- Input and output pulleys are equal, direct drive (GR = 1) is accomplished. The reason it's termed "direct drive" is that both sets of pulleys

The Toyota CVT gearbox model K111, shown in Fig. 18.2, has been employed to assess EMA's actuation performance. With a metal pushing belt, it has two sets of pulley-movable sheaves (one for primary and one for secondary). Figure 18.3 displays the kinematics of the pulley's moving sheave. The secondary pulley's movable sheave is operated by a hydraulic system, while the primary pulley's is activated by a spring with a rigidity of 40 N/mm.

The CVT pulley sheave angle θ_b based on actual measurement then was estimated by trigonometry as 10.74°. While the belt vertical movement length (b) equal to 50 mm, the pulley sheave angle (θ_b) of 10.74°, the one-sided pulley maximum moving distance is found by following equation $\tan\theta_b = \Delta x/b$ and Δx found to be 9.4837 mm. The total pulley sheave movement or the stroke length (L_{stroke} or $2\Delta x$) is equal to

Fig. 18.2 CVT pulley sheave angle [5]

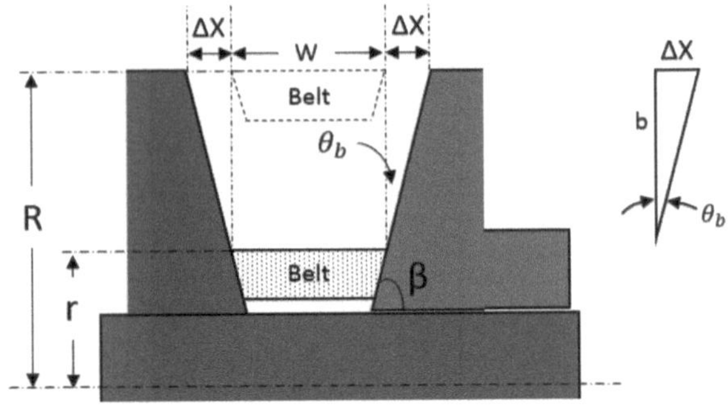

Fig. 18.3 CVT pulley and belt diagram [5]

18.9675 mm. There is about 2.6% different from the actual measurement, which is 19.50 mm. Since simple tools were used for the measurement, the primary cause of the discrepancy is measurement error or human error. Additionally, using the following formula, the pulley sheave radius (R) for different gear ratios was estimated:

$$\begin{aligned} R &= r + \Delta x * tan\beta \\ &= r + \Delta x * tan\left(90^0 - \theta_b\right) \end{aligned} \tag{18.8}$$

where, r is the minimum pulley sheave radius, which is when the belt at pulley bottom which is about 31 mm. In situation when the belt is at the top of the pulley sheave position (where the one-sided moving distance at maximum, $\Delta x = 9.4837$ $\Delta x = 9.4837$ $\Delta x = 9.4837$), the maximum pulley radius is found as 80.99 mm, the gear ratio of the CVT is computed by using the Eq. (1.5). The lower gear ratio or under drive and overdrive gear ratio are:

(i) Drive or Traction or Acceleration:

$$GR_{LOW} = \frac{D_S}{D_P} = \frac{R_S}{R_P} = \frac{80.99}{31} = 2.612$$

(ii) Overdrive or Cruising

$$GR_{OD} = \frac{D_S}{D_P} = \frac{R_S}{R_P} = \frac{31}{80.99} = 0.3827$$

According to Wikipedia (2019d), the calculated gear ratios agree with the manufacturer's actual gear ratios, which are 2.386 and 0.411 for the low and overdrive gear ratios, respectively. The discrepancy results from the pulley being measured by hand using a few tools. According to the agreement between the manufacturer's information and the calculations made using the equation, the gear ratio is negotiable. Nonetheless, the value that the manufacturer supplied is taken into account throughout this investigation.

18.4 Kinetics of Cvt Clamping Force

The GR increases in proportion to the primary and secondary pulleys' continual movement of movable sheaves. The primary pulley's movability is controlled by the spring force. The primary focus of this study is the electromagnetic force generated by the solenoid, which controls the movement of the secondary pulley's sheave. Two sets of solenoid sets with a common plunger were constructed by authors Ataur [7] to develop the pushing and pulling forces of the secondary pulley's movable sheave, as seen in Fig. 18.4.

The force of clamping Fc is the primary pulley's actuation force. Assuming that a spring is utilized to provide the primary and secondary pulley actuation force (Fig. 18.4), Fc, also known as the clamping force for each pulley, is calculated as follows:

$$F_c(t) = k_p \delta_p(t) cos\theta \tag{18.9}$$

where k_p is the spring stiffness that keeps the primary pulley in the highest diameter in N/mm and $\delta_p(t)$ is the instantaneous displacement. In this study, we have replaced the spring of the secondary spring by an electromagnetic actuator. In order to change the gear ratio of the EMA-CVT, the EMA needs to develop the electromagnetic force, which must be, $F_{em}(t) \geq k_p \delta_p(t) cos\theta$ for the changing the gear ratio of the CVT to make the vehicle in traction mode or in cruising mode.

The actuator should be built so that it can change the gear ratio in the range of by producing electromagnetic force that is at least as strong as the CVT's clamping force. The required gear ratio of the vehicle is estimated by using the equation:

$$\mathrm{T_{initial}} = \mathrm{m_c\, g}\left[\mu_\mathrm{R} \frac{(\mathrm{l_f} - \mathrm{f_r\, h})/\mathrm{L_w}}{1 + \mu_\mathrm{R}\, \mathrm{h}/\mathrm{L_w}} + \sin\theta_\mathrm{R}\right](\mathrm{r_w}) \tag{18.10}$$

$$T_{final} = \left(f_r\, W + \frac{1}{2}\rho_a A_f\, C_D\, v_c^2\right) r_w \tag{18.11}$$

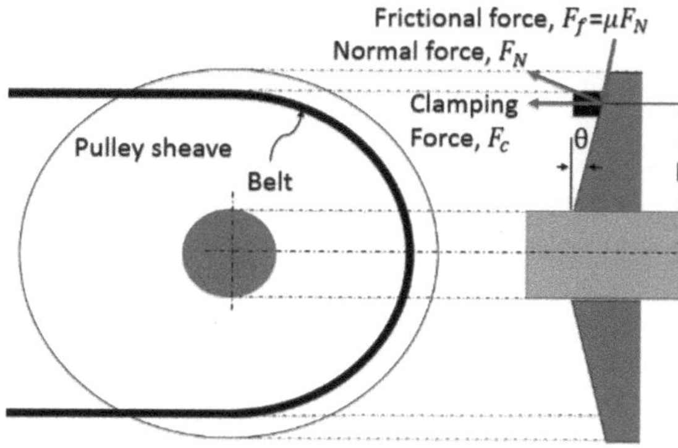

Fig. 18.4 Kinetic analysis of the pulley [6]

A car weighing 1700 kg, with a top speed of 120 km/h or 33 m/s, a wheel radius of 0.295 m, a drag area of 0.633, a rolling motion resistance coefficient of 0.02, and a road friction coefficient of 0.5 are used. Equation (18.10) for the initial traction torque (wheel torque, Tw) and Eq. (18.11) for the gradient 30% yield starting values of 1560 Nm and 1500 Nm, respectively [8, 9].

Traction torque at dynamic mode can be estimated using the equation,

$$T_W(t) = m\frac{\partial v}{\partial t} * r_w + \begin{vmatrix} f_r W & 0 < v \le 120\,km/h \\ W\,Sin\theta & 0 < G \ge 20\%\,,\ \ 0 < v \le 40\,km/h \\ \frac{\rho}{2} C_d A_f v^2 & G = 0\%,\ 40 < v \le 120\,km/h \end{vmatrix} * r_w \tag{18.12}$$

The following formula can be used to predict the gear ratio that the gearbox needs:

$$GR = N_g = \frac{T_w}{T_e N_d \eta_t} \tag{18.13}$$

where Gear ratio is denoted by N_g, driving wheel torque by T_w, engine torque by N.m. and Te, differential speed constant by N.m. and N_d, and transmission efficiency by η_t. The primary and secondary pulleys' movable sheaves have been activated to modify the CVT's gear ratio (GR). As previously mentioned, the primary pulley's movable part is controlled by the spring force. However, the electromagnetic force generated by the solenoid controls the secondary pulley's movable sheave. In this study, the EMA's design requirement is to activate the secondary pulley's movable sheave in order to raise the CVT's gear ratio (GR) to 1.1 or higher. But the maximum gear ratio that was intended to be 5.

Based on the CVT gear ratio and the final drive ratio of 4:1, the secondary pulley's maximum output torque is predicted to be 390 Nm. To retain the pulley sheave in the proper position for the intended gear ratio, the clamping force needed is determined by the input pulley torque.

The clamping force on the secondary can be estimated as,

$$F_{c(s)} = k_p \delta_p(t) cos\theta_b$$

It is considered that the pulley belt angle (θ_b) is 12°. The clamping force of CVT is 210 N for the maximum gear ratio of 5. The spring stiffness, ks is calculated to be 11.72 N/mm, and the primary pulley's spring displacement is 18 mm. The EMA must thus create an actuation force of 300 N, which includes its supporting spring force, in order to ensure smooth operation and quick car acceleration. To achieve the vehicle traction torque of 1560 Nm at a gear ratio of 5.0, the EMA's actuation system, which includes the secondary pulley back spring force, must be developed to maintain the torque at 390 Nm at the secondary pulley.

The CVT's actuation system is composed of the EMA, a set of springs, a power pack, and a plunger (one end of which is attached to the pulley's movable sheave and the other end to a metal disc). Section 18.3 has a thorough treatment of the subject. In order for the EMA to attain the maximum gear ratio of the pulley sets and prevent creep motion during return, it has been supported by a back spring (EMA's supporting spring) with a rigidity of 5.86 N/mm. For the 18 mm displacement to achieve the gear ratio of 5, the predicted back spring force is 105 N. Consequently, the EMA was created in this study to generate an actuation force of 200N for the CVT's seamless operation.

However, the CVT system's primary pulley incorporates a spring with a rigidity of 11.72 N/mm, which is utilized to activate the primary pulley when the EMA's power supply drops or stops. This function is taken into account while the vehicle is cruising with the gear ratio closed to 1.0. The EMA's total actuating force, including the spring force, can be calculated as follows:

$$F_{act.} = F_{em} + k_s \partial_s$$

where, F_{act} is the movable sheave of the secondary pulley which is also called the clamping force (F_c), F_{em} is the electromagnetic force in N, k_s is the spring stiffness of the spring of secondary pulley, N/m, ∂_s is the spring displacement, m.

The total energy store in the EMA coils due to the supply current can be modelled as,

$$\begin{aligned} E_{act} &= E_{d(EMA)} + E_{sd} \\ \text{with } E_{d(EMA)} &= \frac{1}{2\mu} B^2 V_{EMA} + \frac{1}{2}\varepsilon_0 E^2 \end{aligned} \quad (18.14)$$

where, E_{act} is the energy actuation of the EMA system, $E_{d(EMA)}$ energy develops at the EMA due to supply current, I and E_{sd} developed by the spring and $V_{EMA} = \pi r^2 l_c$ is the volume of the solenoid. If the spring of stiffness k_s is added at the other end of the plunger, by simplifying the (18.4), the energy develops by the EMA actuation system can be modelled as,

$$E_{act} = \frac{1}{2}\mu N^2 I^2 V_{EMA} + \frac{1}{2}\varepsilon_0 E^2 + k_s x^2 t_{act} \quad (18.15)$$

The power loss of the EMA due to the current supplied to its coils $I_0{}^2R_{rad}$ associated with the radiation resistance, whereas the power loss $I_0{}^2R_c$ associated with the real resistance is due to Ohmic heating of the EMA. The R_{rad} can be defined by using the equation:

$$R_{rad} = \frac{2\pi}{3\varepsilon_0 c}\left(\frac{l_c}{\lambda}\right)^2$$

Total power loss of the EMA can be defined as,

$$P_{EMA(loss)} = I^2\left[R_c + \frac{2\pi}{3\varepsilon_0 c}\left(\frac{l_c}{\lambda}\right)^2\right] \tag{18.16}$$

The effective energy of the EMA to do the actuation of the CVT's pulley,

$$E_{eff\,(EMA)} = E_{act(EMA)} - E_{l(oh)} - E_{l(rad)} \tag{18.17}$$

By simplifying the Eq. (21.17), the equation can be rewritten as,

$$E_{eff\,(EMA)} = \frac{1}{2}\mu N^2 I^2 \pi r^2 l_c + \frac{1}{2}\varepsilon_0 E^2 - I^2\left[R_c + \frac{2\pi}{3\varepsilon_0 c}\left(\frac{l_c}{\lambda}\right)^2\right]\int_0^t dt \tag{18.18}$$

18.5 EMA-CVT Force Modeling

The electromagnetic force (axial shifting force) required led to the development of the electromagnetic actuator (EMA) with copper coil winding. Electro-motive force (also known as voltage or EMF) in an electric circuit is equal to magneto-motive force (MMF). A-t, or ampere-turns, is its unit. A coil of N turns and I amps produces an MMF of NI Ampere-turns. In order to maintain the vehicle's gear ratio or transmission ratio and modify it as needed, the electromagnetic actuator is utilized. A fuzzy controller has been used to regulate the battery DC power supply to the EMA by using the sensors' signals. The EMA automatically determines the CVT's gear ratio depending on the sensors' responses (such as the slope and speed sensors and/or both.

The magnetic flux density and supply current determine the electromagnetic force, while the coil pack's length is held constant in the magnetic field. One coil or more coils, as illustrated in Fig. 18.5, can be used to measure the magnetic flux density of EMA. Simplifying Gauss's law, Maxwell's equations, the ampere circuital equation, Faraday's law, and Poynting's equation allowed for the creation of the EMA force modeling.

The EMA has considered a thin coaxial cylindrical conducting shells of radii a and c. The current, I is supplied with number of turns N per unit length. According to Gauss's law, the electric field takes the form $E = E_r\ e_r$ where,

$$E_r = \begin{vmatrix} -Q/(2\pi\varepsilon_0 rl) & for\, a \le r \le c \\ 0 & \text{Otherwise} \end{vmatrix} \tag{18.19}$$

Similarly, According to Ampere's circuital law, the initial magnetic field of the EMA due to supply current,

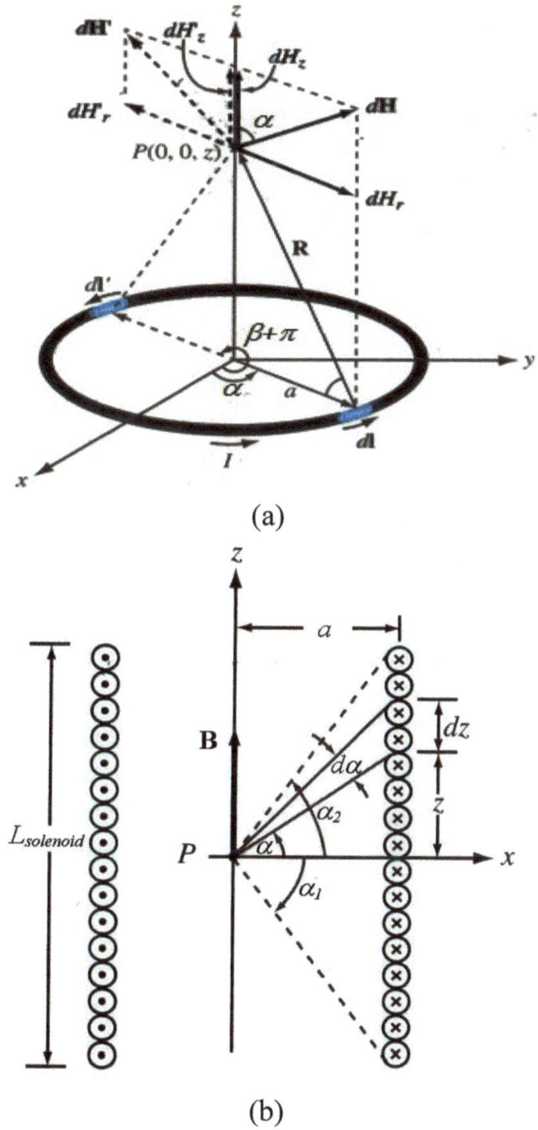

Fig. 18.5 Magnetic flux density **a** single coil, **b** multiple coils [1]

$$B_z = \begin{vmatrix} \mu_0 NI & for\, r \leq c \\ 0 & \text{Otherwise} \end{vmatrix} \tag{18.20}$$

The initial momentum density of the electromagnetic field could be,

$$g_\theta = \begin{vmatrix} \mu_0 NIQ/(2\pi rl) & for\, a \leq r \leq b \\ 0 & \text{Otherwise} \end{vmatrix} \tag{18.21}$$

Any change in the current flowing in the EMA winding generates an inductive electric field, which can be defined as,

$$E_\theta = \begin{vmatrix} -\mu_0 NIr/2 & for\, r \le b \\ -\mu_0 NIb^2/2r & \text{Otherwise} \end{vmatrix} \tag{18.22}$$

However, if a current of uniform density, $j = je_z$ within the coil of the EMA, the net energy flux U into the EMA surface, coaxial within the wire, of radius r and length l_c could be modelled as,

$$U = u2\pi rl_c = \rho j^2 V(r) \tag{18.23}$$

where $V_{EMA}(r) = \pi r^2 l_c$ is the volume enclosed by the surface of the EMA and $r = c - a = D_2 - D_1$. The EMA develops the magnetic force, F_{em} is the function of supplied current and number of coils as stated in Eq. (18.24). Therefore, the electromagnetic force [6]:

$$\begin{aligned} F_{em} &= \frac{E_{eff\,(EMA)}}{d} \\ &= \frac{1}{d}\left[\frac{1}{2\mu}B^2 V_{EMA} + \frac{1}{2}\varepsilon_0 E^2\right] - I^2\left[\frac{R_c}{l_c} + \frac{R_{rad}}{d(r)}\right] \\ \text{with } B &= \mu NI \\ d &= \text{pulley sheave moval distance, } m \end{aligned} \tag{18.24}$$

where, μ is magnetic permeability (degree of magnetization of a material in response to magnetic field), l_c length of the core, *m*, *d(r)* is the distance of the radiation measuring, and t is the time, s.

By applying the principle of virtual work, we can model an equation to estimate the total actuation force of the movable sheave of secondary pulley against the clamping force exerted from the action of the spring of primary pulley,

$$F_{c(s)} = k_{sp}\delta_p(t)cos\theta_b \tag{18.25}$$

The gear ratio of the CVT can be maintained using the EMA, if the total force develops by the EMA with associated the spring can be estimated using Eq. 2.24 and 2.25, as $F_{act.} = F_{em} + k_s\partial_s$. The electromagnetic force, F_{em} can be estimated as,

$$\begin{aligned} F_{em} &= \frac{1}{d}\left[\frac{1}{2\mu}B^2 V_{ema} + \frac{1}{2}\varepsilon_0 E^2 + k_{ss}\delta_{ss}\right] - I^2\left[\frac{R_c}{l_c} + \frac{R_{rad}}{d(r)}\right] \\ \text{with } B &= \mu NI \\ d &= \text{pulley sheave moval distance, } m \end{aligned} \tag{18.26}$$

where, $A_e = \pi\left(\frac{D_2^2-(D_i+2t)^2}{4}\right)$ m^2, A_e is the effective cross-sectional area of the EMA for coil winding, and t is the bobbin's thickness. It is commonly known that the current flowing through the coil (loop) produces a magnetic field. As seen in Fig. 18.5a, an imaginary coil with radius a and a center on the z-axix can be created by applying Ampere's circuital law. One can estimate the coil's current flows as follows:

$$I(r) = \begin{vmatrix} 0 & r < a \\ \{I * (r^2 - a^2)/(c^2 - a^2)\} & a \le r \le c \\ I & c > r \end{vmatrix} \quad (18.27)$$

Hence the magnetic flux density can be estimated,

$$B_\theta(r) = \begin{vmatrix} 0 & r > a \\ [\mu I/2\pi r][(r^2 - a^2)/(c^2 - a^2)] & a \le r \le c \\ \mu I/_{2\pi r} & c > r \end{vmatrix} \quad (18.28)$$

The standard value for the magnetic permeability is 0.008 H/m. It is observed that the solenoid efficiency is 0% when the EMA is powered by a current (I) of 17A and its temperature has reached 100 °C. As a result, the EMA loses energy in the form of electromagnetic radiation and home heat.

Supply current to the EMA to make the vehicle acceleration can be simulated by using the equation:

$$a(t) = \begin{vmatrix} \text{At start} - \text{from} - \text{rest} & I_{max} = 12.5\,A \\ At\,G = 20\% & 7A < I \le 12.5A \\ At\,G = 10\% & 5A < I \le 7A \\ \text{Otherwise} & I = 0\,A \end{vmatrix} \quad (18.29)$$

where, G is the gradient of the road in percentage (%), *a(t)* is the instantaneous acceleration of the vehicle in m/s^2. Total length of wire can be computed as follows:

$$l_c = \frac{\left[D_o^2 - (D_i + 2t)^2\right]}{d_w^2}(l_{ema}) \quad (18.30)$$

where, D_O and D_i is the outer and inner diameter of the solenoid respectively. d_w is the wire diameter and $L_{solenoid}$ is the length of the solenoid. Again,

$$R = \frac{4\rho l_{wire}}{\pi d_w^2} \quad (18.31)$$

where, R is the wire resistance in Ω, ρ is the resistivity of conductor in Ωm, l_{wire} is the total length of the conductor and d_w is the wire diameter:

$$d_w = \sqrt[4]{\frac{\rho L_{solenoid}(D_0^2 - D_i^2)}{R}} \quad (18.32)$$

where, d_w is the diameter of the coil, m

Number of turns of the coils of the EMA can be modelled as,

$$N = \frac{[D_0 - (D_i + 2t)]}{2d_w^2} * \underset{packing}{f} \quad (18.33)$$

where, $f_{packing}$ is the packing factor.

18.5.1 Heat Generation

When there is a process of current flow in an object such a wire, it generates heat. The heat generation at the EMA body due to current flow can be estimated by using the Eq. (18.3):

$$Q_{\Delta energy} = \left[\left(\begin{array}{c} I_{b(d)} \cdot \sum_i \frac{T_{batt(i)} \Delta S_i}{n_i F} + \\ \sum_i I_{b(d)} \eta_i + I_{b(d)}^2 R \end{array}\right) - h_m A_s \left(T_{batt(i)} - T_\infty\right)\right] \quad (18.34)$$

with

$$T_{batt(f)} = T_{batt(i)} + \frac{1}{mc_p}\left[\left(\begin{array}{c} I_{b(d)} \cdot \sum_i T_{batt(i)} \Delta S_i \\ \sum_i I_{b(d)} \eta_i + I_{b(d)}^2 R \end{array}\right) - h_m A_s \left(T_{batt(i)} - T_\infty\right)\right] \quad (18.35)$$

and

$$I_{b(d)} = \frac{\left[\begin{array}{l} m\frac{dv}{dt} + \left(0.005 + {}^1/_{P_{t(p)}}\left(0.01 + 0.09\left({}^V/_{100}\right)^2\right)\right)\frac{m}{g} \\ + \frac{\rho_a C_D}{2} A_f V^2 + \frac{m}{g} \sin\theta \end{array}\right][V]}{V_{m(vol))} \eta_m} \quad (18.36)$$

where, $Q_{\Delta energy}$ is the heat energy developed due to the flow electrical current, I_b is the current level in ampere, T_b is the temperature in °C, T_∞ is the ambient temperature °C, m is the mass of the coil in kg, C_p is the heat transfer coefficient, h_m is the enthalpy, and S is the entropy.

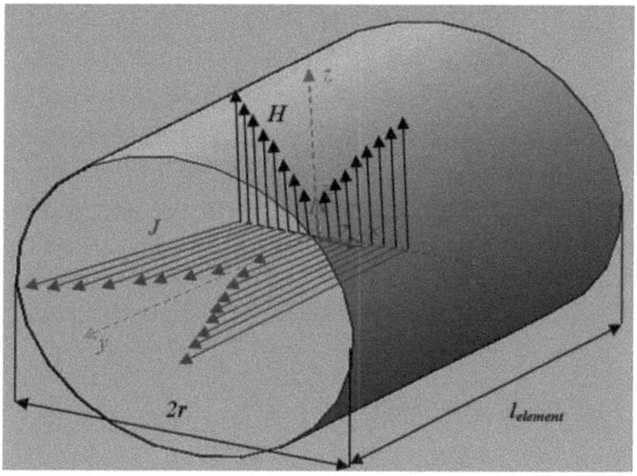

Fig. 18.6 Segment of a copper wire [5]

The heat generation within the copper wire is shown in Fig. 18.6 based on each heat source factor. It is clear from the simulation results that the primary source of heat generation is resistive heat, also known as the Joule effect. Joule heating is a process where heat is internally generated by electrical resistance when an electric current is conducted through it. It is also known as Ohmic heating, electrical resistance heating, direct electrical resistance heating, electro-heating, or electroconductive heating [38]. The amount of current flowing through the wire directly correlates with the amount of heat generated. Additionally, this factor is exactly proportional to the overall resistance of the copper wire. In contrast to the other two components, the heat generation factor brought on by polarization resistance was discovered to be extremely little.

18.5.2 Electromagnetic Energy Loss

Temperature spikes (eddy current loss), irreversible loss (magnetic hysteresis loss), and magnetic flux leaking through the solenoid surface are thought to be the causes of the EMA's electromagnetic energy loss. The first two types of energy loss are referred to as core loss. Each segment loss is identified by taking into account a copper wire element (Fig. 18.6), and the total summation of the segments for a winding loss is then followed. This amount multiplied by the total number of windings shows the EMA's overall loss.

18.5.3 Eddy Current Loss

The currents generated in the magnetic material by the time-varying magnetic induction are what cause the eddy current loss. Eddy currents are produced by the magnetic material's electric conductivity σ (5.8 × 107 S/m at 20 °C for copper), which is induced by the time-varying flux φ that is surrounded by each loop in the material. Moving farther away from the wire element's center causes this generation to accelerate. Over time, the eddy current losses are caused by the magnetic field in the magnetic elements. It can be defined using the equation:

$$\begin{aligned} E_{eddy\,current\,loss} &= N\int_{t_1}^{t_2}\left(\sum_{no.\,segment}\left(\int_{y=0}^{l_{segment}}\int_{\theta_s=0}^{2\pi}\int_{x=z=0}^{2r}\frac{J^2}{\sigma}r.dr.d\theta.dy\right)\right)dt \\ &= N\frac{\pi r^2 I^2 l_{\sin gle\,winding}}{\sigma}(t_2-t_1) \end{aligned} \tag{18.37}$$

where, N is the total number of winding in solenoid, J is the current density, I is the supply current, r is the wire radius, $l_{single\ winding}$ is length of single winding, σ is the electric conductivity and $(t_2–t_1)$ is the time interval of the solenoid operation. Heat is produced as a result of eddy currents being induced within the core. One way to lessen eddy current loss is to laminate the core material. In addition, there is magnetic domain mobility. It is possible to say that the boundaries of the magnetic domains shift as the magnetic field changes because certain magnetic domains expand while others contract. Energy is absorbed by this motion. Magnetic fields are created by the inductance of circulating current eddies. When electromagnetic forces are acting, these fields produce the opposite effect.

18.5.4 Hysteresis Loss

Ferromagnetic materials are known to exhibit hysteresis. A ferromagnetic material's atomic dipoles align themselves with the external magnetic field when it is applied. A portion of the magnetic alignment will remain even after the external field is eliminated. The material's magnetic field H and magnetic induction B are related to the hysteresis loss, which is dependent on the magnetic field's history. For the time interval from t_1 to t_2 hysteresis loss in accordance with the formula [1]:

$$\begin{aligned} E_{hysteresis\,loss} &= N\sum_{no.\,segment}\left(\int_{y=0}^{l_{segment}}\int_{\theta_s=0}^{2\pi}\int_{x=z=0}^{r}\left(\int_{H(t_1)}^{H(t_2)}H\frac{\partial B}{\partial H}dH\right).r.dr.d\theta.dy\right) \\ &= N\frac{\pi r^2(B_{t_2}-B_{t_1})^2 l_{\sin gle\,winding}}{2\mu} \end{aligned} \tag{18.38}$$

where, N is total number of winding in solenoid coil, H is magnetic field intensity, B is the magnetic flux density with respect time, r is the wire radius, $l_{single\ winding}$ is the length of single winding and μ is the magnetic permeability.

18.5.5 Electromagnetic Flux Leaking Loss

When the solenoid surface is not shielded, electromagnetic flux flowing through it causes a considerable loss of electromagnetic field. In this study, outer loop windings are exclusively taken into account when determining leakage losses. The simplified equation can be used to calculate the magnetic field loss resulting from flux leakage of the solenoid;

$$H_{in} = \frac{r_{in} \iint_{S(r_{in}, l_{segment})} J dS_{wire}}{2\pi r^2} \quad (for\ r_{in} \leq r) \tag{18.39}$$

Again,

$$H_{out} = \frac{\iint_{S(r_{out}, l_{segment})} J dS_{wire}}{2\pi r_{out}} \quad (for\ r_{out} > r) \tag{18.40}$$

H_{out} is the main concern for finding the flux leakage through the solenoid surface. The magnetic field H_{in} is effective to develop the desired magnetic force to operate the movable sheave of the pulley while H_{out} is the causes of the magnetic loss. The energy loss of the solenoid due to the flux leaking can be computed by the simplified equation [1, 6, 10]:

$$\begin{aligned} E_{flux\ leakage\ loss} &= N_{out}\left(\sum_{no.\ segment} \frac{\iint_{S(r, l_{segment})} {}^{J}/_{2} dS_{wire}}{2\pi r_{out}}\right) \\ &= \frac{I(h_2 h_1) L_{bobbin}}{4\pi r_{wire} r_{out}} l_{\sin gle\ winding} \end{aligned} \tag{18.42}$$

where, N_{out} is number of winding in the solenoid surface. *J/2* is due to the half circle of the copper wire, I is the supply current. h_2 and h_1 is the solenoid outer and inner diameter respectively, L_{bobbin} is length of the solenoid, $l_{single\ winding}$ is length of single winding r_{wire} is the wire radius and r_{out} is the outer radius where $r_{out} > r_{wire}$.

18.6 Conclusion

The currents generated in the magnetic material by the time-varying magnetic induction are what cause the eddy current loss. In fact, the magnetic material's time-varying flux φ, which is surrounded by each loop, generates an electromotive force that is induced by the magnetic field.

Acknowledgements The project was funded by the Prototype Research Grant Scheme (PRGS) 2016 by the Malaysian Ministry of Higher Education, for which the authors are thankful. The authors also acknowledge the funding provided by the International Islamic University Malaysia for the development of prototype of EMA-CVT and the Eastern Michigan University in Ypsilanti, Michigan, USA, for enabling the finalization of EMA-CVT for commercialization.

References

1. Ulaby, F. T. (2005). *Electromagnetics for engineers*. Pearson International Edition.
2. The Mathworks. (2012). *MATLAB: Fuzzy logic toolbox users' guide*. Natick, MA. http://www.mathworks.com/products/fuzzy-logic/demos.html
3. Rahman, A., Nur, F., Hawlader, M. N. A., & Afroz, R. (2015). Fuzzy controlled evaporative battery thermal management system for EV/HEV. *International Journal of Electric and Hybrid Vehicles, 7*(1), 22–39.
4. Pesgens, M., Vroemen, B., Stouten, B., Veldpaus, F., & Steinbuch, M. (2006). Control of a hydraulically actuated continuously variable transmission. *Vehicle System Dynamics, 44*(5), 387–406.
5. Hassan, A. J. (2020). *Development of intelligent electromagnetic continuously variable transmission* [A PhD Thesis Dissertation]. IIUM Library, International Islamic University Malaysia.
6. Rahman, A., Jaffar, A. H., & Sany, I. I. (2022). IoT-based electromagnetic actuator for CVT: Basic design and prototyping. *International Journal of Powertrains, 11*(1), 38–61.
7. Ataur, R., Nurul, H., Hassan, A. J., Mohiuddin, A. K. M., & Sany, I. I. (2019). Study on the development of electromagnetic seamless two speed gearbox for EV. *International Journal of Recent Engineering and Technology, 7*(6S), 147–152.
8. Wong, J. Y. (2008). *Theory of ground vehicles* (Vol. 1, 4th ed.). Wiley-Interscience, John Wiley & Sons.
9. Ataur, R., Rafia, A., & Jafar, A. H. (2019). Green transportation system, (1st Ed.). IIUM Press, International Islamic University Malaysia, ISBN 978-967-491-021-1.
10. Rahman, A., Sazzad, B. S., & Hossain, A. (2012). Kinematics and non-linear control of electromagnetic actuated CVT system. *Journal of Mechanical, Science and Technology., 26*(7), 2189–2196.
11. Kaur, N., & Singh, A. K. (2016). Ohmic heating: Concept and applications—A review. *Critical Reviews in Food Science and Nutrition, 56*(14), 2338–2351.

Chapter 19
Intelligent System for EMCVT Control

Abstract Automotive engineering's fuzzy inference system (FIS) is expanded to include vehicle dynamics, namely CVT with electromagnetically actuated systems. The EMA-CVT system management's control goal is to use a fuzzy logic control system to govern the flow of electrical power through changes in current level. The load weight that replicated the traction torque at the propeller shaft was detected by a load sensor. The necessary current level to run the EMA in each driving scenario is then estimated using the related load signal. Then, using the Current Control Module (CCM) to regulate the power supply to the EMA, FLC adjusts the CVT gear ratio as necessary. The aggregation method is used to merge the output fuzzy sets produced by the fuzzy inferring (reasoning) system into a single fuzzy set. Active fuzzy rule outputs are then combined to create a final output fuzzy set, which is then defuzzed using the centroid approach to yield the servo motor's crisp output (S) at a 270° angle.

Keywords Fuzzy inference system · CVT gear ratio · Current control module · FLC_Servo motor

19.1 Introduction

The intelligent control system is a practical technological problem that the automotive industry aims to solve by simulating the functioning of an intelligent biological system, such as a person. An intelligent control system essentially takes ideas from people who are skilled at solving control problems by hand. Lotfi Zadeh first proposed fuzzy logic in 1965. It is a rule-based intelligent control system that mimics human decision-making by taking into account all intermediate possibilities between two conceivable values, which are often represented as True and False.

The accuracy of mathematical models that incorporate control parameters for a car with an electromagnetically actuated continuously variable transmission under different driving conditions, which involve a variety of errors and ambiguous phenomena, is called into question. Such a complex system cannot be adequately

A. Rahman et al., *Autonomous Driveline System for Electric Vehicle*, Green Energy and Technology, https://doi.org/10.1007/978-981-96-8418-2_19

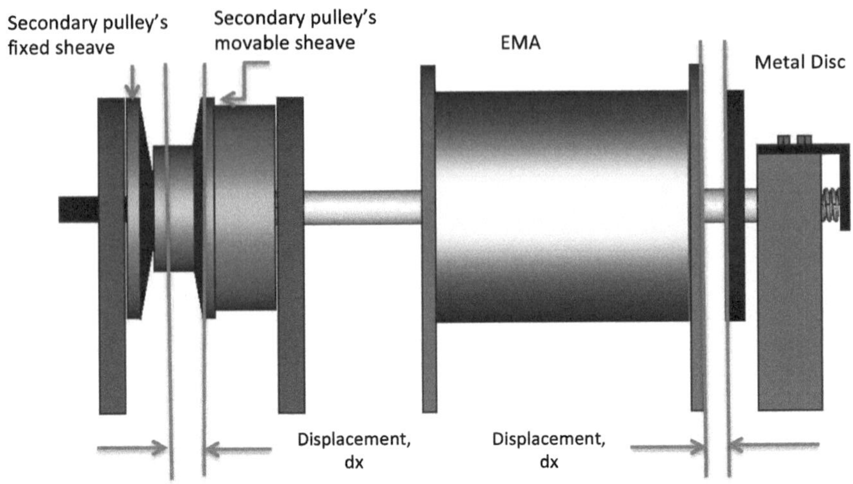

Fig. 19.1 Schematic view of MA-CVT

described by any mathematical model. Although all theoretical models are obligated to make simplifying assumptions, it is possible that they will influence performance rules of a broad variety of the interaction dynamics of the vehicle. It is therefore normal to search for alternative approaches. In this sense, as seen in Fig. 19.1, the fuzzy inference system (FIS) for automotive engineering is expanded to vehicle dynamics, specifically to CVT with electromagnetically actuated system.

The EMA-CVT system management's control goal is to use a fuzzy logic control system to govern the flow of electrical power through changes in current level.

19.2 Power System Design

The EMA must produce the current required to maintain the intended CVT gear ratio based on the clamping force that was determined in the preceding section as well as compiled in Table 19.1. The required maximum current is approximately 9.33 A. A high voltage power source is necessary to create the high current as indicated by Ohm's law because of the EMA's copper coil's low resistance. Two battery modules with a nominal voltage of 51.2 V and a capacity of 50 Ah were connected in series to create a battery pack with a rated capacity of 50 Ah and a voltage of approximately 100 V for the EMA's operation, as shown in Table 19.2.

Brake Traction Load

In order to simulate the transmission torque generated by the wheel after the last gear, a brake system was fitted at the driven shaft of the pulley (driveshaft) because a dynamometer device was available. By dividing the traction torque at the wheel

Table 19.1 EMA current requirement

Condition	θ_R (°)	V (m/s)	F_T (N)	T_W (Nm)	Tout (Nm)	F_C (N)	Current (A)	Power (W)
Initial	11.3	2.78	9,110.80	2,870	536	6,157	9.33	933
Cruising	11.3	4.17	3,489.57	1,099	205	2,358	5.77	577
		13.89	3,551.93	1,119	209	2,400	5.83	583
		33.33	3,878.13	1,222	228	2,621	6.09	609

Table 19.2 Battery pack's specification

Parameters	Unit	Value
Battery electrochemistry	–	LiFePO4
Battery cell's nominal voltage	V	100
Battery cell's capacity	Ah	50

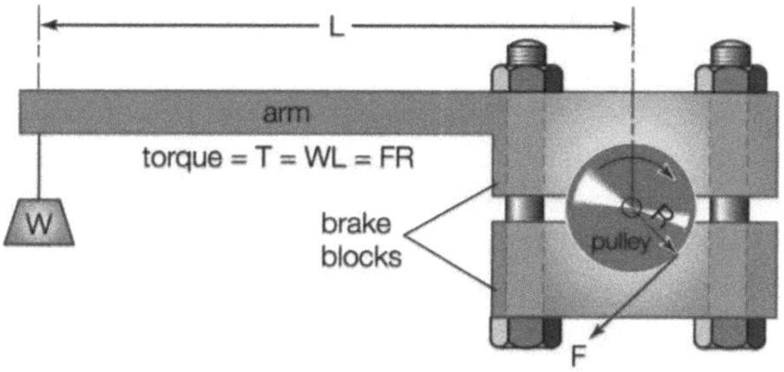

Fig. 19.2 Typical diagram of a Prony-brake dynamometer

by the final drive ratio, the transmission torque at the propeller was determined. As with a Prony-brake dynamometer, a hanging weight was attached in order to measure and generate an equivalent torque at the propeller. A dynamometer measures the mechanical power that a revolving shaft transmits. Dynamometers are basically instruments for measuring torque because power is the result of the product of torque (turning force) and angular speed. Figure 19.3 shows the Prony-brake like torque input that has been used to mimic the traction torque at a car (Fig. 19.2).

19.3 Intelligent Control System for Ema

In Fig. 19.4, the intelligent electromagnetic actuator for continuously variable transmission (i-EMA-CVT) functional block diagram is displayed. As previously stated, the EMA's power supply system will be powered by a lithium battery pack; however, in practical use, the vehicle's alternator will provide electricity. The load weight that replicated the traction torque at the propeller shaft was detected by a load sensor. The necessary current level to run the EMA in each driving scenario is then estimated

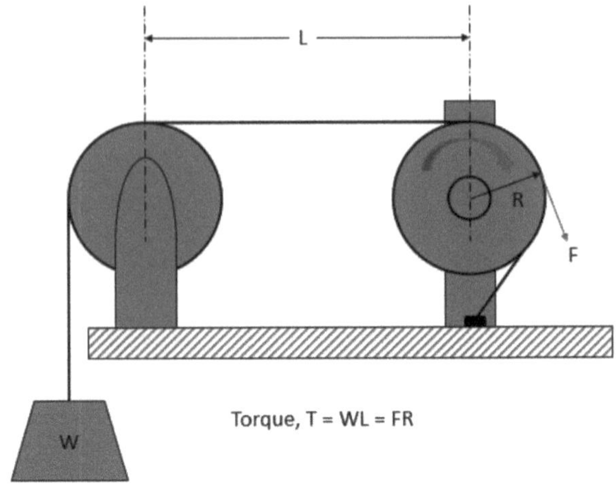

Fig. 19.3 Torque load input diagram for test-bed

using the related load signal. The real current sensor that was sent to the EMA is continuously measured by another current sensor.

When the actual current level differs from the estimated current level, the fuzzy logic controller (FLC) will get the error signal. The FLC then modifies the CVT gear ratio appropriately by regulating the power supply to the EMA via the Current Control Module (CCM), as seen in Fig. 19.4, based on the error, or the difference between the targeted and actual level. The CCM in this project is powered by a servomotor and a buck converter to turn the adjustable knob. The CCM permitted a suitable current level to provide sufficient clamping force in order to generate and sustain the desired gear ratio. Additionally, the CVT is equipped with a failsafe mechanism called the center manual switching system (CMSS) to guard against harm in the event that some of the sensor's malfunction.

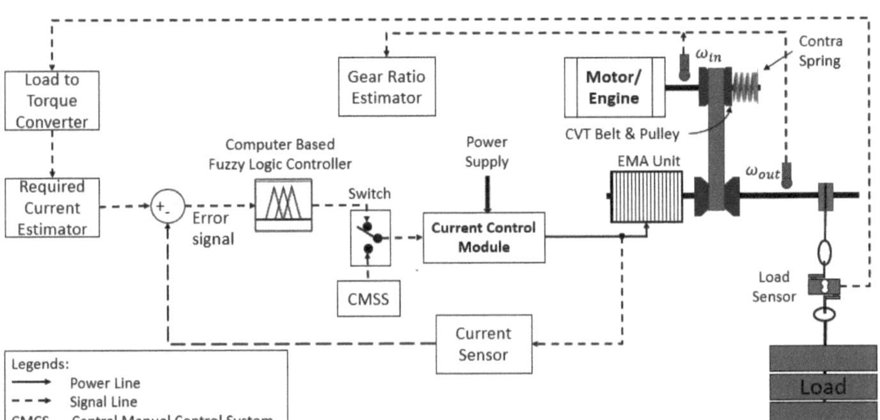

Fig. 19.4 The general diagram of i-EMA-CVT testbed system [1]

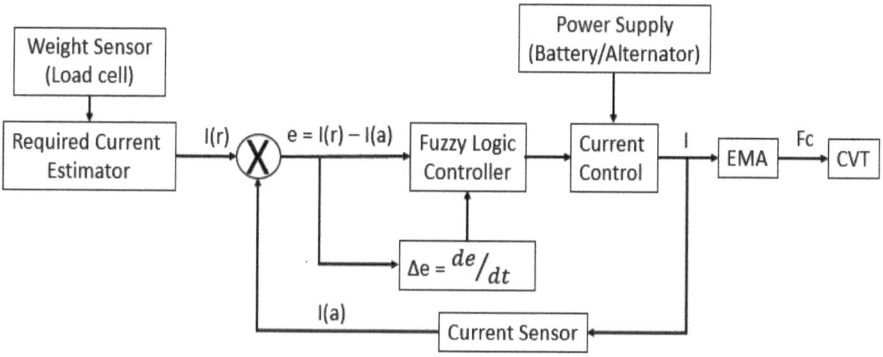

Fig. 19.5 Control system block diagram of the i-EMA-CVT [2]

The drive may switch to manual mode in failsafe mode to protect the CVT from harm and fix the gear ratio equal to first gear. Additionally, there are two-rpm sensors (tachometers) that detect the propeller shaft and traction motor's rotational speed continually and estimate the real gear ratio.

The control block diagram for the EMA-CVT's intelligent control system is displayed in Fig. 19.5. The real data of the current level that was supplied to the EMA for various driving scenarios (shown as Ia) would be continuously gathered by a computer-based controller. The torque produced by the load weight as determined by the load cell was used to estimate the reference or necessary current. The error is computed for each road condition with respect to the reference current level (represented by Ir). The primary input to the fuzzy logic controller, which estimates the appropriate current or power to run the electromagnetic actuator and produce an appropriate gear ratio for driving the vehicle, will be the current error (E) and its derivation (RE).Current control functions as a gate to permit the appropriate amount of current from the alternator to the EMA based on the signal from the fuzzy logic controller. This current controls the position of the plunger or movable sheave.

19.4 Function of Fuzzy Logic Controller

The inputs and outputs of the control process must be accurately identified prior to developing any controller. The controller uses the input variables to decide how best to control the plant or process. Therefore, the process must be significantly impacted by the controller's output variables. This paper introduces a fuzzy logic controller that uses a fuzzy inference system (FIS) to regulate an intelligent electromagnetic actuator CVT.

Fuzzifier, rule base (knowledge base), inference, and defuzzifier are the four primary components of the fuzzy logic controller's fundamental structure, which is depicted in Fig. 19.6 [3]. The system inputs, which are discrete numbers, are

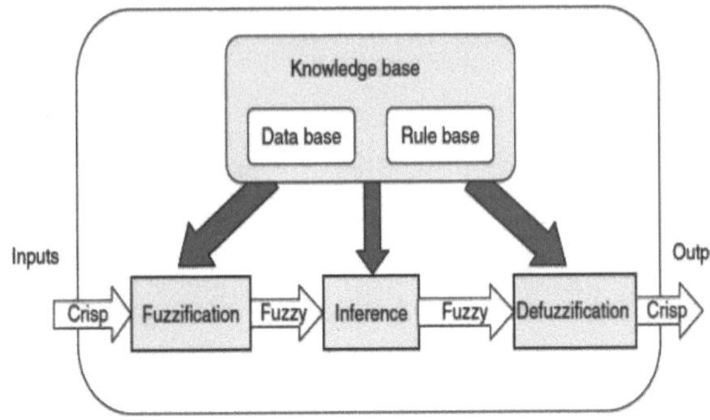

Fig. 19.6 The basic structure of fuzzy logic based controller [4]

converted into fuzzy sets with varying degrees of membership by the fuzzyifier. Anywhere in the range [0, 1] could be this level of membership. The value does not belong to the specified fuzzy set if it is zero, and it fully belongs to the fuzzy set if it is one. The degree of uncertainty that a value belongs in the set is represented by any number between 0 and 1.

The expert's linguistic description of how to achieve appropriate control for a specific application is quantified using fuzzy logic in the knowledge base, which is made up of a database and a rule base. The inference mechanism uses IF–THEN rules and fuzzy inference on the inputs to mimic human reasoning. The fuzzy set created by the inference mechanism is transformed into a precise value or crisp output by the defuzzifier so that the process can use it.

19.5 Implementation of Fuzzy Logic Controller

Servo motor angle (S) was utilized as the output parameter, while electrical current error (E) and rate of current error (RE) were used as input parameters for the fuzzy inference system (FIS), which was used to implement fuzzy values into the system. Figure 19.7 illustrates that E and RE are fuzzy variables of E and RE, respectively. Linguistic variables such as large negative error (LNE), small negative error (SNE), zero error (ZE), small positive error (SPE), and large positive error (LPE) are used to fuzzify these factors. For the gear ratio error (E), large negative rate of error (LNRE), small negative rate of error (SNRE), zero rate of error (ZRE), small positive rate of error (SPRE), and large positive rate of error (LPRE) are used for the rate of position error (RPE) as the input parameters. Likewise, with the servo angle (S) as the output parameter, the linguistic variables large negative servo angle (LNS), small negative servo angle (SNS), zero servo angle (ZS), small positive servo angle (SPS), and large positive servo angle (LPS) are employed.

The fuzzy logic controller uses the inputs from the aforementioned language variables to simulate human decision-making when controlling the EMA CVT

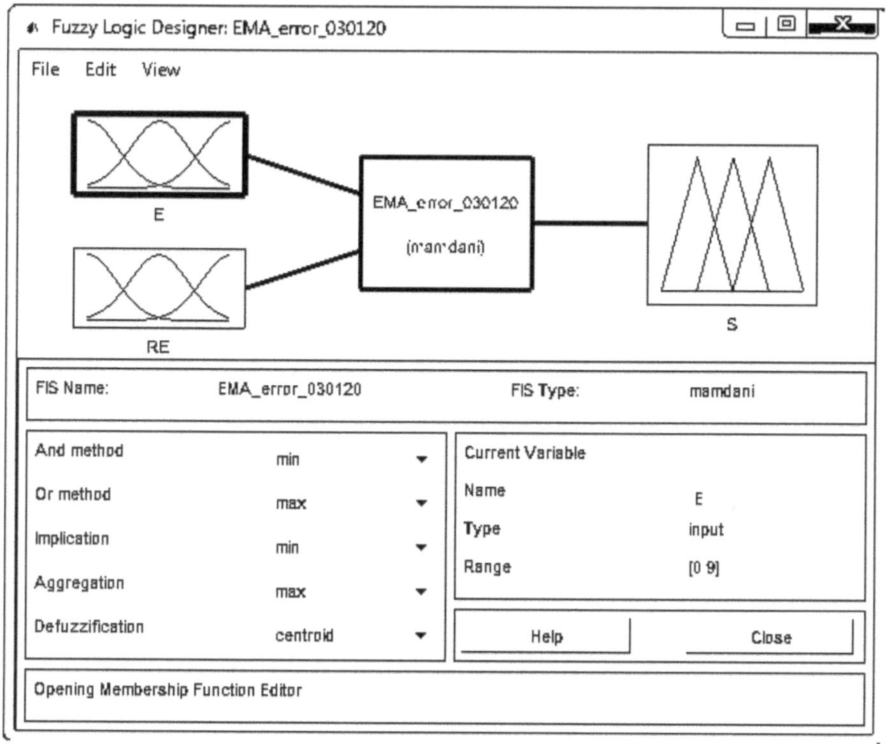

Fig. 19.7 The structure of the fuzzy inference system (FIS) for EMA-CVT [1]

under different driving circumstances. Table 19.3 displays the controller parameters' interference rule.

The center of gravity (Centroid) defuzzification formula method, the aggregation method, and the logical AND have all been applied because they guarantee a linear interpolation of the output between the rules [5]. An individual rule-based inference strategy is employed in this study's Mamdani max–min inference approach, which generates the output when presented with a crisp input.

The system knowledge, expert evaluations, experiment settings, and programmers' effectiveness in transferring their expertise and judgment to the system are the basis for choosing the number of membership functions and the FIS's performance. In conclusion, the form of the membership functions is primarily determined by the specific application. There are numerous membership function types to choose from, depending on the application and expert. As piecewise-linear functions, the triangular shape membership functions are employed in this study for both input and output variables due to their correctness [6, 7].

Moreover, using this fuzzy set for operations and graphical representations is quite easy. As regulation rules, a fuzzy related memory or decision (also known as a decision rule) is created for the two inputs and one output. Twenty-five rules were

Table 19.3 Inference rules of controller parameters

Rule	Input variable		Output variable
	E	RE	I
1	LNE	LNRE	LNS
2	LNE	SNRE	LNS
3	LNE	ZRE	LNS
4	LNE	SPRE	LNS
5	LNE	LPRE	LNS
6	SNE	LNRE	LNS
7	SNE-	SNRE	SNS
8	SNE	ZRE	SNS
9	SNE	SPRE	SNS
10	SNE	LPRE	SNS
11	ZE	LNRE	SNS
12	ZE	SNRE	SNS
–	–	–	–
20	SPE	LPRE	SPS
21	LPE	LNRE	LPS
22	LPE	SNRE	LPS
23	LPE	ZRE	LPS
24	LPE	SPRE	LPS
25	LPE	LPRE	LPS

created utilizing five linguistic variables for each of the two inputs, as Appendix A illustrates. The table displays some of the fuzzy rules that were developed. To determine a fuzzy output once the inputs have been fuzzified, the FIS consults a collection of user-defined if–then rules. Figure 19.8 depicts the schematic of a fuzzy logic expert system for flow rate modeling.

The expert's thoughts regarding the control decision-making process inside the framework created by the selection of fuzzy controller inputs and outputs are expressed in the language provided by the aforementioned linguistic variables and values. In the fuzzy logic system, fuzzification is the first step that transforms each piece of input data into a degree of membership in one or more membership functions. Thus, the fuzzification block compares the input data with the rule conditions to ascertain the degree to which each rule's condition corresponds to that specific input structure. For every rule, there is a single output function. A single output function is created by combining the output functions of many rules, and this output function is subsequently defuzzed into a single numeric result.

The servo motor, gear ratio error (E), and rate of gear ratio error (RE) are fuzzified using the functions shown below. These formulas are computed using measurement values, where i1 stands for the first input variable (E), i2 for the second input variable (RE), and o for the output variable (S).

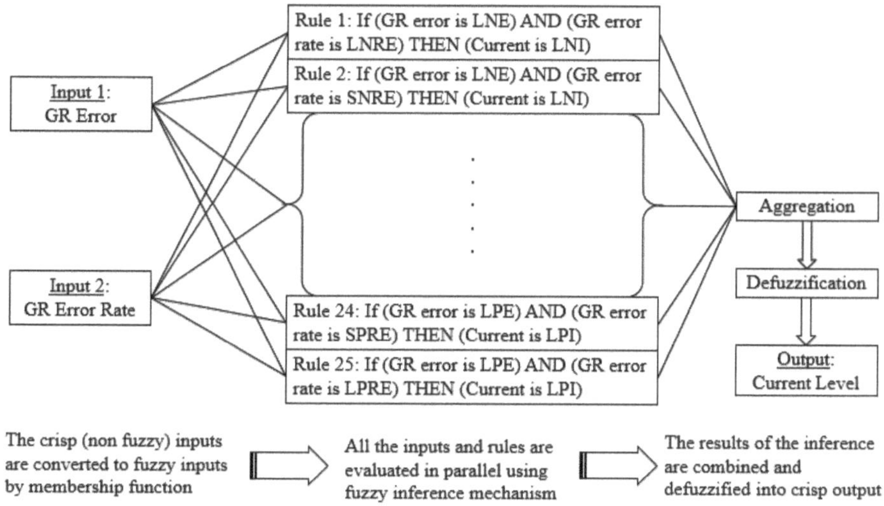

Fig. 19.8 Schematic representation of fuzzy inference system [8]

$$E(i_1) = \begin{vmatrix} i_1 & -9.6 \leq i_1 \leq 9.6 \\ 0 & otherwise \end{vmatrix} \tag{19.1}$$

$$\mathrm{RE}(i_2) = \begin{vmatrix} i_2 & -1 \leq i_2 \geq 1 \\ 0 & otherwis \end{vmatrix} \tag{19.2}$$

$$S(0) = \begin{vmatrix} i_1 & 0 \leq i_1 \leq 360 \\ 0 & otherwise \end{vmatrix} \tag{19.3}$$

The MATLAB Fuzzy Toolbox is used to create the triangular fuzzy sets for the fuzzy variables, the servo motor angle (S), rate of current error (RE), and current error (E). These figures display the membership values that were derived from the aforementioned formulas. In the same way, the degrees of E, RE, and S are measured in centimeters (cm) from −9.6 to 9.6, A/s from −1 to 1, and 0 to 360, respectively. Figures 19.9, 19.10 and 19.11 demonstrate how membership functions can translate numerical variables into language. Before the data was sent into the fuzzy logic controller, it was normalized for the rate of error (RE). A variable is typically normalized when its values fall between 0 and 1.

Each element of X is mapped to a value between 0 and 1, and the membership function for a fuzzy set A on the universe of discourse X is defined as μA: X → [0, 1]. In other words, it converts the input into a 0–1 scale that indicates the strength of a membership in a set. This value, also known as the membership value or degree of membership, quantifies the grade of membership of the element in X to the fuzzy set A. According to Wikipedia (2018), a value of 1 denotes complete membership in the fuzzy set, whereas a value of 0 implies non-membership. The following formulas

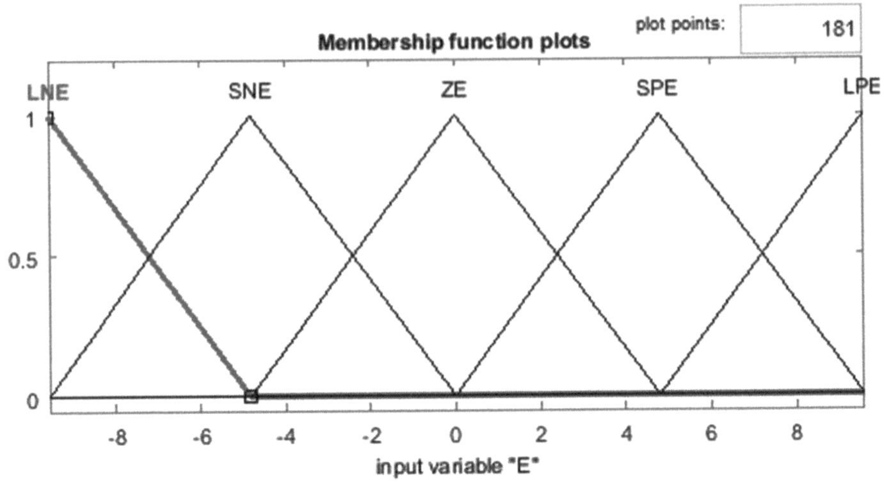

Fig. 19.9 Membership functions of fuzzy input variable 'E'

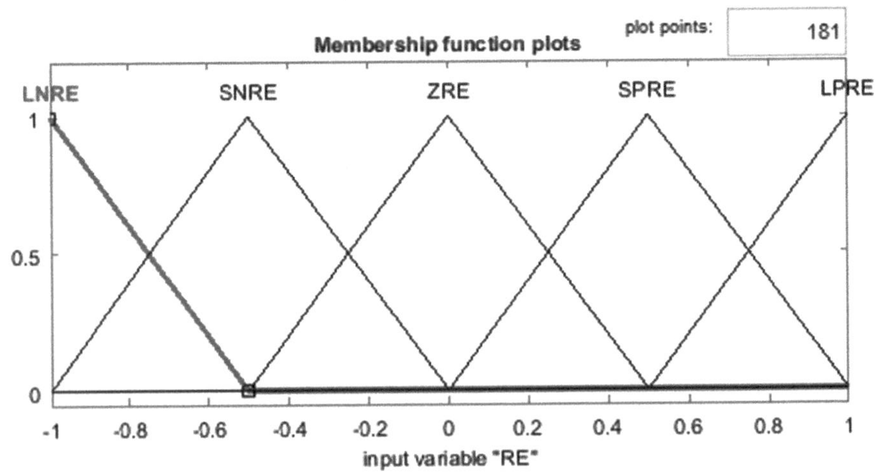

Fig. 19.10 Membership functions of fuzzy input variable 'RE'

use a triangle membership function specified to calculate fuzzy membership values: a lower limit a, an upper limit b, and a centre value m, where a < m < b, as shown in Eq. (19.4).

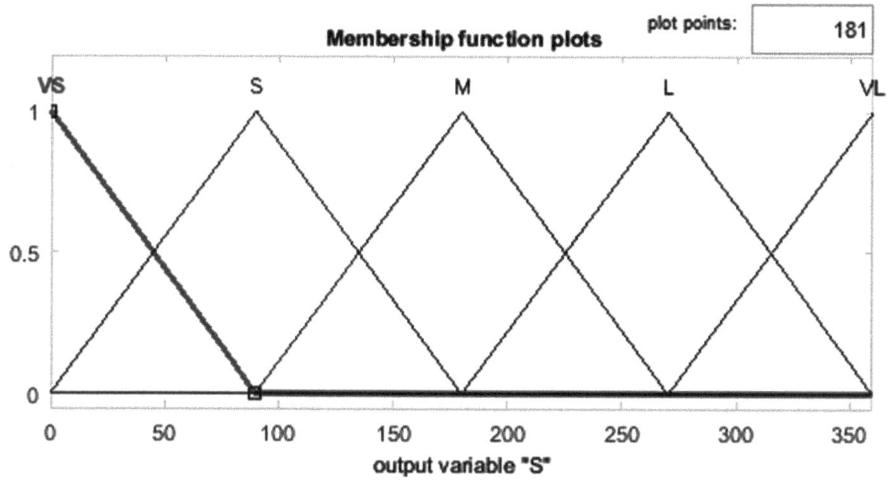

Fig. 19.11 Membership functions of fuzzy output variable 'S'

$$\mu_A(X) = \begin{Bmatrix} 0 & a \geq x \geq b \\ \dfrac{x-a}{m-a} & a < x \leq m \\ \dfrac{b-x}{b-m} & m < x < b \end{Bmatrix} \tag{19.4}$$

In Eq. (19.4), x is the input or output variable, a, m, and b are the values limit for the membership function to describe the input and output variables.

The input and output variables' membership function limits are listed in Tables 19.4, 19.5 and 19.6. It is observed that the values of membership functions vary with time, for instance, as the value of current error (E) changes. For example, it can be observed that as E shifts from −9.6 to 9.6, various membership functions will acquire on zero and nonzero values, reflecting the extent to which the linguistic value accurately depicts the present value of E. At E = −9.6, for instance, it is certain that the position error is "large negative error (LNE)". As PE approaches −5, it becomes less certain that it is "LNE" and more certain that it is "SNE", for example.

Additionally, the membership function linked to the linguistic value "LNE" will be represented in the following using the notation µLNE, with a comparable notation for additional values.

Two steps are typically involved in the inference process: (1) linguistic variables of all the rules are compared to the input data to identify which rules apply to the current situation, and (2) conclusions (i.e., what control actions to take) are arrived at using the rules that have been identified as applicable at the moment. Using a fuzzy set (or sets) to express the certainty that the input to the plant should take on different values, the results are described.

Table 19.4 The limit values for each membership function of input 'E'

Linguistic variables	Limit values		
	a	m	b
Large negative error	–	−9.6	−4.8
Small negative error	−9.6	−4.8	0
Zero error	−4.8	0	4.8
Small positive error	0	4.8	9.6
Large positive error	4.8	9.6	–

Table 19.5 The limit values for each membership function of input RE

Linguistic variables	Limit values		
	a	m	b
Large negative rate error	–	−1	−0.5
Small negative rate error	−1	−0.5	0
Zero rate error	−0.5	0	0.5
Small positive rate error	0	0.5	1
Large positive rate error	0.5	1	–

Table 19.6 The limit values for each membership function of Output S

Linguistic variables	Limit values		
	a	m	b
Very small	–	0	90
Small	0	90	180
Medium	90	180	270
Large	180	270	360
Very large	270	360	–

The rules that are used to determine the appropriate current level through the change in servo motor angle to the plant (current control module) are taken into consideration when drawing conclusions. Each rule's recommendations are taken into consideration separately in order to achieve this. Subsequently, the current level of inputs to the plant is determined by combining the suggestions from all the regulations. During this phase, each rule's truth degrees (μ) were calculated using the min and then the max between working rules.

By employing the aggregation method, the output fuzzy sets produced by the fuzzy inferring (reasoning) system are merged into a single fuzzy set. One of the defuzzification techniques is then applied to the combined output fuzzy set in order to calculate a final, sharp output value. The mean of maximum (MOM), smallest of maximum (SOM), centroid of area (COA), bisector of area (BOA), and largest of maximum (LOM) are the five general techniques for defuzzifying a fuzzy set A of a universe of discourse Z. The MATLAB Fuzzy Toolbox was used to create the

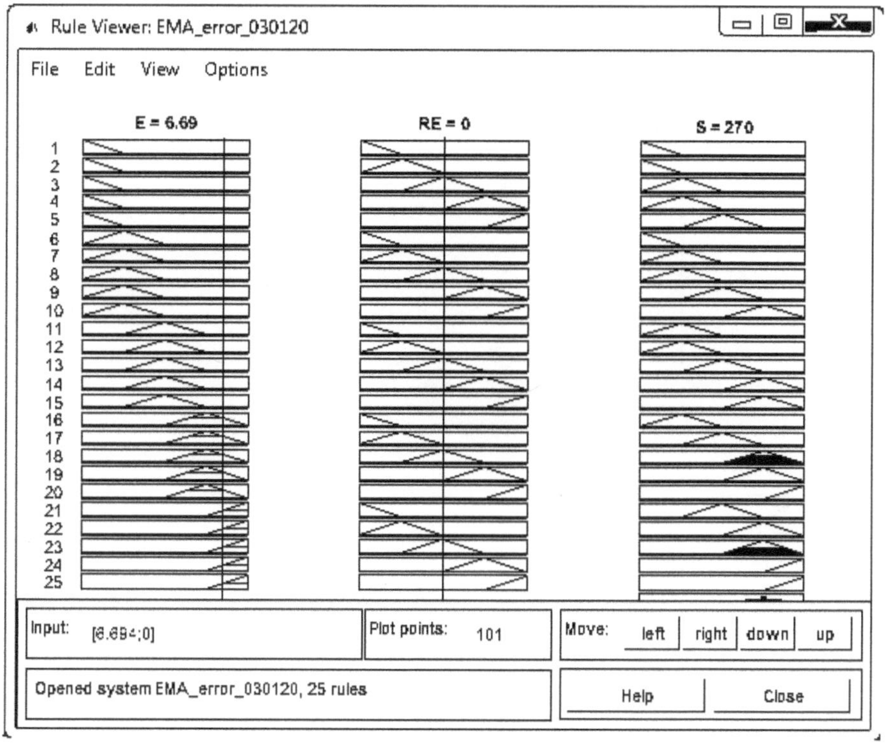

Fig. 19.12 Graphical representation of the fuzzy logic controller operation [1]

fuzzy control system for simulation in this work, as seen in Fig. 19.12. To confirm the output value, adjust the input values by clicking and dragging the input vertical lines, and then observe the interactive changes.

To identify the fuzzy output electrical current level (I) to EMA through change in servo motor angle (S), for instance, all 25-five fuzzy rules are evaluated concurrently if the error is 6.89A and the rate of error is 0 A/s. Unfortunately, some of the rules are still in use because they do not generate any fuzzy sets as their output, and the 'fuzzy and' function was utilized in the fuzzy rules' antecedent component. As seen in Fig. 19.13, the outputs of active fuzzy rules are then combined to create a final output fuzzy set, which is subsequently defuzzed using the centroid approach to yield the servo motor's crisp output (S) at a 270° angle.

Figure 19.13 illustrates how the fuzzy control surface is created using MATLAB. It could be used to illustrate the dynamic, time-varying operation of fuzzy logic expert systems. In this mesh plot, the controller output servo angle (S) on the output side is represented by the current error (E) and rate of current error (RE) on the input side. For every potential PE and RPE input, this control surface shows the range of defuzzified values. To determine whether the rules and membership functions are appropriate and whether changes are required to improve the output, the plot is

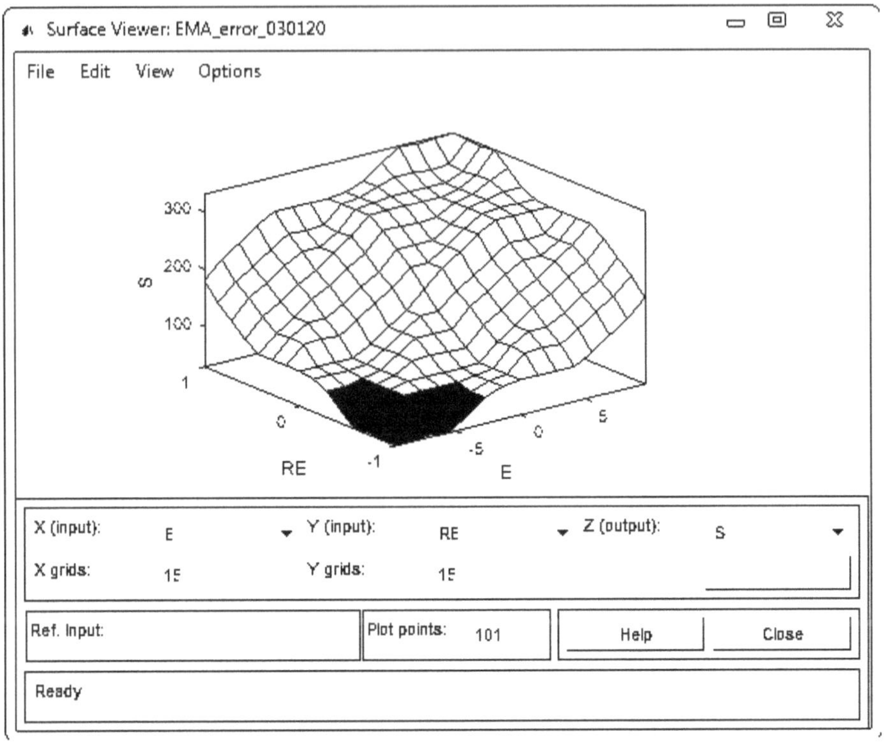

Fig. 19.13 Control surface of the fuzzy rules [1]

utilized. The rule foundation for the fuzzy sets is adjusted as needed to achieve the appropriate output curves. The effects of position error and position error rate factors on the flow rate are also shown in the surface map. It demonstrates that, as predicted, there is a corresponding rise in flow rate through the change in valve position when the vehicle position error and rate of position error grow positively. When the position error and rate of position error both reach their greatest levels, the flow rate achieves the apex.

19.6 Fuzzy Logic Controller for Ema-Cvt

The EMA-CVT testbed's general schematic of its electrical and mechanical systems. Figure 19.14 displays a detailed schematic of the testbed's electronics control system. It described the communication between input sensors and output actuators and the main controller board. This project's primary controller board is the Arduino Uno, an open source microcontroller board built by Arduino.cc and based on the Microchip ATmega328P microprocessor. When a product is open source, the creator grants

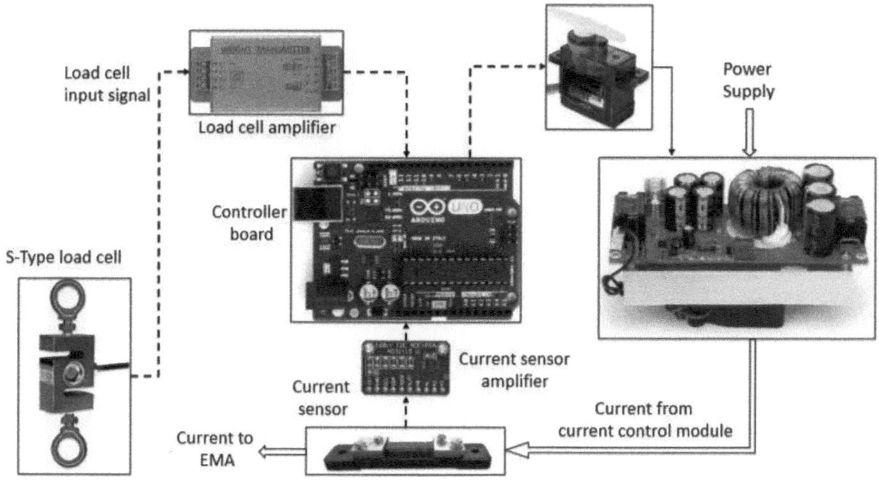

Fig. 19.14 Electronics control system block diagram for the EMA2SGB testbed [1]

authorization for the usage of its design documents, source code, or content. One of the key benefits of the Arduino board series, which helped make them well-known microcontrollers, is that the inventor produced and made available a library for programming them in C.

With a clock speed of 16 MHz, the Arduino Uno board features 14 digital input/output (I/O) pins (six of which may output PWM), 6 analog I/O pins, and the Arduino IDE (Integrated Development Environment) for programming over a type B USB connector. It can be powered by an external 9-V battery or by the USB cable, although it can handle voltages ranging from 7 to 20 V.

In essence, the load cell and current sensor provide the main controller board with two input signals. The traction torque generated at the propeller shaft is being replicated by the load cell. To accurately amplify the sensor output signal into a standard current and voltage signal output that the Arduino Uno controller board can accept, the load cell requires an amplifier. The DY510 load cell amplifier model, which has an accuracy of 0.05%, was utilized in this project. As was covered in the previous part, the Arduino Uno microcontroller then mathematically transformed the load cell signal to the torque and the corresponding needed current.

The current sensor is the second input to the controller; it measures the current flowing from the electromagnetic actuator to the current control module. Because the current that needs to be measured is large (more than 9 A) and continuous, the current flow was measured using a shunt type current sensor. The shunt enables the measurement of huge current values that are too great for a specific ammeter to measure directly. In this instance, the amplifier is connected in parallel with a separate shunt, which is a resistor with a very low but precisely known resistance, so that all of the current to be measured passes through the shunt. In order to get a

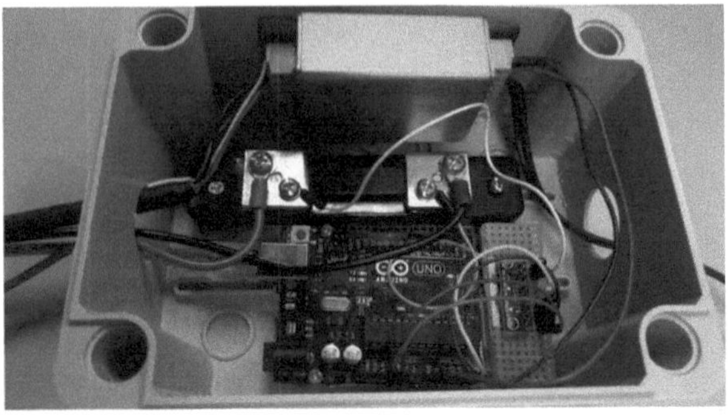

Fig. 19.15 EMA electronic control system [1]

noticeable voltage drop, the resistance is selected to be sufficiently low to low but precisely known resistance that is connected in parallel with the amplifier.

The resistance is adjusted so that the resulting voltage drop is detectable but not so large as to cause the circuit to malfunction. Since the current passing through the shunt determines the voltage across it, the measured voltage can be scaled to show the current value directly (Wikipedia, 2019c). Between the Arduino Uno microcontroller and the shunt resistor or current sensor, an analog-to-digital converter with amplifier (model ADS1115) is used to generate a current signal that the microcontroller can understand.

The C and C++ programming languages can be used to create the mathematical algorithm known as a fuzzy logic controller. Using the Arduino Integrated Development Environment (IDE) to compile and upload the sketch is necessary for the ATMega Microcontroller (or Arduino board) to apply the Fuzzy Logic Algorithm as illustrated in Fig. 19.15. The IDE is a cross-platform tool that is written in C and C++ functions and is available for Windows, macOS, and Linux. Importing and adding the fuzzy library to the /libraries directory is another requirement for the fuzzy logic code. In this work, the Robotics Research Group (RRG) of the State University of Piauí (UESPI-Teresina) developed the embedded Fuzzy Logic Library (eFLL) library [3].

19.7 Conclusion

In order to drive the CVT, traction torque is needed. Clamping force is calculated using this traction torque as a basis. Analysis of electromagnetic force and energy conservation at various angles has been done in order to optimize the design parameters for the creation of electromagnetic actuators. To ensure smooth operation of the EMA-CVT system under dynamic conditions, a fuzzy logic controller has been incorporated to regulate the actuator coil's power consumption at any given time. The current control module produced enough clamping force to generate and sustain the desired gear ratio at the right current level. To further minimize harm in the event that a sensor malfunctions, the CVT is equipped with a failsafe mode called the center manual switching system (CMSS).

In order to prevent damage to the CVT and fix the gear ratio equal to first gear, the drive may switch to manual mode during failsafe mode.

Acknowledgements The project was funded by the Prototype Research Grant Scheme (PRGS) 2016 by the Malaysian Ministry of Higher Education, for which the authors are thankful. The authors also acknowledge the funding provided by the International Islamic University Malaysia for the development of prototype of EMA-CVT and the Eastern Michigan University in Ypsilanti, Michigan, USA, for enabling the finalization of EMA-CVT for commercialization.

References

1. Hassan, A. J. (2020). *Development of intelligent electromagnetic continuously variable transmission* [A PhD Thesis Dissertation]. IIUM Library, International Islamic University Malaysia
2. Rahman, A., Jaffar, A. H., & Ihsan, S. I. (2022). IoT-based electromagnetic actuator for CVT: Basic design and prototyping. *International Journal of Powertrains, 11*(1), 38–61.
3. Altas, I. H., & Sharaf, A. M. (2007). A generalized direct approach for designing fuzzy logic controllers in matlab. *Simulink Gui Environment, 1*(4).
4. Alves, A. (2019). *Embedded fuzzy logic library (eFLL)*. Retrieved October 12, 2019, from https://www.arduinolibraries.info/libraries/e-fll
5. Çarman, K. (2008). Prediction of soil compaction under pneumatic tires a using fuzzy logic approach. *Journal of Terramechanics, 45*(4), 103–108.
6. Barua, A., Mudunuri, L. S., & Kosheleva, O. (2014). Why trapezoidal and triangular membership functions work so well: Towards a theoretical explanation. *Journal of Uncertain Systems, 8*(3), 164–168.
7. Hashmi, S. (2014). Sensor-controlled intelligent vehicle system: Demands and needs for a global automotive landscape. In *Comprehensive materials processing* (p. 480). Newnes.
8. Altab, H., & Rahman, A. (2012). Intelligent air-cushion tracked vehicle performance investigation: Neural-networks. *International Journal of Heavy Vehicle System, 19*(4), 407–426.

Chapter 20
Design Optimization of EMCVT

Abstract The purpose of the single plunger type solenoid actuator is to control the movable sheave of the secondary pulley, which ensures that the vehicle driveline has the proper gear ratio. A coil is used in the solenoid actuator to generate an electromagnetic field when current is supplied on it. The solenoid actuator uses PWM signals to reduce power consumption, and temperature rises with a maximum current supply for gear ratio 2.6 and a minimum for gear ratio 1.2. Both with and without an intelligent controller, the full-scale EMA has been parametrically studied to actuate the CVT system of a passenger car. The electromagnetic force, the iron core's absolute permeability of approximately 80 × 10^{-3}N/A2, the number of turns (163), the effective area (625.15 mm^2), and the gap (5 mm) were the EMA characteristics that were noted.

Keywords EMA design optimization · Vehicle propulsion · Pulse width modulation · Performance of EMA to actuate CVT

20.1 Introduction

The value of creativity of EMA-CVT is identified by conducting the literature review presented in Chap. 19. By chipping away weight, EMA-CVT intends to address the issues with the current transmission by lowering acceleration time, transmission loss, gear shifting issues, power consumption from the heavy gearbox, and removing reliance on engine rotation. Furthermore, on a hill climb where hydraulically operated CVTs are not noticeable, the EMA-CVT provides smooth operation to develop driving torque. To manage CVT operations, it is crucial to optimize the EMA design.

This study's primary focus is on the intelligent controlling theory of EMA-CVT operational management, which aims to automate EMA-CVT operation. The EMA design optimization based on the examination of laboratory-scale metal bobbin EMA experimental results has been described in this chapter. In order to create this high-end EMA, the EMA Lab Scale was created based on this study. According to the article, the primary goal of the EMA laboratory scale was to optimize the design

A. Rahman et al., *Autonomous Driveline System for Electric Vehicle*, Green Energy and Technology, https://doi.org/10.1007/978-981-96-8418-2_20

of high-end EMA by understanding how electromagnetic force and heat generation are developed by the supplied current, how long it takes for the secondary pulley's movable sheave to move, and how much electromagnetic force is needed to move the movable pulley against the primary pulley's clamping force in order to set the gear ratio required.

20.2 Optimization Process

In order to modify the intended gear ratio for the vehicle's propulsion, the single plunger type solenoid actuator is intended to operate the secondary pulley's movable sheave. The solenoid actuator is made up of an iron core (bobbin) and a disk that form a magnetic circuit to direct the mechanical movements of the plungers, as well as a coil that uses the electromagnetic field created by the applied voltage to draw the plunger horizontally from positions A and B for the desired gear ratio of the CVT.

Determining design parameters such as enameled wire diameter and coil bobbin shape that result in the lowest power consumption under specified saturation temperature and attraction force is the goal of optimal design for solenoid actuators. In Fig. 20.1, the optimization procedure is displayed. The average length of one winding and the number of coil turns can be found when the coil resistance for power consumption has been established. The magneto-motive force, magnetic flux density, temperature rise, coil change resistance, and attraction force at saturation temperature are all measured.

The conditions under which the coil would have good contact with the metal component were used to determine the temperature gradient. The temperature increases after applying battery power with a maximum current supply for the gear ratio of 2.6 and a minimum for the gear ratio of 1.2. The solenoid actuator employs PWM signals to minimize power usage. To demonstrate the suggested idea, a smaller EMA was created and built. The results of the analysis were then utilized to create a larger EMA that would run the CVT. Figure 20.2 depicts the schematic diagram of a laboratory-scale electromagnetic force generator (EMA), which uses an iron core to boost magnetic energy storage and generate a larger electromagnetic force.

Based on the input pulley torque, the amount of clamping force required to maintain the pulley sheave at correct position is calculated using the equation as:

$$F_c(t) = k_p \delta_p(t) cos\theta \tag{20.1}$$

where k_p is the spring stiffness that keeps the primary pulley in the highest diameter in N/mm and $\delta_p(t)$ is the instantaneous displacement. In this study, we have replaced the spring of the secondary spring by an electromagnetic actuator. In order to change the gear ratio of the EMA-CVT, the EMA needs to develop the electromagnetic force F_{em}, which must be,

$$F_{em}(t) \geq k_p \delta_p(t) cos\theta \tag{20.2}$$

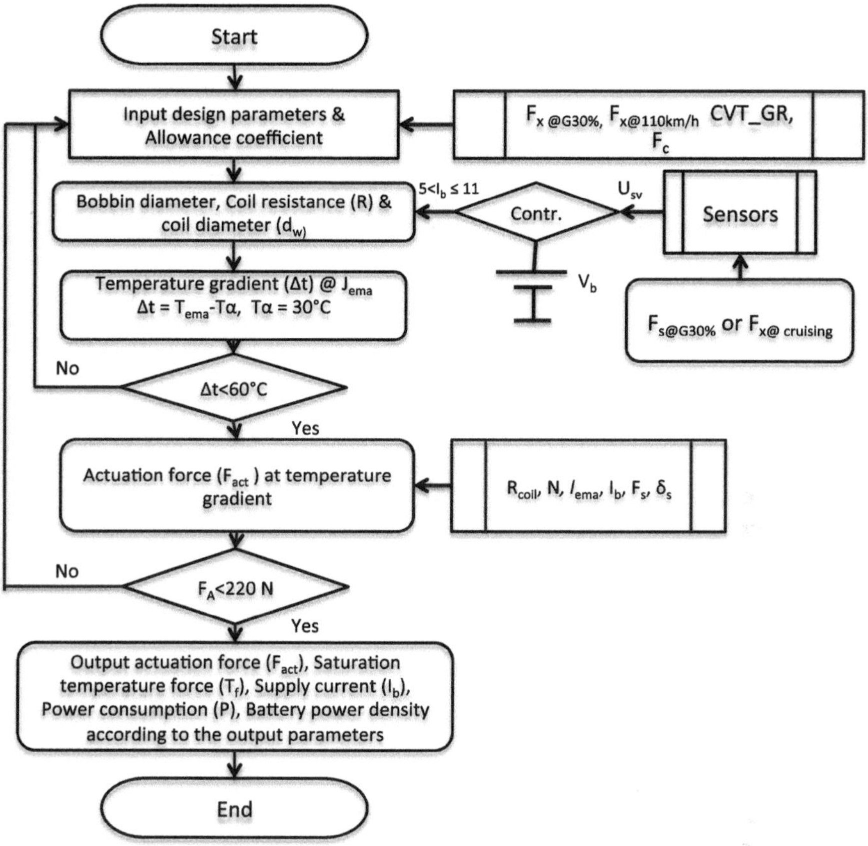

Fig. 20.1 Optimization process of EMA design [1]

The pulley radius is 0.0855 m, the belt friction coefficient is 0.5, and the pulley belt angle is 10.74°. As a result, the clamping forces for slope rising and the initial condition are roughly 100 and 217 Nm, respectively. Estimation are made for the vehicle's tractive force $Fx(t)$, wheel torque Tw, clamping force FC, and power under different road conditions. This figure serves as the foundation for the actuator's design in order to generate an electromagnetic force (F_{em}) that is more than or equal to the clamping force (F_c) by connecting the spring force, F_{act},

$$F_{act.} = F_{em} + k_s \partial_s \tag{20.3}$$

Table 20.1 displays the specification for laboratory-scale EMA. In the laboratory, the bobbin Cast iron is used to make EMA, and copper wire is wound around it to boost the electric energy density (*Eem*) and provide a stronger electromagnetic force. A mover or stopper is used to attach a non-magnetic piston made of aluminum alloy, which serves as an actuator for the pulley's movable sheave. The lab scale's goal to

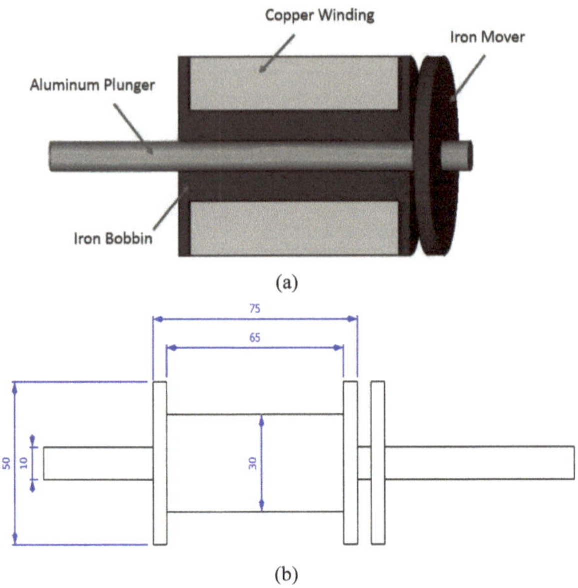

Fig. 20.2 Laboratory scale EMA **a** cut view of new electromagnetic actuator **b** iron core bobbin dimensions for laboratory-scale (all dimensions are in mm) [1]

Table 20.1 Specification of lab scale EMA

Parameters	Values
Coil diameter (dw), mm	1.45
Cross sectional area of wire, mm^2	1.65
Length of solenoid (Lsol), mm	75
Solenoid minimum diameter (hi), mm	30
Solenoid maximum diameter (ho), mm	50
Packing factor $f_{pf} = \frac{N_{actual}}{N_{theoretial}}$	0.525
Number of turns (N) $N = \frac{F_{ema} * L_{coil}}{\mu I^2 L_{sol}}$, turns	163
Length of coil (Lcoil), $L_{coil} = \frac{L_{sol}}{d_w^2}\left[D_0^2 - (D_i + t)^2\right](f_{pf})$, m	30
Mass of solenoid, $m_{sol} = r_{cu} V_{sol}$, kg ρcu = 8940 kg/m at room temperature	2.64

investigate the actuation properties of the full-scale EMA-CVT of a passenger car, EMA was created. A 10 A current was used in the initial experiment to move a 5 N object on a wood surface, and a 5 A current was used to move a 5 N object on a typical road surface.

The main goal of this laboratory-scale electromagnetic actuator (EMA) development is to optimize the design of the full-scale EMA-CVT for passenger cars by

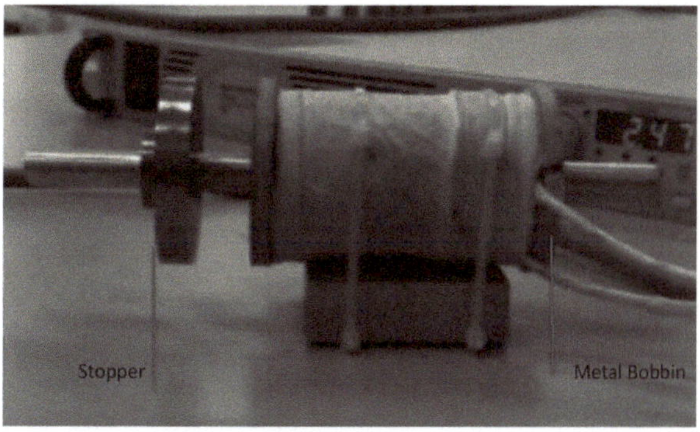

Fig. 20.3 Prototype laboratory scale EMA [2]

doing parametric studies of the electromagnetic actuator (EMA) in both simulation and experimentation. Chapter 21 presents a detailed description of the laboratory EMA as depicted in Fig. 20.3. The EMA's parametric study has been carried out.

The laboratory scale EMA's performance has been parametrically studied in order to actuate the passenger car's CVT system both with and without an intelligent controller. As illustrated in Fig. 20.4, the EMA parameters were as follows: electromagnetic force, the iron core's absolute permeability of approximately 80×10^{-3}N/A2, the number of turns (163), the effective area (0.000625146 m^2), and the gap (5 mm).

Simulation Study on EMA Optimization

The area that perpendicular to the coil turn is considered as the effective area of the actuator. The effective area of the core of EMA, could be modelled as:

$$A_e = \pi\left(\frac{D_2^2 - (D_i + t)^2}{4}\right) \tag{20.4}$$

where D_O (h_1) is outer radius and D_i (h_2) is the inner radius of the actuator bobbin.

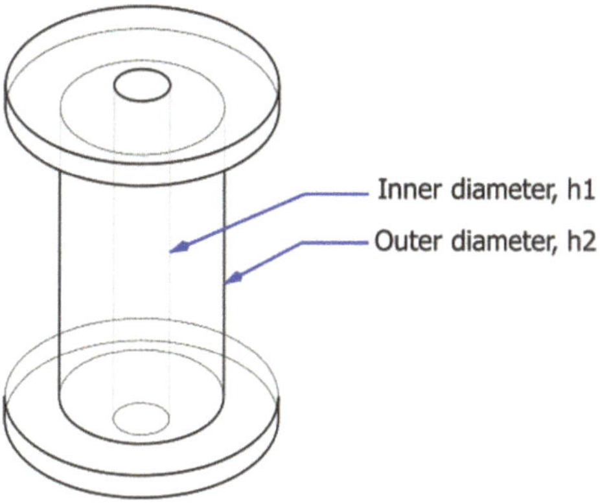

Fig. 20.4 Effective area of actuator bobbin

The simulation of the electromagnetic force of the EMA, clamping force of the secondary pulley, is conducted using the equation by varying the current supplied using the Eq. [3]:

$$F_{em} = \frac{1}{d}\left[\frac{1}{2\mu}B^2 V_{ema} + \frac{1}{2}\varepsilon_0 E^2 + k_{ss}\delta_{ss}\right] - I^2\left[\frac{R_c}{l_c} + \frac{R_{rad}}{d(r)}\right] \quad (20.5)$$
$$with\ B = \mu NI$$
$$d = pulley\ sheave\ moval\ distance, m$$

20.3 Performance of Intelligent Controller

Figure 20.5 depicts the EMA simulation diagram, which includes the input load signal generator, fuzzy logic controller, and EMA model. To observe the generated electromagnetic force in relation to the provided current, the load was varied between 0 and 60 N. The ANFIS changed the fuzzy logic controller's membership functions, allowing the fuzzy system to learn how much current is required to deliver the necessary clamping force (Fem). Figure 20.6 depicts the force produced by the actuator in response to the load and delivered current. Depending on the function, the force produced increases and decreases in response to the supplied current.

$F = \oint (i^2)$. Simulation of the force generated by the actuator corresponding the load as well as the supplied current is shown in Fig. 20.7.

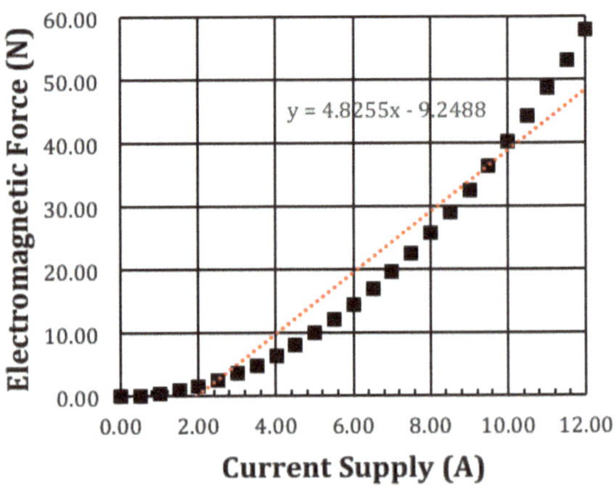

Fig. 20.5 A typical simulation result of laboratory-scale EMA force generated with respect to the supply current [3]

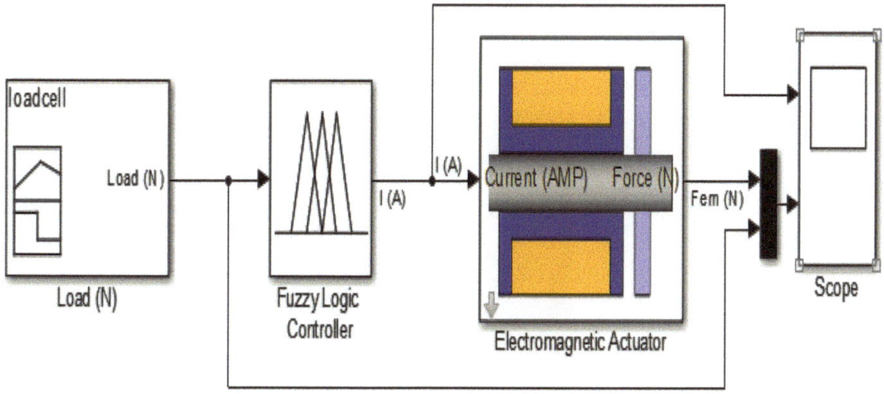

Fig. 20.6 Lab.scale EMA simulation block diagram in Matlab

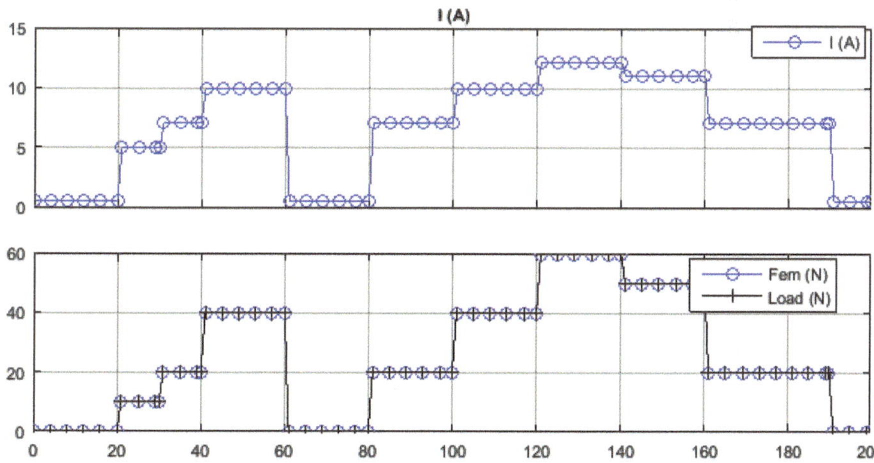

Fig. 20.7 Simulation result for supply current, load and generated F_{em}

20.4 Experimental Study on Ema Optimization

Figure 20.8 depicts an experimental setup for measuring the provided current as it responds to weight (load). Bending beam load cell strain gauge technology was used to measure physical qualities such as compression, tension, bending, and shear and convert them to an electrical signal. The computer-based controller then transformed the signal to an equivalent weight value. The weight serves as the basis for the controller to send a signal to the programmable power supply and generate the appropriate current flow.

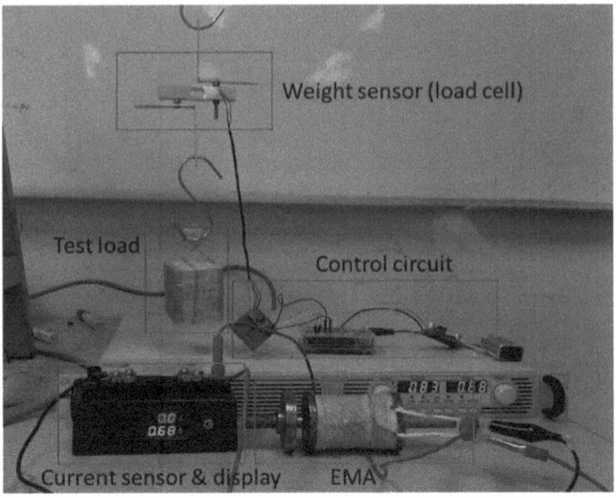

Fig. 20.8 Electromagnetic actuator experiment setup block diagram [4, 5]

Fuzzy Intelligent System

Figure 20.8 shows an experimental setup for measuring the provided current in response to weight (load). Bending beam load cell strain gauge technology was used to measure physical qualities such as compression, tension, bending, and shear and convert them to an electrical signal. The computer-based controller then transformed the signal to an equivalent weight value. The weight serves as the basis for the controller to send a signal to the programmable power supply and generate the appropriate current flow.

The experimental result of EMA The provided current is determined by the input load at the load cell. The current is proportional to both the load and the actuator's electromagnetic force. Matlab simulations of the relationship between clamping force, electromagnetic force, and current supply were performed using FIS methods. The created laboratory scale EMA was investigated in both modeling and experimentation.

The main conclusions drawn from this study are as follows:

i. *Theoretically*

The laboratory scale EMA can generate electromagnetic force in the range of 0 N to 60 N by delivering current up to 10.50 amps. The FIC can generate according to the input load.

ii. *Experimentally*

The laboratory-scale EMA required a maximum current of 12.2 A to generate an electromagnetic force of 60 N. These data were analyzed and used as the basis for developing a full-size actuator capable of driving a genuine car's CVT system.

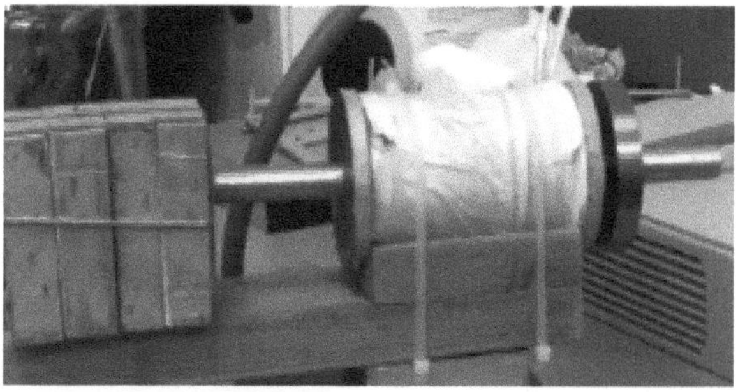

Fig. 20.9 Pushing force experimental setup for lab-scale actuator [2]

Pushing Force

The goal of this experiment is to optimize the design requirements for the upgraded EMA. The experimental setup of the laboratory scale EMA, illustrated in Fig. 20.9, was designed to measure the current required to push the load, which symbolizes the electromagnetic force generated by the EMA as the current source is varied. The 530-g load (containing a set of metal parts weighing 378 g plus 152 g of mover and plunger) was situated at the end of the plunger. The current supply was increased from 0 to 11.5 amps. Figure 20.9 shows that a maximum current of 11.0A was required to move the 530 g (5.2 Newton) load on the surface of wood.

Clamping Force

The goal of determining the EMA's hanging capacity (holding capacity of the EMA against the pulley clamping) is to optimize the clamping force required to clamp the CVT's movable sheave against the primary pulley in order to maintain the desired gear ratio. For example, the movable sheave on the secondary pulley of a full-scale CVT requires 374 N to achieve a gear ratio of around 5:1. The experimental setup was also created to test the current required to supply the laboratory scale. EMA optimizes the required current for the EMA solenoid to provide 374 N to clamp the secondary pulley against pushing force, as illustrated in Fig. 20.10. The 530 g load was hung from the plunger of the small scale EMA.

At first, the current supply was set to a high value of 10 A and subsequently decreased until the actuator could no longer withstand the load. To hold the 5.2 Newton load hanging, a minimum current of 4.1 A is necessary, which is regarded the clamping force that the EMA creates with a 4.1 A supply current. The EMA for the full-scale of the passenger car's CVT can be optimized to be 30 times larger than the laboratory scale EMA while keeping the current supply constant, or to be 10 times larger than the laboratory scale EMA while increasing the current supply by around 12 A. However, by attaching a spring with a stiffness of 40 N/cm, the needed clamping force can be created more than once by adjusting the current supply to the EMA between 7 and 11.5 A.

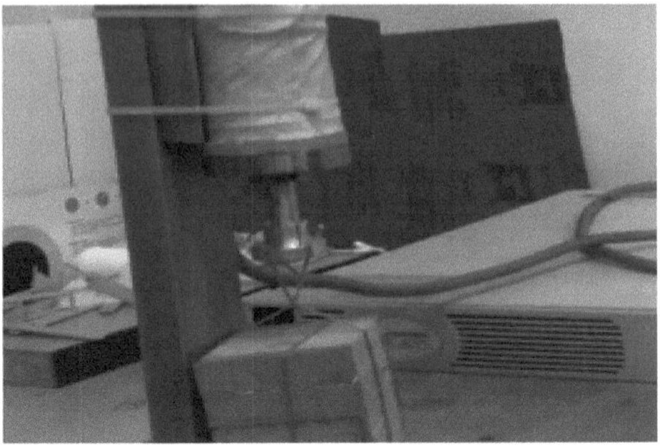

Fig. 20.10 Hanging force experimental setup for lab-scale actuator [2]

Holding Force

Figure 20.11 depicts the performance of the EMA laboratory scale when working with a load, which is regarded the driving power of the movable pulley of the secondary pulley to accelerate or cruise the vehicle. EMA performance has been evaluated using load displacement on a hardwood surface. The force required to move a 5.2 N load to a hardwood surface at different offsets develops EMA as the current supply changes. The results reveal that a larger current supply affects the force development of the EMA, resulting in a longer acceleration time. It was discovered that the higher quantity of current of 11.0 A supply to the EMA causes heat, which reduces the EMA force due to the development of resistance.

Supply Current and Gap Analysis

First, the effective gap across which the actuator could create pulling force and the related needed current were calculated and summarized in Table 20.2. Experiments revealed that below 4.0A, there is no sufficient pulling force to pull the mover. The

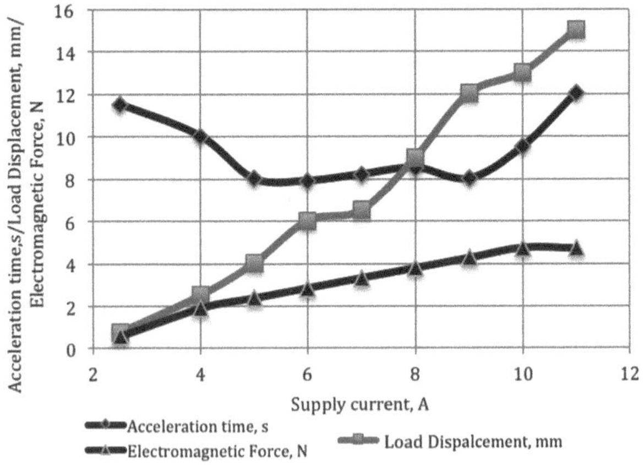

Fig. 20.11 Performance of laboratory scale EMA [1]

Table 20.2 Performance of the laboratory scale EMA

Supply current	Gap (load displacement)	Applied load	EMA force	Time taken	Temperature gradient[a]
(A)	(mm)	(N)	(N)	(s)	(°C)
3	1.2	0.473	0.56	11.5	1.23
4	2.5	0.473	1.892	10	2.74
5	4	0.473	2.365	8	3.42
6	6	0.473	2.838	7.9	4.87
7	6.5	0.473	3.311	8.2	6.88
8	9	0.473	3.784	8.5	9.31
9	12	0.473	4.257	8	11.09
10	13	0.473	4.73	9.5	16.26
11	15	0.473	5.203	12	24.85

[a] Temperature gradient, $\delta T = T_f - T\alpha$, T_f is the temperature develops at EMA due to supply current, $T\alpha$ ambient temperature. The expected temperature gradient is 20 °C

actuator began to exert significant force at 4.0 A, attracting the mover at a maximum distance of 2.5 mm. With a current supply of 11 A, the actuator can pull the mover 15 mm. The more the supplied current, the greater the gap distance that the magnetic force can govern.

In the previous paragraph, it is indicated that the temperature gradient of the EMA reaches 20 °C for supply current 11.0 A. The temperature gradient of 20 °C indicates a temperature of 50 °C at the solenoid, compared to Malaysia's typical outdoor temperature of 30 °C. The EMA temperature, which develops as a result of ohm heat, has a substantial impact on electromagnetic force development, increasing acceleration time and delaying gear changes. As a result, the optimal current supply to the EMA should be 10 A. However, for short-term operation, the EMA might be supplied with 11.0 A of current to achieve a greater gear ratio at the lowest acceleration time. As a result, assuming the temperature remains below 500C, the vehicle's EMA-CVT will be able to reduce transmission loss by around 10%.

Figures 20.12, 20.13, 20.14, 20.15 and 20.16 show the variation of electromagnetic force development with the variation of current supply. Result shows that the electromagnetic force of the EMA decreases with increasing the current supply 11 A or more, since electromagnetic force,

$$F_{em} = \frac{N\mu i^2 l}{2\pi r} \tag{20.6}$$

where l is the conductor length, m, μ is the absolute permeability of the iron core, N/A^2, and r is the radius of the EMA. The absolute permeability is calculated as $\mu = 80 \times 10^{-3}$ N/A^2. A high current supply generates higher electromagnetic force instantly. However, it begins to diminish immediately due to excessive heat generation, as

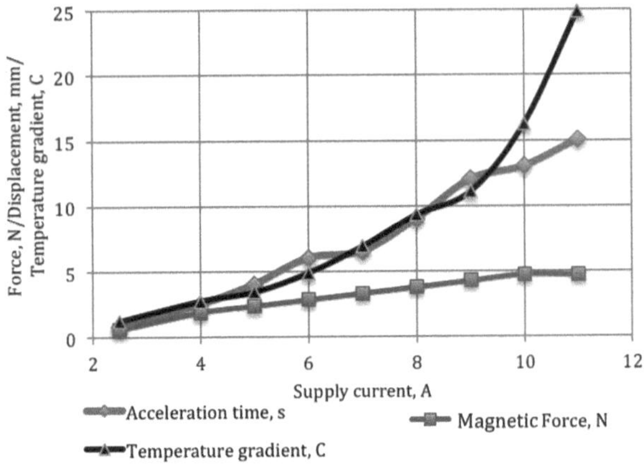

Fig. 20.12 Temperature effect on EMA force and acceleration time

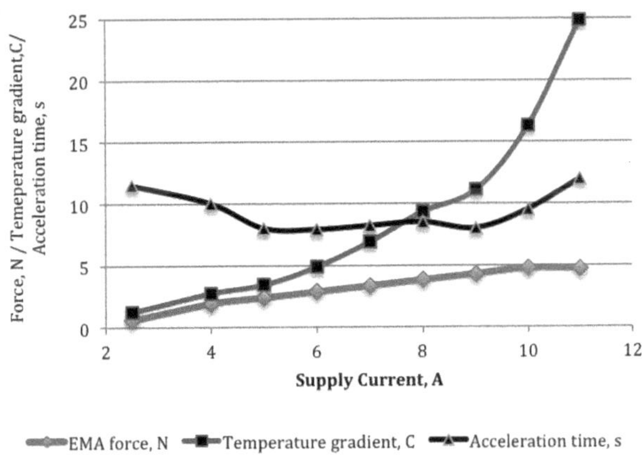

Fig. 20.13 Heat generation effect on acceleration time and EMA force [1]

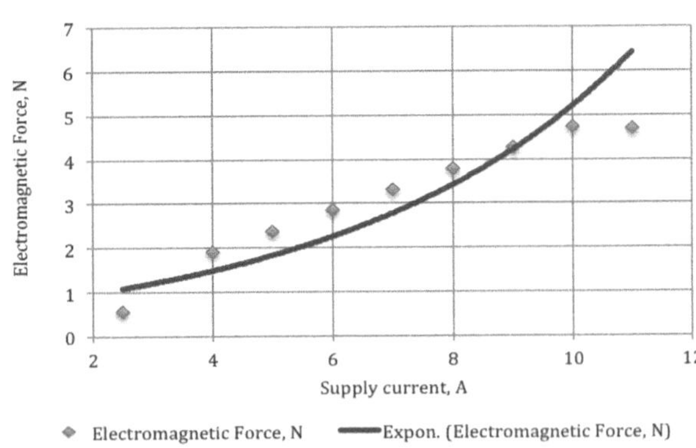

Fig. 20.14 Electromagnetic force with current supply

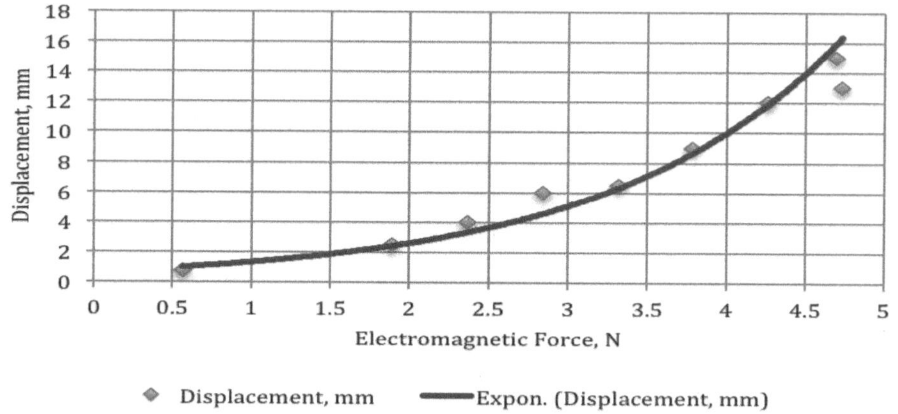

Fig. 20.15 Electromagnetic force causes the load displacement [3]

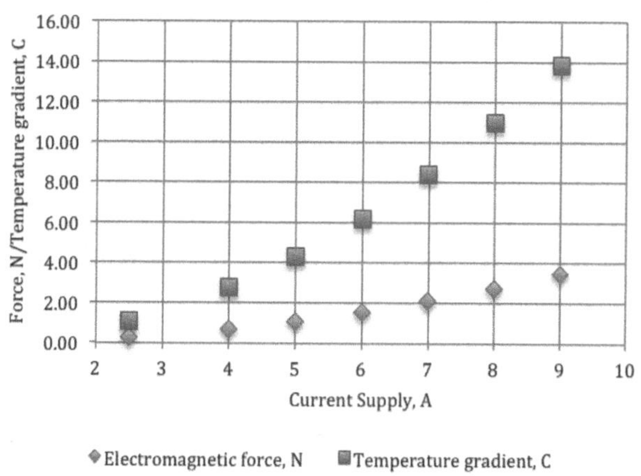

Fig. 20.16 Electromagnetic shifting force and temperature gradient [3]

illustrated in Fig. 20.15. As a result, the optimal current delivery to the EMA of a 163-turn copper coil with a diameter of 1.45 mm is 11.0 A.

20.5 Validation of Laboratory Scale EMA

To validate the concept of laboratory scale EMA design, the electromagnetic actuator was tested on a model of an electromagnetic two-speed gearbox (EMA2-SGB), as seen in Fig. 20.17. It is constructed of two-speed gear sets (gear ratio), two laboratory scale EMAs, a fuzzy controller, and speed sensors. The first gear is used to accelerate the vehicle, while the second gear is used to maintain the vehicle's moderate or cruising pace. The actuator was functioning well, efficiently moving the gear from

Fig. 20.17 Laboratory scale EMA-2SGB [4, 5]

one to the other. Shifting time is the time interval between gear changes that occur without interruption. The fuzzy controller is responsible for switching the power supply to the EMA.

The laboratory scale metal bobbin EMA can generate electromagnetic force to move the gear in 1–2 s with a current supply in the range of 7–8. Table 20.3 displays the time it takes to shift from neutral to first gear or second gear, as well as from first gear to second gear and second gear to first gear, under heavy-duty operation. Gear shifting time for second gear engagement is 25% longer than for first gear, which could be attributed to the vehicle's speed fluctuation.

Figures 20.18 and 20.19 illustrate the validation of the laboratory scale EMA, which was performed using the identical EMA in EMA-2SGB. The strong agreement between electromagnetic force on load displacement on a wooden surface and gear shifting supports the development of EMA for mechanical system actuation. The effective use of mechanical systems, such as load shift, gravity load, and automatic transmission, with EMA electromagnetic force implies that the design of EMA with the hollow metal bobbin and alloy is significant and sustainable control for the CVT of the passenger vehicle. This was the purpose of creating the EMA laboratory scale,

Table 20.3 Shifting performance of EMA-2SGB

Gear shifting	Gear shifting distance (mm)	EMA current required (A)	Gear shifting time (ms)
1st gear engagement for neutral position	7	7	75.5
2nd gear engagement from 1st gear	14	8.5	150.5
1st gear engagement from 2nd gear	14	8.5	120.25

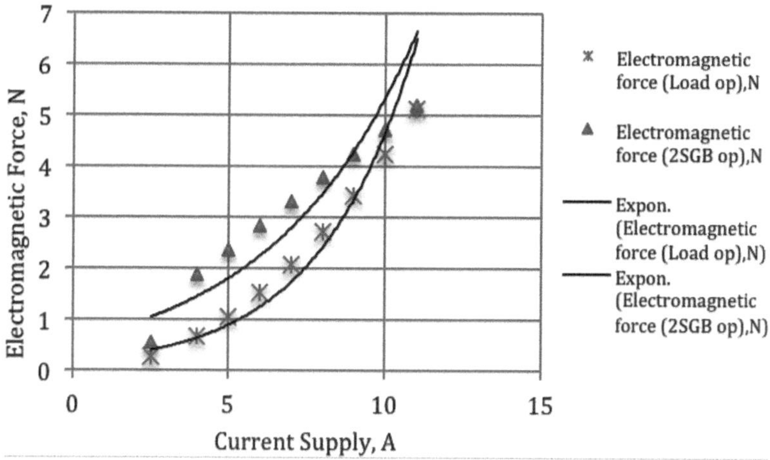

Fig. 20.18 Variation of laboratory scale EMA performance [3]

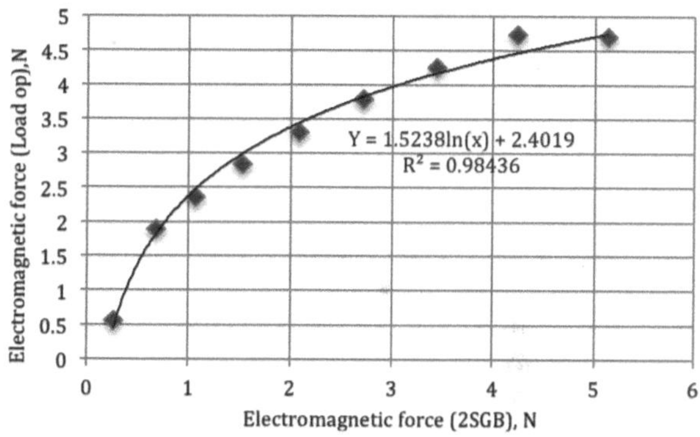

Fig. 20.19 Validation of laboratory scale performance [3]

which explains the innovative principle behind the development of upmarket EMA for passenger car CVT operation.

20.6 Parametric Study of Upscale EMA-CVT

The estimated parameters of the traction torque, clamping force for the CVT pulley, length and diameter of the EMA bobbin, bobbin materials, diameter of the EMA coils, total coil length, number of coil turns, and electrical current required to develop the electromagnetic force by the EMA. The EMA is designed to provide electromagnetic force that is equivalent to or greater than the CVT's clamping force. The clamping force on the movable sheave of the primary pulley is determined by the primary's stiffness, which is 117.4 N/cm.

The clamping force is necessary to hold the secondary pulley sheave against the force exerted by the primary pulley in order to maintain the number of gear ratio. The gear ratio is necessary to provide traction torque at the driving wheel's tire-road interface. For example, the traction torque required for the 1700 kg car on 25% gradient 1221 Nm and the initial start-up 2870 Nm using the equation;

$$T_W(t) = m\frac{\partial v}{\partial t} * r_w + \left| \begin{array}{ll} f_r W & 0 < v \leq 120\, km/h \\ W\, Sin\theta & 0 < G \geq 20\%\,,\ \ 0 < v \leq 40\, km/h \\ \frac{\rho}{2} C_d A_f v^2 & G = 0\%,\ 40 < v \leq 120\, km/h \end{array} \right| * r_w \tag{20.7}$$

The parameters evaluated are vehicle maximum speed of 120 km/h, traveling on a dry asphalt surface with a coefficient of road adhesion (p) of 0.85, wheel radius of 0.315 m, air density of 1.225 kg/m3, drag area of 0.58, and rolling motion resistance coefficient of 0.013. The simulation results are shown in Table 20.4.

The transmission torque of the CVT pulley at initial condition and during motion is computed as 660 Nm and 87 Nm, respectively, using a maximum gear ratio of 5:1 based on a Toyota Wish automobile with CVT (Wikipedia, 2017b) [6].

Design Parameter Study of the Upscale EMA

The length of the EMA was chosen based on the amount of space available in the Toyota Wish Car transmission. This analysis takes into account an electrical motor's maximum torque of 260 Nm, voltage of 150 V, power capacity of 15 kW, and differential speed ratio of 4.0. The gear ratio of the EMA-CVT has been estimated based on the traction force required for the vehicle to propel on the road surface, whether traction or cruising. For example, the traction force necessary to start the car from rest is 8111 N, with a wheel torque of 2670 N.m.

Table 20.4 EMA-CVT clamping force

Condition	θ (°)	V (km/h)	V (m/s)	F_{TR} (N)	T_W (Nm)	T_{out} (Nm)	T_m (Nm)	GR	F_C (N)
Initial	11.3	10	2.78	8111	2670	660	260	2.7	217
		90	25	494	163	47	38	0.8	69
Cruising	0	100	27.77	628	207	60	60	1.0	75
		120	33.33	878	290	83	100	1.2	100

- F_{TR} = vehicle required traction force, N
- T_w = vehicle driving wheel torque
- T_{out} = output torque of the secondary pulley
- T_m = torque of the motor
- $F_c = F_{em} + F_s$

where, F_c = clamping force, F_{EM} = Electromagnetic force, F_S = Spring force

The EMA-CVT's gear ratio is estimated to be 2.7 based on a maximum motor torque of 260 Nm. The EMA-CVT will be capable of producing 660 Nm, resulting in a wheel torque of 2670 Nm. The clamping force for the secondary pulley's moveable sheave should be 217 N. The clamping force is defined as the sum of the EMA force and the spring force. Figure 20.19 depicts how the EMA force and spring force combine to make up the total force. For example, the EMA force must generate 146 N and the spring force 74 N to achieve the EMA-CVT's total clamping force of 2.6 in first gear. The force developed by the EMA is determined by the current supplied to the EMA's coil. While, the spring force, $F_s\ (x) = k_s\ \Delta x$ where Δx is the spring displacement and k_s is the stiffness of the spring which is considered as 70 N/cm.

Table 20.5 summarizes the minimum and highest current requirements for generating electromagnetic force (clamping force) from 100 to 220 N using different numbers of copper wire. According to the results, the required current might be achieved by wrapping the copper wire between 3000 and 4000 turns. Given the supply current required to operate the EMA and the physical space of the testbed, 3080 coil turns were chosen so that the EMA could create the aforementioned electromagnetic force with a supply current ranging from 6 A to 9.33 A. The packing factor used in the calculation was 90.7%, as reported in (Wikipedia, 2019b) [6] for a factory coil winding.

Design Parameters of Upscale EMA-CVT

The required electromagnetic force generated by the coil might be used to estimate the appropriate copper wire diameter. The coil wire diameter and current limit could be compared to the table provided by the sources listed below [7]; Wikipedia, 2019a). Equation 20.5 is used to calculate the desired amount of current with varying the number of windings for the electromagnetic force in the range of 160 N to 220 N (initial or maximum), where the absolute permeability of the iron core, μ (80×10^{-3} N/A2), is the air gap or stroke length separating the two ferromagnetic bodies, and is the effective air-gap face area of the actuator. The bobbin or housing of the actuator, depicted in Fig. 20.20, was chosen based on the limited space in the Toyota Wish automobile. Different winding numbers of 1000, 1500, 2000, 2500, 3000, and 3500 are examined to optimize the current supply to the EMA coil if it is fully scaled up for the passenger car's CVT operation.

The relationship between the number of copper coil turns, current supply, and generated electromagnetic force for a vehicle's needed traction force. It showed the

Table 20.5 Current requirement for different number of turns

Electromagnetic force, N	Number of turns	Current (Ampere)	
		Minimum	Maximum
220	1000	25.18	31.58
	2000	12.59	15.79
	3000	8.39	10.53
	4000	6.30	7.89

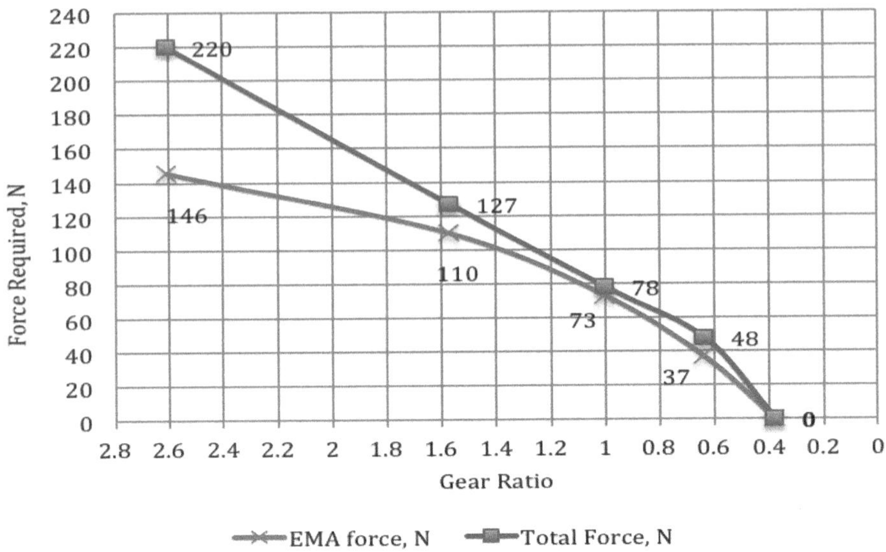

Fig. 20.20 Clamping force respect to gear ratio [4, 5]

maximum and minimum needed electromagnetic force for different winding and current levels. When designing the testbed space, a turn count of 3000 to 3500 could be considered. With 3080 revolutions, the electromagnetic actuator used 6.09 to 9.33 amps of current to provide an electromagnetic force of 100 to 146 N, which is capable of developing the vehicle's traction force between 2.6 and 6.2 kN. Wire with a diameter of 1.6–1.9 mm was sufficient to support the current required, as illustrated in Figs. 20.21 and 20.22.

Table 20.6 shows that the CVT required 217 clamping force to create a gear ratio of 2.6 for traction torque of 2670 Nm at the vehicle's driving wheel. According to the EMA, an electromagnetic force of 146 N is required to generate a clamping force of 217 N. Furthermore, Fig. 20.21 shows that increasing the supply current by 10 or more causes an increase in the temperature gradient, which reduces electromagnetic force development if the EMA is formed with a copper coil with a diameter ranging from 1.8 to 2.0 mm.

If the upscale EMA is made with a copper coil diameter of 1.85 mm, the maximum number of EMA coil turns should be optimized at 3100 to achieve an electromagnetic force of 146–155 N for a supply current of 10 A, as shown in Fig. 4.19b. If the EMA coils have more than 3100 turns, the supply current should increase to 11 for 3000 turns and 11.5 for 2800 turns. Both EMAs will struggle as the temperature gradient increases.

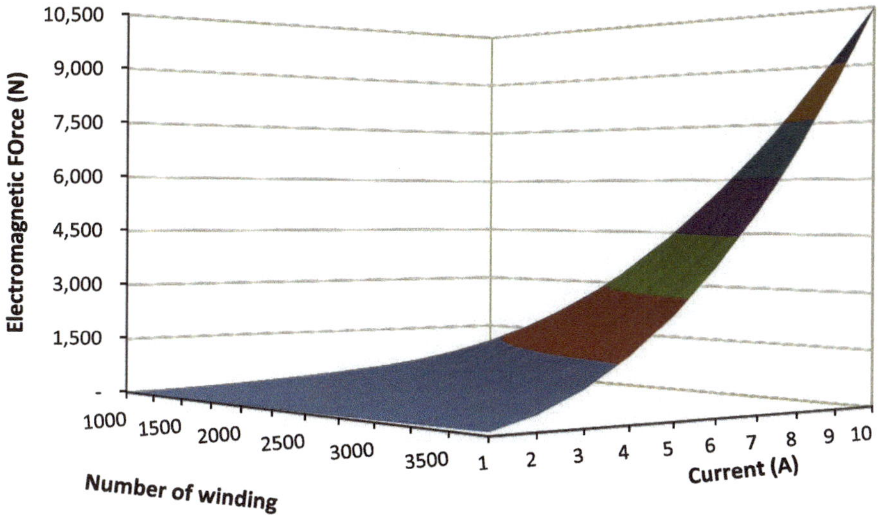

Fig. 20.21 Coil wire diameter VS number of windings and current supply [1]

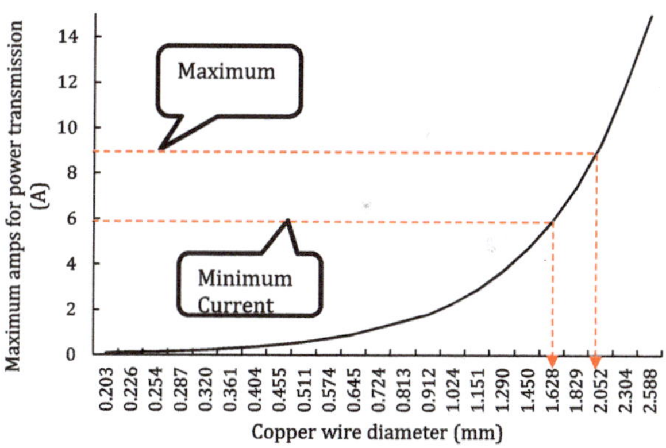

Fig. 20.22 Copper wire maximum amps for different diameter based on chart of [7]

Figure 20.23 depicts the voltage and current relationship for activating the EMA to generate the electromagnetic force required to clam the movable sheave of the secondary pulley while maintaining a constant voltage. In the previous discussion, it was stated that the maximum current of 11.0 A must be supplied to the EMA in order to obtain an electromagnetic force of 146 N. This section demonstrates how to supply current to the EMA's coil using a 100 V battery.

Table 20.6 Design parameter of the upscale EMA

D_O (m)	Di (m)	*fp*	dw (m)	Lema (m)	L_c (m)	I_s (A)	N (Turns)	Fema (N)
0.15	0.08	0.65	0.0018	0.22	710.59	10.5	3100	166.99
0.13	0.08	0.65	0.0018	0.22	463.43	10.5	2800	108.13
0.12	0.08	0.65	0.0018	0.22	353.09	10.5	2100	77.68
0.1	0.08	0.65	0.0018	0.22	158.89	10.5	1530	33.10
0	0.08	0.65	0.0018	0.22	0	0	0	0

Notifications:
D_O & D_i = outer and inner diameter of the EMA, m
f_p = packing factor
d_w = diameter of the wire, m L_{ema} = length of the EMA, m
L_c = length of the coils, m
N = Number of turns of coils

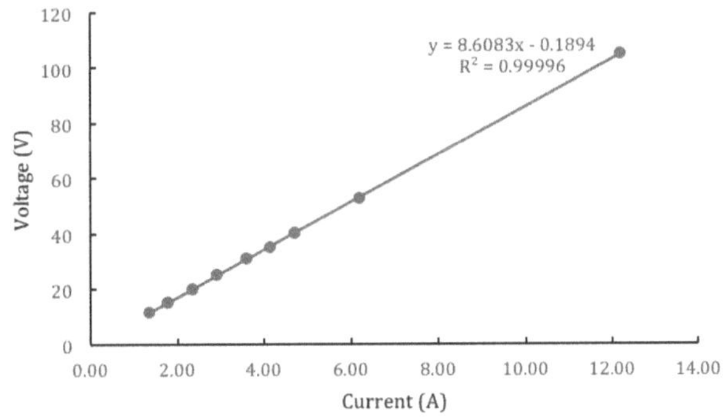

Fig. 20.23 EMA's voltage requirement to meet the current supply [1]

20.7 Conclusion

Prior to the prototype of full-size EMA-CVT, modeling and experimental tests on laboratory-scale EMA were performed to validate the suggested concept. The primary goal of the EMA laboratory scale was to optimize the design of high-end EMAs by understanding the development of electromagnetic force and heat generation due to the supplied current, the acceleration time of the movement of the movable sheave of the secondary pulley, and the electromagnetic force required to move the movable pulley against the clamping force of the primary pulley to set the gear ratio as required. The results were studied for parameter optimization when building the actual size EMA to run the CVT.

Acknowledgements The authors are grateful to Malaysia's Ministry of Higher Education for financing this project to construct an upscale EMA-CVT prototype under the Prototype Research Grant Scheme (PRGS) in 2016. The authors are also appreciative to the Faculty of Engineering, International Islamic University Malaysia, for prototyping, and GameAbove Engineering and

Technology, Eastern Michigan University, Ypsilanti, Michigan, USA, for enabling continuous advancement of EMA-CVT.

References

1. Hassan, A. J. (2020). *Development of intelligent electromagnetic continuously variable transmission* [A PhD Thesis Dissertation]. IIUM Library, International Islamic University Malaysia.
2. Jaafar, A. H., & Rahman, M. A. (2019). Study on the intelligent control system of EMA-CVT. *International Journal of Recent Technology and Engineering (IJRTE), 7*(65), 48–53.
3. Rahman, A., Sazzad, B. S., Mohiuddin, A. K. M., Mahbubur, R., & Hossain, A. (2014). Energy efficient electromagnetic actuator for CVT system. *Journal of Mechanical Science and Technology, 28*(4), 1153–1160.
4. Rahman, A., Jaffar, A. H., & Sany, I. I. (2022). IoT-based electromagnetic actuator for CVT: Basic design and prototyping. *International Journal of Powertrains, 11*(1), 38–61.
5. Rahman, A., Nurul, H., & Sany, I. I. (2022). A fuzzy logic controlled two speed electromagnetic gearbox for electric vehicle. *International Journal of Technology, Universitas Indonesia, 13*(2), 167–179.
6. Mathworks. (2012). *MATLAB: Fuzzy logic toolbox users' guide*. Natick, MA. http://www.mathworks.com/products/fuzzy-logic/demos.html
7. Lund Instrument Engineering Inc. (2018). *Wire gauge and current limits including skin depth and strength*. Retrieved February 20, 2018, from https://www.powerstream.com/Wire_Size.htm

Chapter 21
Performance of EMA-CVT

Abstract The EMA-CVT technology is made to efficiently control dynamic torque in automobiles, especially in difficult situations like a 20% road grade. With a motor torque of 490 Nm and an overall gear ratio of 11.7, the system can maintain a dynamic tractive force of 6200 N with a generated electromagnetic force of 200 N. The current supply may be precisely controlled between 6 and 11 amps by the fuzzy logic controller, which maximizes the power from a battery rated at 150 V and 144 Ah. With this modification, the system can produce clamping forces of up to 6.2 kN, guaranteeing sufficient torque delivery to the vehicle's drive shaft, which was between 1221 and 2870 Nm. The overall goal of the EMA-CVT is to improve vehicle efficiency and performance by using electromagnetic technology and sophisticated control systems to react dynamically to driving situations. A smart controller benefits are provided by EMA-CVT for automotive transmission systems. Its greater performance over conventional gearboxes are demonstrated by its capacity to accelerate quickly from 0 to 60 km/h in under 220 ms. Driver convenience and safety are improved by this rapid response time, which is accomplished without the need for human intervention. A smoother driving experience is also a result of the lack of the mechanical rattling noise that comes with traditional gear changes. Additionally, the intelligent control system maximizes energy efficiency, lowering pollutants and fuel consumption while preserving improved vehicle performance, particularly in traction. All things considered, the EMA-CVT is a promising development in vehicle gearbox technology that combines quiet, speed, and energy saving.

Keywords EMA-CVT · Fuzzy logic controller · EMA power requirement

21.1 Introduction

This study focuses on the design and development of an Electromechanical Actuator (EMA) for the CVT of the Toyota Wish car. The design methodology and optimization processes for the EMA are detailed in Chaps. 19 and 20, respectively. This chapter presents both theoretical and experimental performance evaluations of the EMA in

A. Rahman et al., *Autonomous Driveline System for Electric Vehicle*, Green Energy and Technology, https://doi.org/10.1007/978-981-96-8418-2_21

its role of actuating the CVT. The analysis involves simulating the desired traction torque of the vehicle across various operating conditions using analytical methods, including Matlab Simulink and fuzzy logic. The estimation of the clamping force of the CVT pulley takes into account the vehicle's traction force. The EMA has been designed to generate an electromagnetic force that matches or exceeds the necessary clamping force for optimal performance.

This chapter discusses a simplified model of the rear-wheel drive system of the Toyota Wish, aimed at visualizing the functioning of the developed Electromagnetic Actuator Continuously Variable Transmission (EMA-CVT). By supplying current from a 100 V battery pack, the EMA is activated. The chapter also explores the dynamic behavior of the electromagnetic actuating system using a Fuzzy Logic approach. To validate the simulated performance of the EMA, experimental results were obtained from a newly developed test bed, which equipped equivalence of road loads. The experimental findings for the EMA are compared and verified against the fuzzy simulated results, demonstrating the effectiveness of the simulation in capturing the system's behavior.

21.2 Theoretical Performance of EMA-CVT

The performance investigation of the EMA-CVT system for the Toyota Wish has been theoretically assessed to optimize its traction capabilities on both inclined and flat surfaces. Key design parameters for the Toyota Wish include:

- **Mass**: 1700 kg
- **Wheelbase**: 2800 mm
- **Center of Gravity (CG) Height**: 500 mm
- **Differential Speed Constant**: 3.5
- **Tire Radius**: 330 mm
- **CG Position (CGx)**: 1.27 cm behind the front axle

Configured as a front-wheel drive (FWD) vehicle, the study focuses on determining the maximum gear ratio needed to achieve optimal traction force. The vehicle is powered by an electric motor with a voltage rating of 150 V and a peak power output of 38 kW.

Simulations were conducted using MATLAB Simulink, allowing for detailed analysis of how the EMA-CVT affects traction performance under various conditions. Figures 21.1 and 21.2 provides insights into these simulation results, highlighting the effectiveness of the EMA-CVT in enhancing the traction capabilities of the Toyota Wish, especially during steep climbs and on flat roads. This approach underscores the potential benefits of integrating advanced control systems in modern vehicle designs.

The simulation results for the EMA-CVT on a 20% grade reveal key insights into its performance under varying current supplies. When adjusting the current between

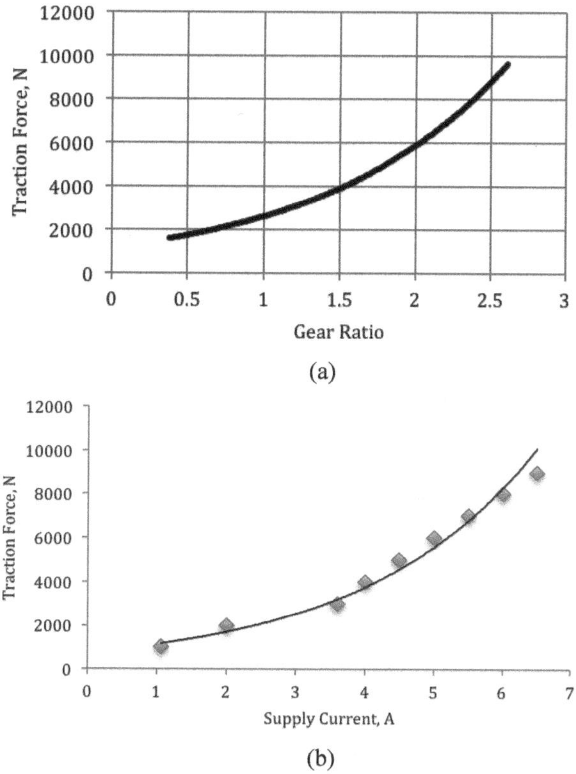

Fig. 21.1 Vehicle traction force for the variation of **a** gear ratio **b** supply current [5]

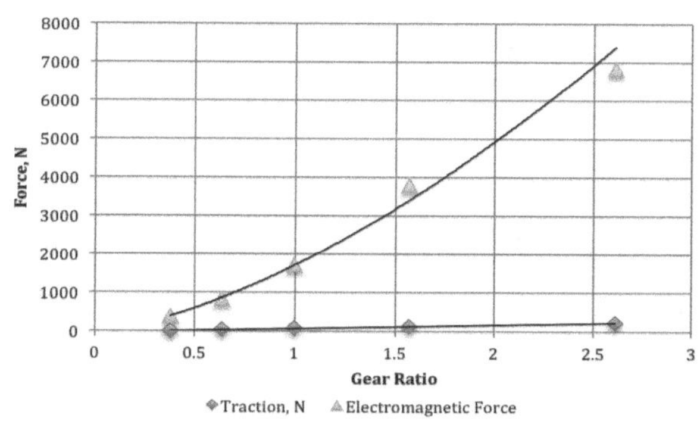

Fig. 21.2 Electromagnetic force, wheel torque versus current [5]

6 and 11 amps, the system generates electromagnetic forces that translate to clamping forces capable of producing 2.6 kN of traction, particularly with a gear ratio of 2.6.

As illustrated in Fig. 21.1, the relationship between the supply current and the electromagnetic force, field, and energy is critical. Initially, increasing the current effectively boosts the electromagnetic field and energy output. However, it's important to note that after reaching a certain threshold, the electromagnetic field stabilizes due to hysteresis and heat generation.

This plateau indicates that further increases in current yield diminishing returns, resulting in excessive heat generation that not only represents a loss of energy but can also pose risks to system integrity. Therefore, optimizing current supply is essential to balance performance and efficiency, avoiding unnecessary heat production while maximizing the system's operational capabilities.

Figure 21.2 shows the electromagnetic force development with supply current, which is considered in this study the gear ratio setting and the clamping force of the secondary pulley by pushing the movable pulley sheave of the secondary pulley. Simulated result shows that the electromagnetic force required to develop 220 to claim the secondary pulley against the primary pulley to set the gear ratio of 2.6 and the wheel torque 2,170 Nm (equivalent traction force of 6.2 kN).

Figure 21.3 illustrates that when a power of 1000 W is delivered at a supply current of 11.5 A, the EMA effectively generates a clamping force resulting in a traction torque of 2170 Nm with a gear ratio of 2.3. As discussed in Chap. 4, it's clear that supplying a current of 12 A to the EMA does not lead to excessive heat generation, making it the optimal current for efficient operation. The EMA-CVT's primary goal is to produce electromagnetic force that maintains the vehicle's dynamic torque within a range of 1221–2870 Nm. Additionally, with 3100 turns, the EMA can develop electromagnetic forces of 200 N, contributing to a dynamic traction force of 6200 N at a gear ratio of 2.6. This configuration requires a motor torque of 490 Nm and a final drive ratio of 4.5. The power supplied to the EMA is sourced from the vehicle's main battery pack, which operates at 150 V with a capacity of 144 Ah. To optimize performance and responsiveness, the current supplied to the EMA will be managed by a Fuzzy Logic Controller, enabling precise adjustments based on real-time vehicle dynamics and performance requirements. This intelligent control mechanism aims to enhance efficiency and overall system reliability.

21.3 FIS Controlled EMA-CVT

Matlab Simulink is used for analysis of the EMA-CVT and the suggested fuzzy logic control system that were covered in Chap. 20. The servo motor angle, which modifies the current control module as illustrated in Fig. 21.4, is the output of the fuzzy logic controller, which has two inputs: the electrical current error and its rate of error. Every simulation component block comes from the Matlab Simulink library or can be made by the user using the subsystem. A subsystem is a collection of blocks that may be created using Matlab® and substituted with a single subsystem block. Blocks could be grouped into a subsystem or a small number of subsystems to simplify the model as it grows in size and complexity (Mathworks, 2019).

The signal that regulates the servo motor's degree, which mechanically turns the current control module's knob, is the fuzzy logic controller's output. The Simulink

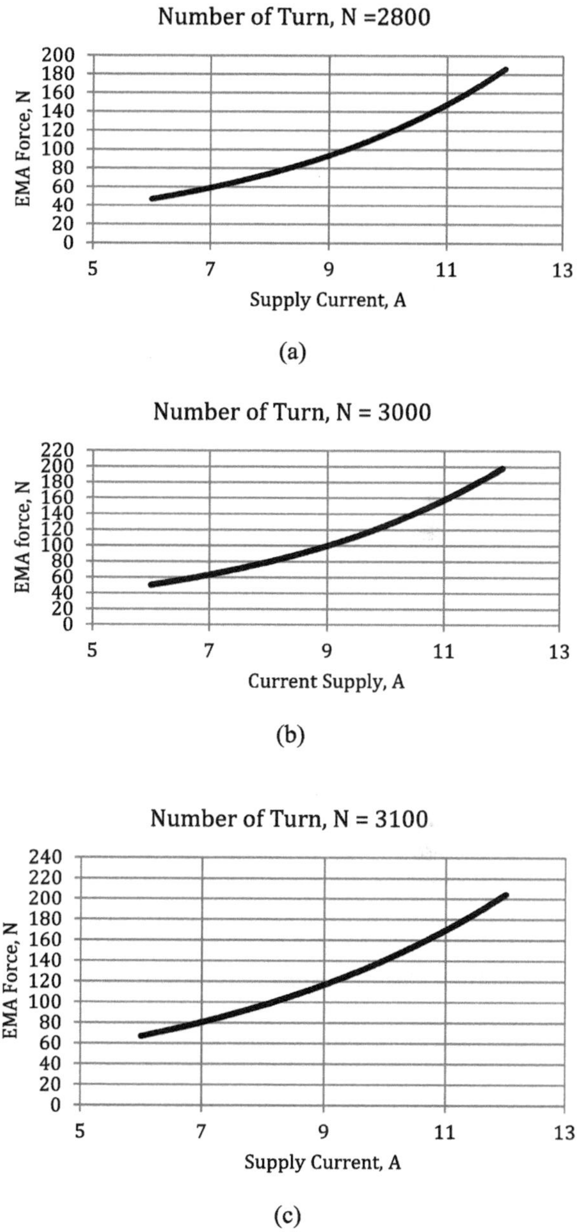

Fig. 21.3 EMA force development with the variation of supply current [1]

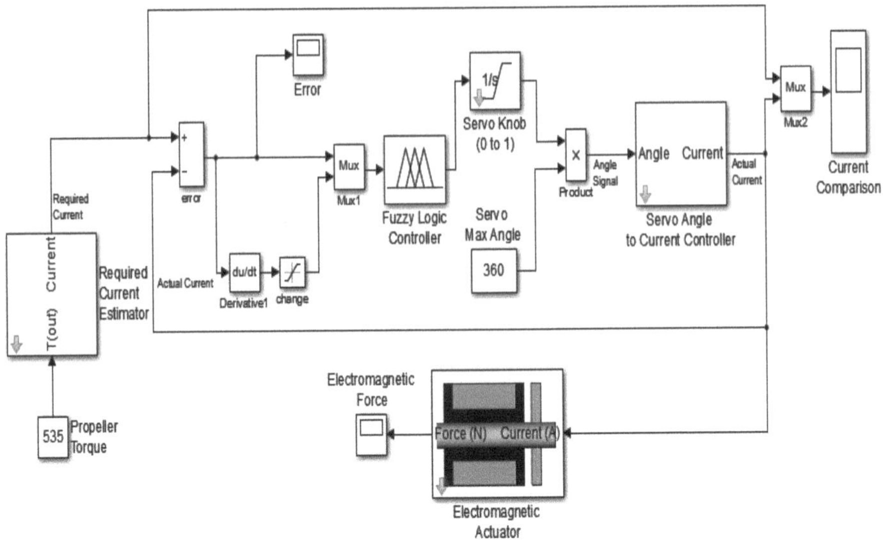

Fig. 21.4 Simulink block diagram of EMA-CVT fuzzy logic controller [1]

block for the servo to current controller translated the servo motor's angle (in degrees) into the predetermined current required to run the EMA. The block includes the Simulink subsystem block that models the equation for a line, y = mx + b, where m is the slope obtained by dividing the maximum required current by the maximum servo angle, y is the current level in ampere, x is the servo angle in degrees, and b is the point where the line crosses the vertical y-axis.

The output of the servo to current controller, which models the servo motor that control the knob of the current control module, is then fed back to the fuzzy logic controller loop as an actual current level. The actual current has been compared with the required current to produce error signal for the fuzzy logic controller. Besides, the actual current also being fed to the electromagnetic actuator block to generate electromagnetic force. The reference and actual current is plot together with generated electromagnetic force for further analysis.

The torque generated at the propeller shaft during the car's first motion was roughly 535.8 Nm. Based on the torque, the current estimator model block determined the necessary current level to run the actuator. As seen in Fig. 21.5, the fuzzy logic controller will then operate the current control module to generate the current level appropriately. As seen in Fig. 21.6, the settling time that is, the amount of time it takes for the current to rise from zero to the desired current of 9.33 is roughly 2.85, which is acceptable when shifting the automobile from standstill. The electromagnetic force that generated is about 6.2 kN is enough to clamp the CVT pulley sheave to create the gear ratio during car initial movement as shown in Fig. 21.7.

The torque generated at the propeller shaft at a cruising speed of 120 km/h was approximately 228 Nm. The current estimator model block used to determine the necessary current level to run the electromagnetic actuator (EMA). As seen in

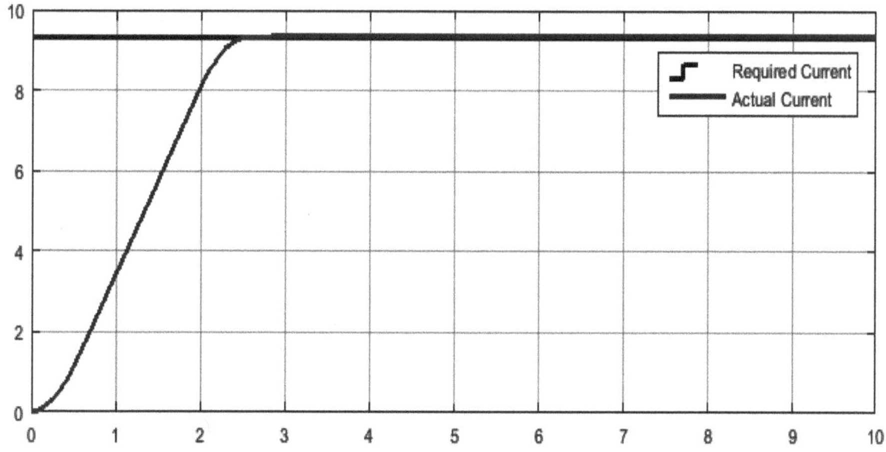

Fig. 21.5 Simulation result for required and actual current for T_{out} 535.8 Nm)

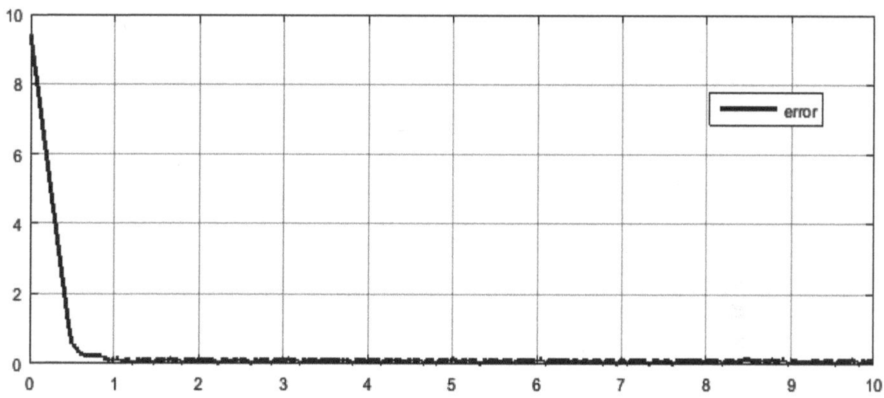

Fig. 21.6 Error response (for T_{out} 535.8 Nm)

Fig. 21.8, the fuzzy logic controller then operates the current control module to generate the current level appropriately. As seen in Fig. 21.9, the settling time that is, the time it takes for the current to rise from zero to the desired current of 6.09 A is roughly 2.85 s. This is acceptable because it simulates the vehicle to accelerate from 0 to 120 km/h. The produces about 2.2 kN of electromagnetic force, which is sufficient to clamp the CVT pulley sheave and produce.

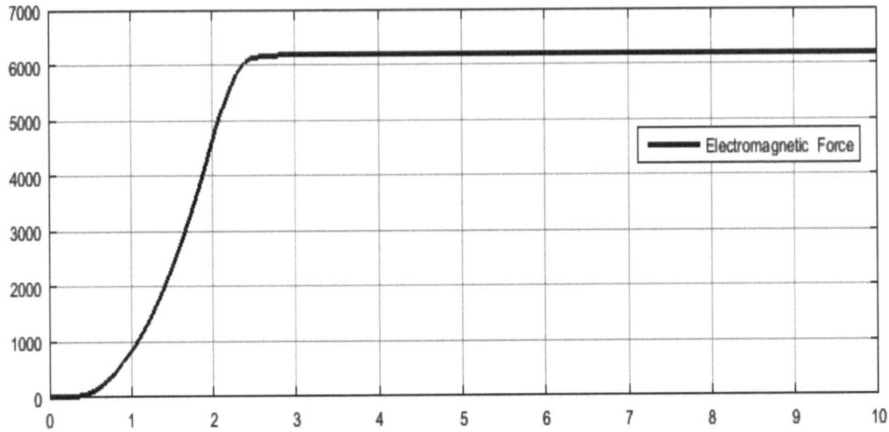

Fig. 21.7 Generated electromagnetic force (for T_{out} 535.8 Nm)

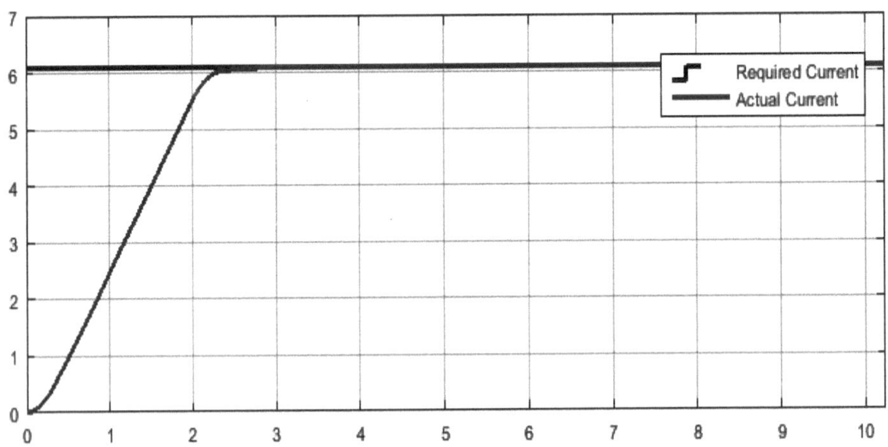

Fig. 21.8 Simulation of required/targeted and actual current (for T_{out} 228 Nm)

21.4 EMA-CVT Performance: Theoritical

This section presents the electromagnetic force and electrical current requirements for the EMA-CVT of a Toyota Wish passenger car for the EPA Highway Fuel Economy Test Cycle (HWFET) and the EPA Urban Dynamometer Driving Schedule (UDDS) cycles. A car weighing 1700 kg, having a top speed of 120 km/h (33 m/s), a wheel radius of 0.295 m, a drag area of 0.633, a rolling motion resistance coefficient of 0.02, and a road friction coefficient of 0.5 were all taken into consideration. Vehicle speed and road grades are the system's input parameters, and the traction torque related to the clamping force needed to run the EMA-CVT system is modeled in Matlab simulation as shown in Fig. 21.4. The wheel radius size and the final gear ratio were

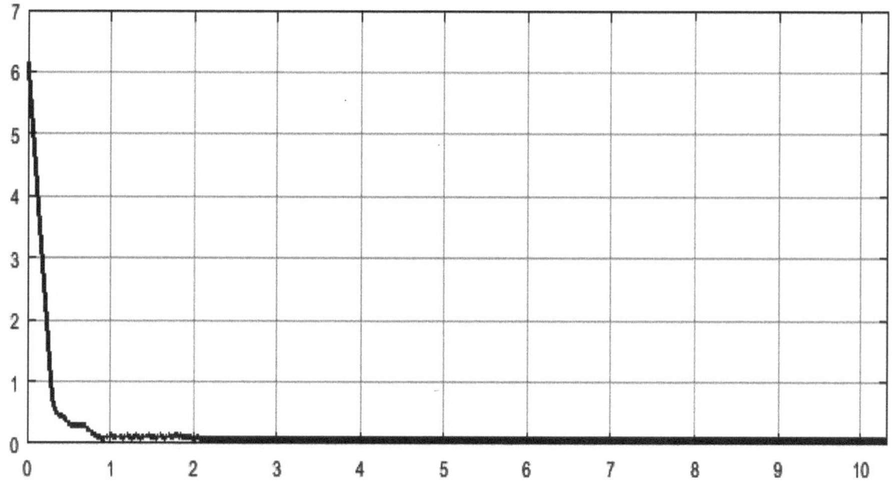

Fig. 21.9 Simulation of the controller error response (for T_{out} 228 Nm)

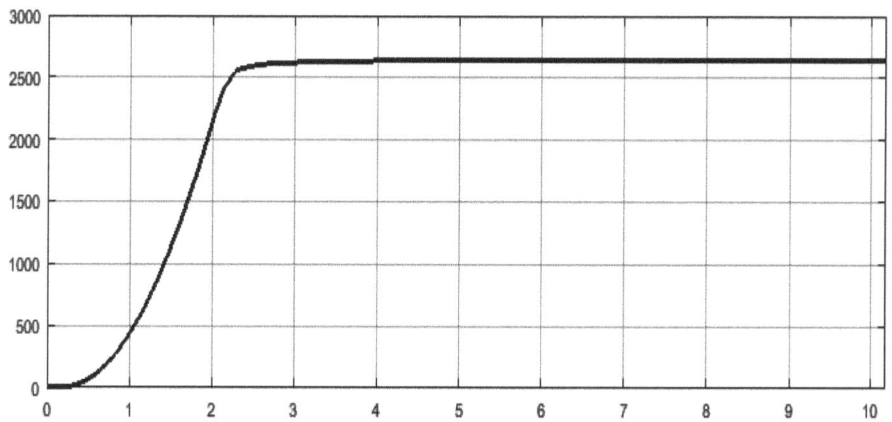

Fig. 21.10 Simulation of electromagnetic force generated (for T_{out} of 228 Nm) [1]

multiplied by the traction force to determine the traction torque at the drive shaft (or the CVT's output shaft).

According to predetermined driving dynamics known as the EPA Highway Fuel Economy Test Cycle (HWFET) and EPA Urban Dynamometer Driving Schedule (UDDS), the necessary traction was designed (United States Environmental Protection Agency, 2018). The UDDS cycle replicates a 12.07-km urban circuit with numerous stops. The average speed is 31.5 km/h, with a maximum speed of 91.25 km/h. As seen in Fig. 21.11, the cycle is divided into two phases: the first phase starts cold and lasts for 505 s, or around 5.78 km at an average speed of 41.2 km/h. The second phase runs for 867 s. Moreover, the US EPA created the HWFET cycles, which are displayed in Fig. 21.12, as a chassis dynamometer driving schedule to assess the fuel efficiency of light-duty cars. With an average speed of 77.7 km/h at 765 s, it replicates a 16.45 km highway trips.

The necessary clamping force and generated electromagnetic force for the UDDS (Fig. 21.15) and HWFET (Fig. 21.16) driving cycles were compared in Figs. 21.13 and 21.14. The CVT pulley sheave can retain its position because the electromagnetic forces produced are at least as strong as the necessary clamping force. The current level error, or the discrepancy between the intended current and the generated current level, is displayed in Figs. 21.17 and 21.18. The controller needed around 2.1 and 2.2 s to adjust the current level from zero to the desired level when zoomed in.

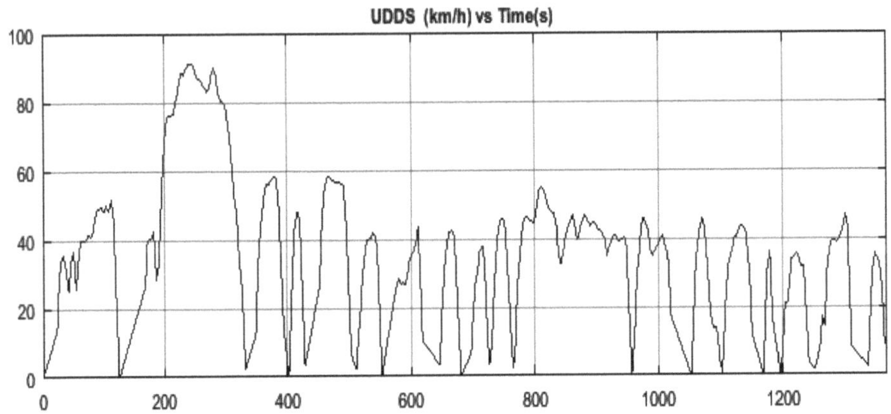

Fig. 21.11 Urban dynamometer driving schedule (UDDS)

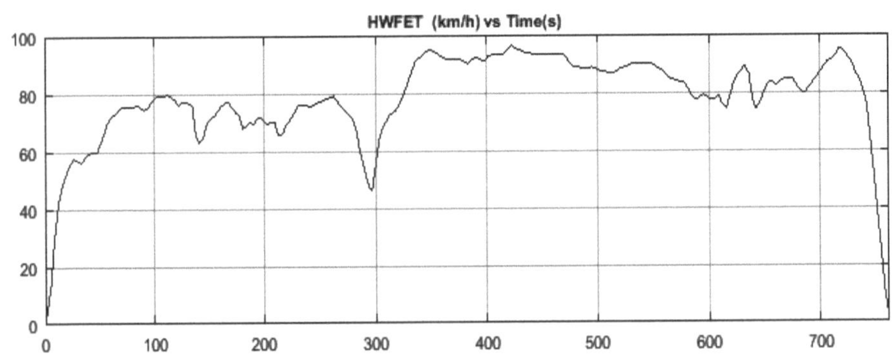

Fig. 21.12 Highway fuel economy test cycle (HWFET)

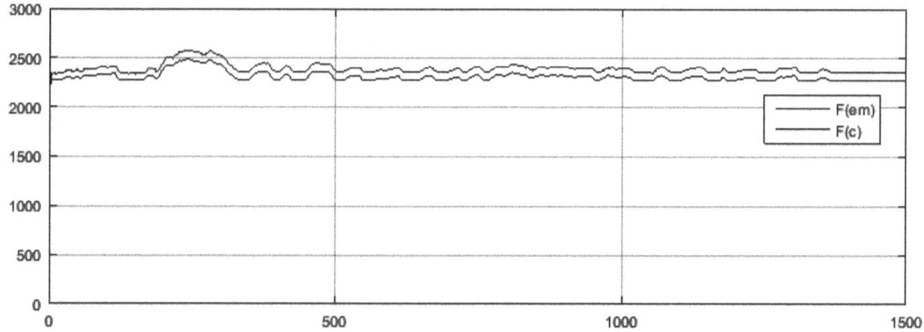

Fig. 21.13 Simulated clamping force and electromagnetic force (UDDS cycle)

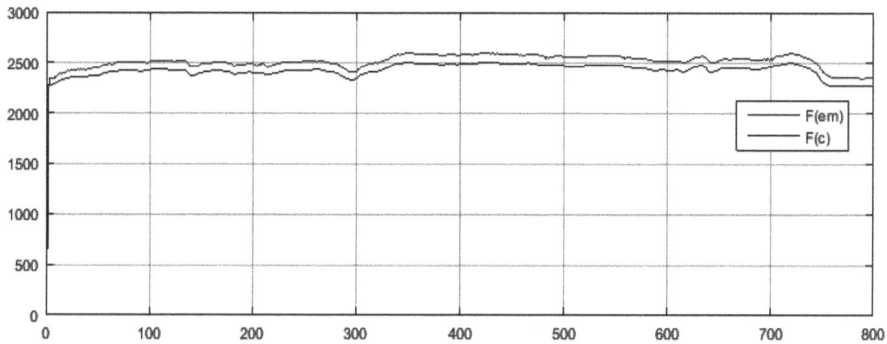

Fig. 21.14 Simulated clamping and electromagnetic force for HWFET cycle

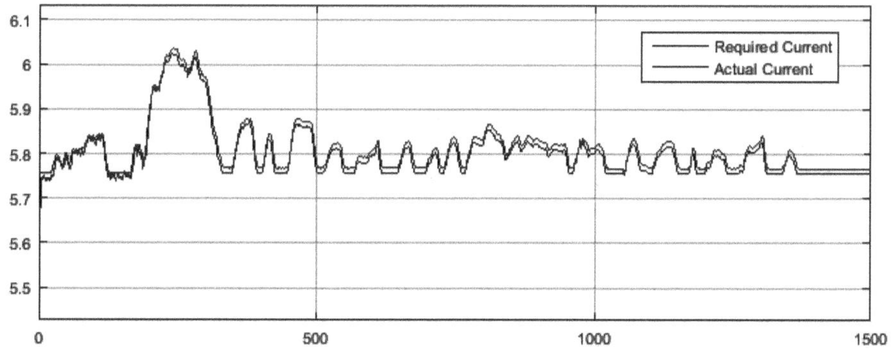

Fig. 21.15 Simulation required and actual current for UDDS cycle

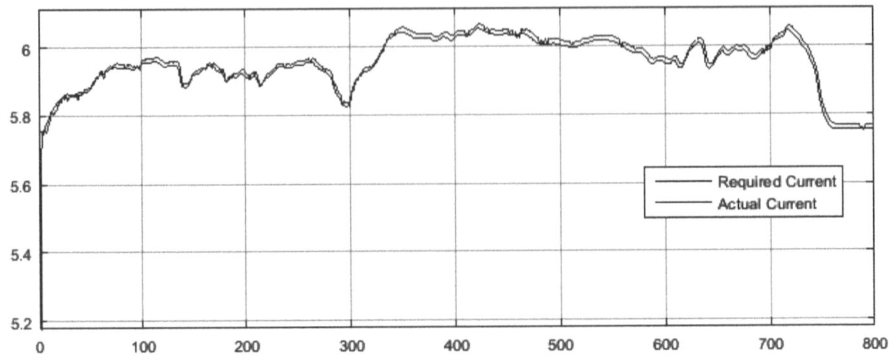

Fig. 21.16 Simulated required and actual current for HWFET cycle

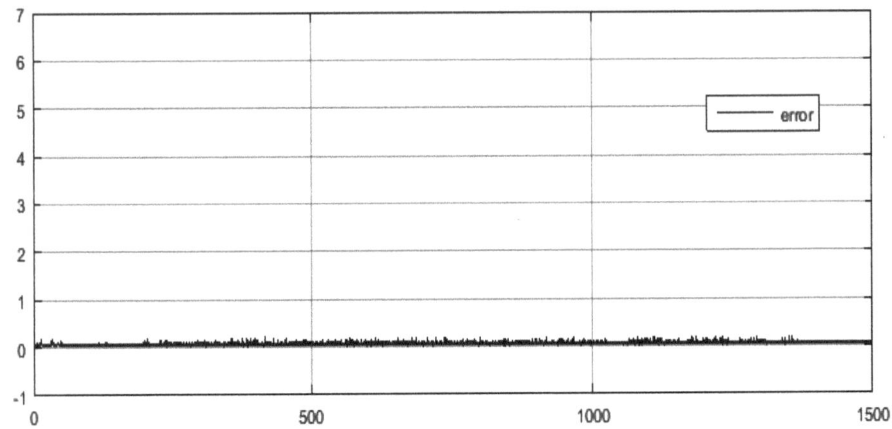

Fig. 21.17 Simulated controller error for UDDS cycle

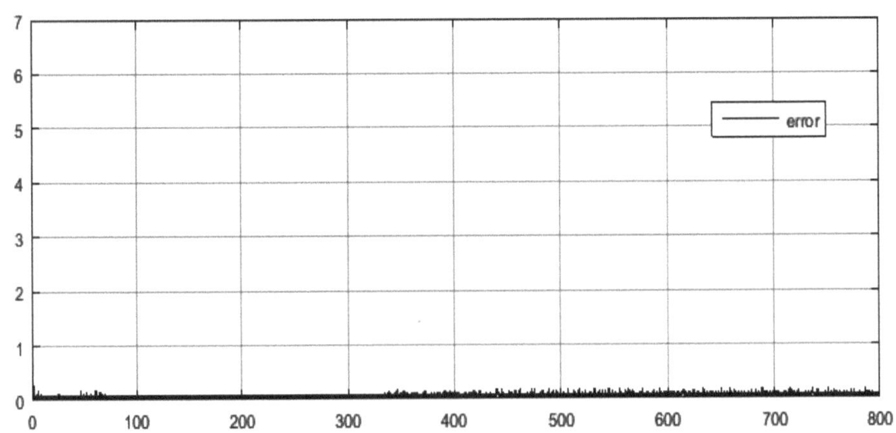

Fig. 21.18 Simulation controller error for HWFET cycle

21.5 EMA-CVT Prformance: Experimentally

The original Toyota Wish CVT (model K311) included a spring at the primary pulley with a rigidity of 117 N/cm and a hydraulic pumping system connected to the secondary pulley to regulate the CVT's gear ratio as needed. The engine powers the hydraulic pumping. The CVT has been disassembled from its initial configuration. Its main pulley spring hasn't changed, though. The only thing that has been removed to connect with the proposed EMA is the secondary pulley's hydraulic actuation system.

An electromagnetic actuated CVT testbed comprised of a traction motor, a battery pack that powers the EMA to generate the desired electromagnetic force the equivalent of clamping force and an electromagnetic actuator that maintains the belt tension and adjusts the gear ratio was created based on design specifications and theoretical analysis. In Fig. 21.19, the system's functional diagram is displayed. Using an electronic motor, the EMA-CVT driving torque for the primary pulley has been produced.

By activating the coil to engage the precise gear ratio, this solenoid actuator enables the push mechanism. The load weight that replicated the traction torque at the propeller shaft was detected by a load sensor. The current or power level estimator then receives the accompanying load signal and uses it to determine the appropriate electrical current that should be delivered to the EMA in order to generate sufficient force in every scenario. Additionally, the vehicle's real gear ratio is measured by two rpm sensors (tachometers) that continuously detect the motor and propeller shaft's rotating speeds.

As seen in Fig. 21.20, the electromagnetic actuator was developed using 3100 wire turns on a 220 mm steel bobbin. It can generate an electromagnetic force of 200 N to 300 N with a supplied current of 5 to 12.0 A, respectively. Steel was used in the construction of the hollow bobbin to provide a stronger electromagnetic force. A 3100-turn copper coil with a 1.8-mm radius has been coiled between the bobbin's flanges. An aluminum alloy plunger passes through the hollow bobbin that has one end attached to the secondary pulley's movable sheave and the other end connected

Fig. 21.19 Test bed for EMA laboratory testing [1]

to an iron disk and a 40 N/cm stiff spring. The full-scale EMA specification is shown in Table 21.1.

The EMA testing instrumentation system, which includes a Toyota Wish CVT, load cell, rpm sensor, a 50 V, 100 Ah battery pack, and an electromagnetic force monitoring laptop, is depicted in Fig. 21.21. The testbed was constructed on a sturdy table with a thick wooden top and was strengthened with aluminum plate that was one centimeter thick. To maintain stability, the primary frame, which measures 180 cm by 80 cm, is constructed from thick hollow steel. A 1.5-inch aluminum bar was used to mount the CVT set in place, and a bolt was used to secure it to the table surface from below.

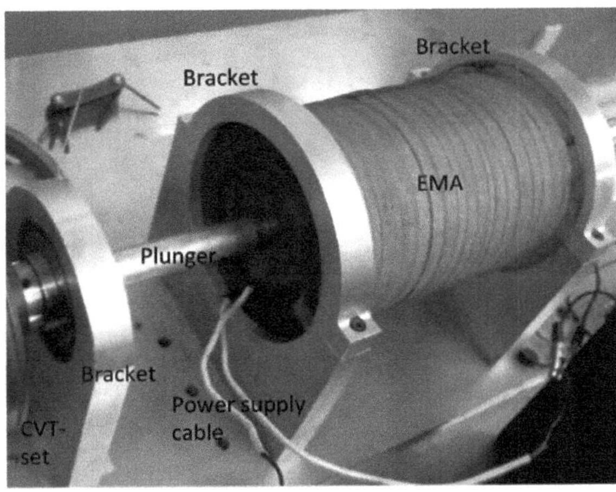

Fig. 21.20 EMA–CVT for passenger car [2]

Table 21.1 Full-scale EMA specifications

Parameters	Values
Coil diameter (dw), mm	180
Cross sectional area of wire, mm^2	1.65
Length of solenoid (Lsol), mm	220
Solenoid Minimum diameter (hi), mm	80
Solenoid Maximum diameter (ho), mm	180
Packing factor, $f_{pf} = \frac{N_{actual}}{N_{theoretial}}$	0.525
Number of turns (N), $N = \frac{F_{ema} * L_{coil}}{\mu I^2 L_{sol}}$, turns	3100
Length of coil (Lcoil), $L_{coil} = \frac{L_{sol}}{d_w^2}\left[D_0^2 - (D_i + t)^2\right](f_{pf})$, m	761

(a)

(b)

Fig. 21.21 EMA force testing without spring loaded [2]

To guarantee the accurate and seamless movement of the primary and secondary pulleys' movable sheave, all of the related original bearings were utilized. Additionally, all of the mounting holes were created using a CNC milling machine to provide greater testbed precision. The driven side of the CVT was attached to the propeller shaft, and the driver side was connected to the AC traction motor. The EMA was positioned on the opposite side of the driving pulley. By pushing the plunger back and forth, the EMA function regulates the position of the pulley's movable sheave. Additionally, an AC traction motor was chosen to drive the CVT system. The pulley gear ratio and torque in the secondary transmission shaft can be used to calculate the torque required to propel the CVT from the electric motor. The traction torque at the wheel was divided by the final drive ratio to determine the transmission torque at the propeller. To regulate the power delivery to the EMA, a fuzzy controller is coupled to the load cell and rpm sensors.

A hanging load, similar to that found in a Prony-brake dynamometer, was attached in order to estimate and generate an equivalent torque at the propeller shaft. A dynamometer is a tool used to measure the mechanical power that a rotating shaft transmits. Dynamometers are basically torque-measuring instruments since power is the product of torque (turning force) and angular speed. By squeezing brake blocks against the wheel and tightening the bolts until the friction torque (FR) equals the torque (WL), a Prony-brake creates mechanical friction on the edge of a rotating pulley. According to the formula m = Tp/gl, a hanging mass of 58 kg is needed to generate 228 Nm of torque at the propeller shaft in order to simulate a 1700 kg car traveling at 120 km/h (33.33 m/s) on an 11.3 degree slope. In order to generate 228 Nm of torque at the propeller shaft, a 58 kg load was suspended at the brake hand. The load cell simultaneously measured the hanging load, which was then converted into an input signal for the fuzzy logic controller.

21.6 EMA-CVT Performance Interpretation

After the EMA-CVT test bed system was successfully fabricated, an actual laboratory test was carried out. The system is operated from a variety of perspectives in accordance with the motor and EMA system's controlled current flow. Different loads result in different torques, which in turn produce varied electromagnetic forces. The only parameter that can be controlled to produce electromagnetic forces is voltage, which controls current. However, producing a significant electromagnetic force requires a high current rating, which the battery power pack provides. A digital clip-on meter is used to measure the supply voltage and the associated current or power in combination.

The load cell was positioned on the same central axis as the EMA's plunger, and a plunger with a disc plate was attached to the load cell in order to measure the attractive force of the plunger produced by the core's magnetization. The linear stage was then shifted using an iron disc plate at intervals of 1 mm from the plunger placement position. The laser displacement sensor was used to verify the actual shift displacement, and a metal ruler was used to measure the disc plate's distance from its initial location. The lithium-ion (LiFePO4) battery was used to apply voltage to the EMA windings when the distance between the magnetic poles was changed.

The data gathering system was used to measure the attractive force from the load cell. The spring was removed in order to measure only the electromagnetic or attractive force in order to compare the experiment's results with the study of the electromagnetic field. To make sure it had an impact on the attractive force that the EMA created as a result of the current supply, the thermocouple was used to measure the temperature that developed as a result of the coil's current supply.

Resistance of the coil at 300C can be estimated as, Rs = 30 degree*0.00393 per degree*1.015 ohms = 0.1197 Ohms and the temperature increment can be estimated as,

$$\Delta T = T_s - T_i = \frac{D_{rs} V_s}{\lambda R_s A_s} \tag{21.1}$$

The temperature rises in the solenoid coil due to the applied voltage, ΔT, V_{RMS} the duty ratio D_{rs} of the applied voltage, resistance at the saturation temperature, heat dissipation coefficient, λ on the coil surface, and the surface area, A_s of the enameled wire.

To measure the electromagnetic force using the EMA while adjusting the current supply, Fig. 21.22 has been set up. The laptop has recorded the observed electromagnetic force by connecting the load-cell. The driving wheel's rotational speed is measured with a tachometer. A digital stopwatch was used to measure the journey time, and a tachometer was used to measure the rpm. The system's acquired data is analyzed and graphically portrayed.

The gear ratios and the matching rpm of the primary and secondary pulleys are displayed in Figs. 21.23 and 21.24. The gear ratio should, in theory, range from 0.820 to 2.369. Both the low gear ratio and the maximum gear ratio indicate that the system reaction is satisfactory in the initial stage. The secondary pulley rpm trend should really be different, but this type of divergence is caused by mechanical constraints such as misalignment or slippage between the belt and pulley, among others. As a result, this experiment was able to approximate but not quite reach the intended gear ratio.

The core of the contemporary CVT technology is electromagnetic force, which precisely positions the pulley shave to provide the required gear ratio. The gear ratio was followed by its linear shift in relation to power. However, the electromagnetic force trend exhibits its fluctuating behavior as a result of slip in the belt and pulley during the gear ratio shift. It takes a sufficient amount of electromagnetic force to grasp the belt and pulley perfectly. By referring to the current or power sent to the coil, the EMA's performance is determined by its capacity to use electromagnetic force to push or pull the pulleys per unit of time.

The power consumption for the needed electromagnetic force to engage the CVT sheave towards the vehicle propulsion is depicted in Fig. 21.25, which is based on experimental data. By covering the solenoid with aluminum foil, magnetic flux leakage may be minimized and the electromagnetic force used to actuate the solenoid operation can be further increased. Significant electromagnetic force loss results from numerous flux leaks from unshielded solenoid housing. By covering the solenoid with aluminum foil, magnetic flux leakage may be minimized and the electromagnetic force used to actuate the solenoid operation can be further increased. The experimental investigation of the EMA to create the electromagnetic force to modify the CVT's gear ratio for the vehicle propulsion based on the road surface.

The motor rpm sensors' signal has been used to maintain the power supply for the corresponding gear ratio. The fuzzy controller has developed the actuation force or clamping force of the CVT by controlling the power flow to the EMA by customizing the sensor signal. For the vehicle to have traction on a 25% gradient with a gear ratio of 2.6, the EMA needs a maximum current supply of 11 A. The relationship between the generated electromagnetic force and supply current for various gear ratios is further depicted in Fig. 21.27. Because of the air gap between the metal mover and the EMA, a different amount of force is produced with the same amount of current for the first gear (high ratio) (Fig. 21.26).

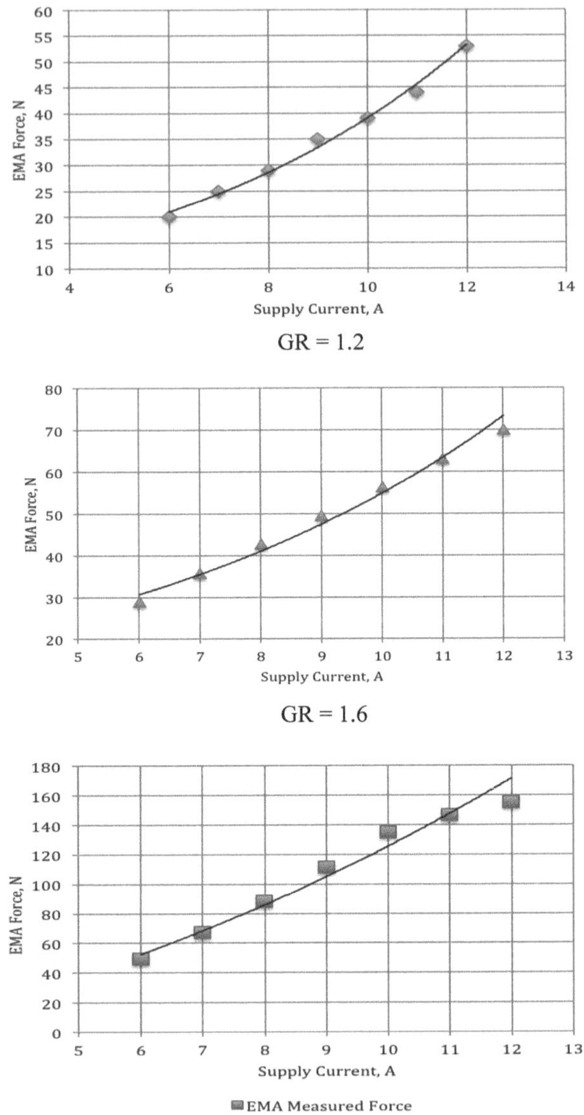

Fig. 21.22 EMA force measurement experimentally [1]

21.7 EMA-CVT Validation

Many researchers from various fields examined and developed the linked systems so that they could be adapted and operated by electrical power as global automakers moved toward automobile electrification. Rahman (2014) developed a quarter-scale CVT that is activated by an electromagnetic system. According to the experimental results, the car accelerates 60% less than the hydraulic CVT in the Supply Current, A.

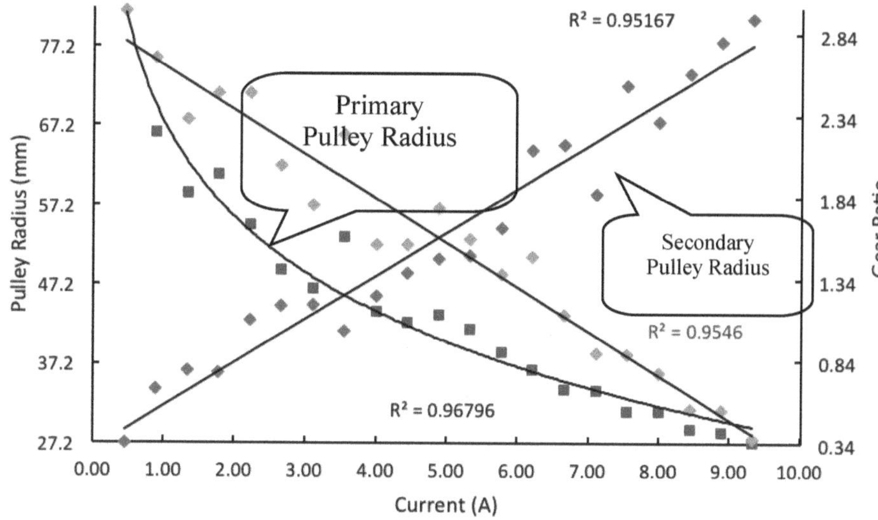

Fig. 21.23 Pulley RPM, gear ratio versus supply current [7]

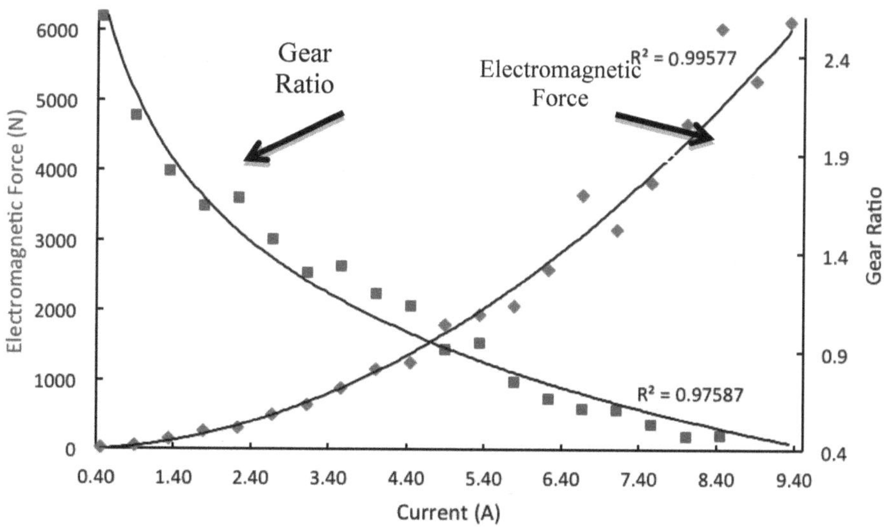

Fig. 21.24 Electromagnetic force, gear ratio versus supply current [7]

However, the plastic solenoid bobbin that the copper coil was wrapped around proved unsuitable for a practical usage due to its inability to withstand the heat produced by the coil's constant current flow and the moving plunger. This project focuses on creating a novel actuator that operates by electromagnetic principles rather than a plastic bobbin solenoid-based EMA.

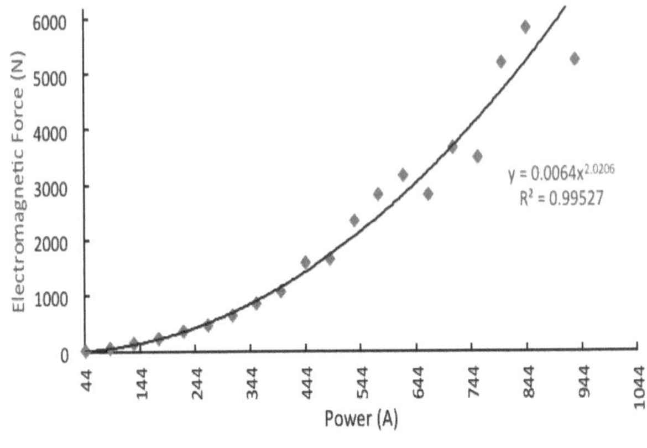

Fig. 21.25 Electromagnetic force versus power consumption [7]

To visualize the car's performance under different road conditions, a simulation using Microsoft Excel and Matlab has been created based on the theoretical model of EMA-CVT that has been constructed from a variety of literature. A laboratory-scale EMA was created to validate the idea before a real-size EMA that could run the actual CVT was created. The correlation between the experimental and simulated electromagnetic force development is displayed in Fig. 21.28. The correlation between the simulation and the experiment for the EMA's generated force is displayed in Fig. 21.29. According to the correlation plot, an extremely high square of the regression coefficient, R^2 value of 0.91 was attained, getting very close to the optimal accuracy value of 1.0.

Additionally, Fig. 21.29 shows how the EMA performs with different power supplies. The range of the movable sheave's reacting (actuation) time is noted by referring to the current that is provided to the coil. Manually adjusting an endless number of gear ratios is challenging. This sophisticated control system would be a superior option to make sure of it. More effort is required to clamp the CVT pulley sheave when starting and on an incline in order to increase the car's GR and torque. The experiment shows that the acceleration time is longer for low rpm (during starting or climbing slope conditions) than for other rpm types. The reason for this is the high torque needed to move the vehicle in the beginning and high torque means high clamping force to be overcome by the actuator.

It takes less time to accelerate once the car has reached its maximum moment of inertia. The lower gear ratio travel time is longer than usual in order to counteract the stall effect. This minimal acceleration time, or travel time in other terms, results from the solenoid's quick response capabilities. This is essentially the main justification for the solenoid's employment in this current transmission system.

Fig. 21.26 Experimental study of EMA performance [7]

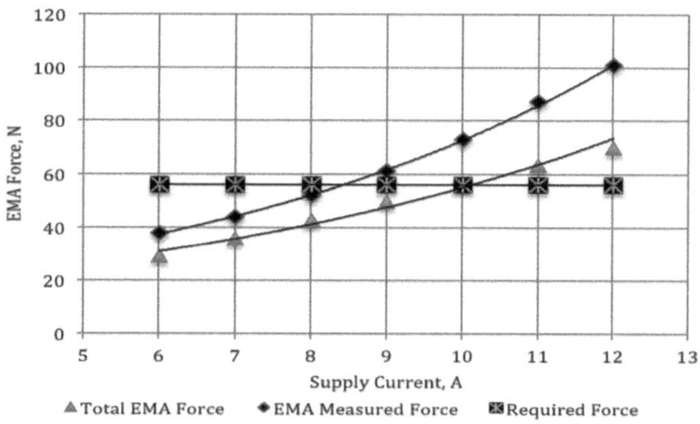

In case of gear ratio, GR = 1.2

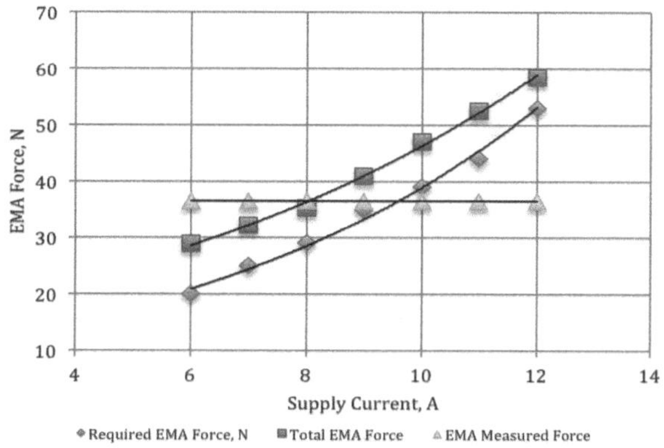

In case of gear ratio, GR = 1.6

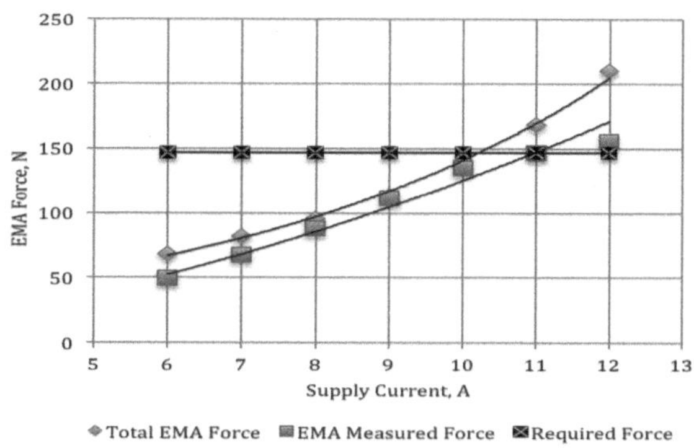

In case of gear ratio, GR = 2.62

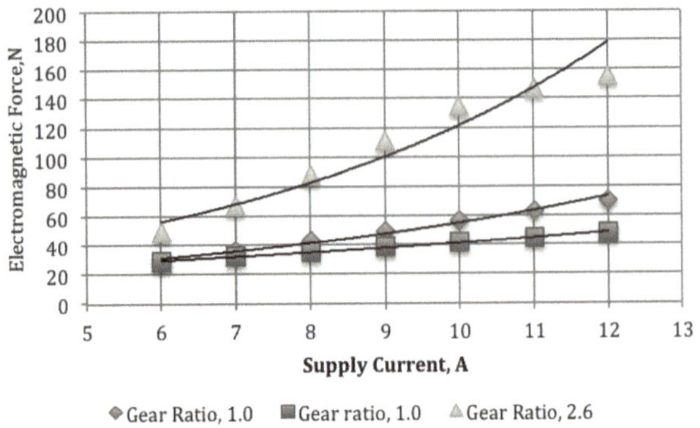

Fig. 21.27 Electromagnetic force for different gear ratio [1]

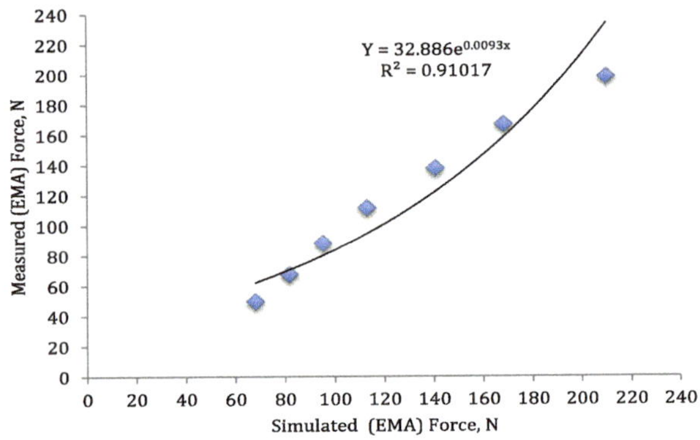

Fig. 21.28 Variation of experimental result and simulation result

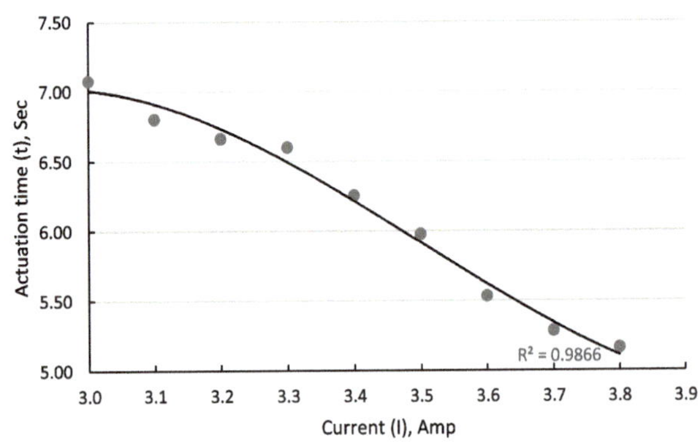

Fig. 21.29 Actuation time versus current supply [1]

21.8 Conclusion

The core of the contemporary CVT technology is electromagnetic force, which precisely positions the pulley shave to provide the required gear ratio. The gear ratio was followed by its linear shift in relation to power. However, the electromagnetic force trend exhibits its fluctuating behavior as a result of slip in the belt and pulley during the gear ratio shift. It takes a sufficient amount of electromagnetic force to grasp the belt and pulley perfectly. Given its quick acceleration (0 to 60 km/h) and ability to set the gear ratio as needed in 220 ms without the need for human intervention, rattling noise, and energy savings, the intelligent control system of EMA-CVT may be a preferable option for car transmissions.

Acknowledgements The authors are grateful to the Ministry of Higher Education of Malaysia for funding this project under the Prototype Research Grant Scheme (PRGS) 2016. The authors are also grateful to the Research Management Center (RMC) of the International Islamic University Malaysia for funding this project to develop an EMA-CVT laboratory scale concept prototype.

References

1. Hassan, A. J. (2020). *Development of intelligent electromagnetic continuously variable transmission* [A PhD Thesis Dissertation]. IIUM Library, International Islamic University Malaysia.
2. Rahman, A., Jaffar, A. H., & Sany, I. I. (2022). IoT-based electromagnetic actuator for CVT: Basic design and prototyping. *International Journal of Powertrains, 11*(1), 38–61.
3. Jaafar, A. H., & Rahman, M. A. (2019). Study on the intelligent control system of EMA-CVT. *International Journal of Recent Technology and Engineering (IJRTE), 7*(65), 48–53.
4. Bdran, S., Saifullah, S., & Shuyuan, M. (2012). An overview on control concepts of push-belt CVT. *International Journal of Engineering and Technology, 4*(4), 392–395.
5. Rahman, A., Sazzad, B. S., Mohiuddin, A. K. M., Mahbubur, R., & Altab, H. (2014). Energy efficient electromagnetic actuator for CVT system. *Journal of Mechanical Science and Technology, 28*(4), 1153–1160.
6. Ulaby, F. T. (2005). *Electromagnetics for engineers*. Pearson International Edition.
7. Ye, M., Liu, Y. G., & Cheng, Y. (2016). Modeling and ratio control of an electromechanical continuously variable transmission. *International Journal of Automotive Technology, 17*(2), 225–235.
8. Cholis, N., Ariyono, S., & Priyandoko, G. (2015). Design of single acting pulley actuator (SAPA) continuously variable transmission (CVT). *Energy Procedia, 68*, 389–397.
9. Erjavec, J. (2005). *Automatic transmission* (1st ed.). Thomson Delmar Learning.
10. Richer, E., & Hurmuzulu, Y. (2001). A high performance pneumatic force actuator system part 1—Nonlinear mathematical model. *ASME Journal of Dynamic Systems, Measurement, and Control, 122*(3), 416–425.
11. Rosli, M. R. B. (2015). Multi-robot tethering using camera. *Journal of Chemical Information and Modeling*, Universiti Teknologi PETRONAS. https://doi.org/10.1017/CBO9781107415324.004
12. Torotrak Group. (2007). Application of Torotrak IVT to an Optare bus produces 19 % fuel economy improvement. Retrieved April 22, 2017, from http://www.torotrak.com/application-of-torotrak-ivt-to-an-optare-bus-produces-19-fuel-economy-improvement/